The tendency of a living organism to move to a more favourable environment is a natural but complex reaction, involving the integration of sometimes conflicting environmental stimuli as well as a co-ordinated mechanical response. The response of motile, single cell organisms to environmental stimuli provides a useful model for understanding, first of all, how the environment is monitored and sensed and, secondly, how his this information is processed to result in an integrated and co-ordinated response. This volume looks at a large number of well-studied examples of the chemotactic response, in prokaryotes and eukaryotes, and casts new light on how cells process information and react to their environment. This fundamental response is of great importance in understanding one of the important characteristics of most living organisms.

The volume will be of interest not only to microbiologists but also cell and developmental biologists with an interest in cell behaviour, cell motility and sensory transduction.

BIOLOGY OF THE CHEMOTACTIC RESPONSE

SYMPOSIA OF THE
SOCIETY FOR GENERAL MICROBIOLOGY*

1 THE NATURE OF THE BACTERIAL SURFACE
2 THE NATURE OF VIRUS MULTIPLICATION
3 ADAPTATION IN MICRO-ORGANISMS
4 AUTOTROPHIC MICRO-ORGANISMS
5 MECHANISMS OF MICROBIAL PATHOGENICITY
6 BACTERIAL ANATOMY
7 MICROBIAL ECOLOGY
8 THE STRATEGY OF CHEMOTHERAPY
9 VIRUS GROWTH AND VARIATION
10 MICROBIAL GENETICS
11 MICROBIAL REACTION TO ENVIRONMENT
12 MICROBIAL CLASSIFICATION
13 SYMBIOTIC ASSOCIATIONS
14 MICROBIAL BEHAVIOUR 'IN VIVO' AND 'IN VITRO'
15 FUNCTION AND STRUCTURE IN MICRO-ORGANISMS
16 BIOCHEMICAL STUDIES OF ANTIMICROBIAL DRUGS
17 AIRBORNE MICROBES
18 THE MOLECULAR BIOLOGY OF VIRUSES
19 MICROBIAL GROWTH
20 ORGANIZATION AND CONTROL IN PROKARYOTIC AND EUKARYOTIC CELLS
21 MICROBES AND BIOLOGICAL PRODUCTIVITY
22 MICROBIAL PATHOGENICITY IN MAN AND ANIMALS
23 MICROBIAL DIFFERENTIATION
24 EVOLUTION IN THE MICROBIAL WORLD
25 CONTROL PROCESSES IN VIRUS MULTIPLICATION
26 THE SURVIVAL OF VEGETATIVE MICROBES
27 MICROBIAL ENERGETICS
28 RELATIONS BETWEEN STRUCTURE AND FUNCTION IN THE PROKARYOTIC CELL
29 MICROBIAL TECHNOLOGY: CURRENT STATE, FUTURE PROSPECTS
30 THE EUKARYOTIC MICROBIAL CELL
31 GENETICS AS A TOOL IN MICROBIOLOGY
32 MOLECULAR AND CELLULAR ASPECTS OF MICROBIAL EVOLUTION
33 VIRUS PERSISTENCE
34 MICROBES IN THE NATURAL ENVIRONMENTS
35 INTERFERONS: FROM MOLECULAR BIOLOGY TO CLINICAL APPLICATIONS
36 THE MICROBE 1984 PART I: VIRUSES
              PART II: PROKARYOTES AND EUKARYOTES
37 VIRUSES AND CANCER
38 THE SCIENTIFIC BASIS OF ANTIMICROBIAL CHEMOTHERAPY
39 REGULATION OF GENE EXPRESSION – 25 YEARS ON
40 MOLECULAR BASIS OF VIRUS DISEASE
41 ECOLOGY OF MICROBIAL COMMUNITIES
42 THE NITROGEN AND SULPHUR CYCLES
43 TRANSPOSITION
44 MICROBIAL PRODUCTS: NEW APPROACHES
45 CONTROL OF VIRUS DISEASES

*Published by the Cambridge University Press, except for the first Symposium, which was published by Blackwell's Scientific Publications Limited.

BIOLOGY OF THE CHEMOTACTIC RESPONSE

EDITED BY

J. P. ARMITAGE AND J. M. LACKIE

FORTY-SIXTH SYMPOSIUM OF
THE SOCIETY FOR
GENERAL MICROBIOLOGY
JOINTLY ORGANISED
WITH THE BRITISH
SOCIETY FOR CELL
BIOLOGY
HELD AT
THE UNIVERSITY OF YORK
DECEMBER 1990

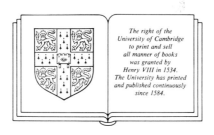

Published for the Society for General Microbiology

CAMBRIDGE UNIVERSITY PRESS

CAMBRIDGE
NEW YORK   PORT CHESTER
MELBOURNE   SYDNEY

Published by the Press Syndicate of the University of Cambridge
The Pitt Building, Trumpington Street, Cambridge CB2 1RP
40 West 20th Street, New York, NY 10011–4211, USA
10 Stamford Road, Oakleigh, Melbourne 3166, Australia

© The Society for General Microbiology Limited 1990

First published 1990

Printed in Great Britain by The Bath Press, Bath, Avon

British Library cataloguing in publication data

Society for General Microbiology, *Symposium: 46th: 1990: University of York*
Biology of the Chemotactic response.
1. Organisms. Cells. Chemotaxis
I. Title II. Armitage, J. P. III. Lackie, J. IV. Series
574.8764

Library of Congress cataloguing in publication data

Society for General Microbiology. Symposium (46th: 1990: University of York)
    Biology of the chemotactic response/
edited by J. P. Armitage and J. Lackie:
Forty-sixth Symposium of the Society for General Microbiology, held at the University of York, December 1990.
    p. cm.
    "Published for the Society for General Microbiology."
    ISBN 0-521-40313-8 (hardback)
    1. Chemotaxis—Congresses. I. Armitage, J. P. (Judith P.)
II. Lackie, J. M.   III. Society for General Microbiology.
IV. Title.
    [DNLM: 1. Chemotaxis—congresses.   QH 647 S678b 1990]
QR187.4.S63   1990
589.9'0876042—dc20
DNLM/DLC
for Library of Congress   90-15040
ISBN 0 521 40313 8 hardback   CIP

CONTRIBUTORS

ARMITAGE, J. P. Microbiology Unit, Biochemistry Department, University of Oxford, Oxford OX1 3QU, UK

BACON, K. B. Institute of Dermatology, St Thomas's Hospital, London SE1 7EH, UK

BERG, H. C. Department of Cellular and Developmental Biology, Rowland Institute for Science, Cambridge MA 02142, USA

BLOCK, S. M. Department of Cellular and Developmental Biology, Rowland Institute for Science, Cambridge MA 02142, USA

BOURGOIN, S. Unité de recherche 'Inflammation et Immunologie/Rhumatologie' Center de recherche du CHUL, 2705 Boulevard Laurier Ste Foy, Québec G1V 4G2, Canada

CAMP, R. D. R. Institute of Dermatology, St Thomas's Hospital, London SE1 7EH, UK

DINGWALL, A. Beckman Center, Department of Development Biology, Stanford University School of Medicine, Stanford, CA94305-542-7, USA

DUNN, G. A. MRC Muscle and Cell Motility Unit, Biophysics Section, King's College, London WC2B 5RL, UK

ELY, B. Department of Microbiology, University of South Carolina, Columbia, SC 29208, USA

EUROPE-FINNER, G. N. Department of Biochemistry, University of Oxford, Oxford OX1 3QU, UK

FINCHAM, N. J. Institute of Dermatology, St Thomas's Hospital, London SE1 7EH, UK

GAMMON, B. Department of Biochemistry, University of Oxford, Oxford OX1 3QU, UK

GAUDRY, M. Unité de recherche 'Inflammation et Immunologie-Rhumatologie' Center de recherche du CHUL, 2705 Boulevard Laurier Ste Foy, Québec G1V 4G2, Canada

HAVELKA, W. A. Microbiology Unit, Biochemistry Department, University of Oxford, Oxford OX1 3QU, UK

LACKIE, J. M. Yamanouchi Research Institute (UK), Littlemore Hospital, Oxford OX4 4XN, UK

LIU, G. Department of Biochemistry, University of Oxford, Oxford OX1 3QU, UK

MCBRIDE, M. J. Department of Molecular and Cell Biology, University of California, Berkeley, CA 94720, USA

MCCLEARY, W. R. Department of Molecular and Cell Biology, University of California, Berkeley, CA 94720, USA

MCCOLL, S. R. Unité de recherche 'Inflammation et Immunologie-Rhumatologie' Center de recherche du CHUL, 2705 Boulevard Laurier Ste Foy, Québec G1V 4G2, Canada

MACNAB, R. M. Department of Molecular Biophysics and Biochemistry, Yale University, New Haven, Connecticut 06511, USA

MCNALLY, D. Laboratory of Molecular Biology, Department of Biological Sciences and Microbiology and Immunology, University of Illinois at Chicago, PO Box 4348 m/c 067, Chicago, Illinois 60680, USA

MARWAN, W. Max-Planck Institut für Biochemie, 8033 Martinsried, W. Germany

MATSUMURA, P. Laboratory of Molecular Biology, Departments of Biological Sciences and Microbiology and Immunology, University of Illinois at Chicago, PO Box 4348 m/c 067, Chicago, Illinois 60680, USA

NACCACHE, P. H. Unité de recherche 'Inflammation et Immunologie – Rhumatologie', Center de recherche du CHUL, 2705 Boulevard Laurier Ste Foy, Québec GIV 4G2, Canada

NEWELL, P. C. Department of Biochemistry, University of Oxford, Oxford OX1 3QU, UK

O'CONNOR, K. A. Department of Molecular and Cell Biology, University of California, Berkeley, CA 94720, USA

OESTERHELT, D. Max-Planck-Institut für Biochemie, 8033 Martinsried, W. Germany

PURCELL, E. M. Department of Physics, Harvard University, Cambridge MA 02138, USA

QUINN, D. G. Institute of Dermatology, St Thomas's Hospital, London SE1 7EH, UK

ROMAN, S. Laboratory of Molecular Biology, Departments of Biological Sciences and Microbiology and Immunology, University of Illinois at Chicago, PO Box 4348 m/c 067, Chicago, Illinois 60680, USA

SEGALL, J. E. Department of Anatomy and Structural Biology, Albert Einstein College of Medicine, 1300 Morris Park Avenue, Bronx, NY 10461, USA

SHAPIRO, L. Beckman Center, Department of Development Biology, Stanford University School of Medicine, Stanford, California 94305-542-7, USA

SCHNITZER, M. J. Department of Physics, Harvard University, Cambridge MA 02138, USA

SOCKETT, R. E. Microbiology Unit, Biochemistry Department, University of Oxford, Oxford OX1 3QU, UK

TRANQUILLO, R. T. Department of Chemical Engineering and Materials Science, University of Minnesota, Minneapolis, MN 55455, USA

VAN HOUTEN, J. Department of Zoology, University of Vermont, Burlington, VT 05405, USA

VOLZ, K. Laboratory of Molecular Biology, Departments of Biological Sciences and Microbiology and Immunology, University of Illinois at Chicago, PO Box 4348 m/c 067, Chicago, Illinois 60680, USA

WILKINSON, P. C. Department of Bacteriology and Immunology, University of Glasgow, Western Infirmary, Glasgow G11 6NT, UK

WOOD, C. A. Department of Biochemistry, University of Oxford, Oxford OX1 3QU, UK

ZUSMAN, D. R. Department of Molecular and Cell Biology, University of California, Berkeley, CA 94720, USA

CONTENTS

	page
Contributors	vii
Editors' preface	xiii

G. A. DUNN
Conceptual problems with kinesis and taxis — 1

M. J. SCHNITZER, S. M. BLOCK, H. C. BERG AND E. M. PURCELL
Strategies for chemotaxis — 15

R. T. TRANQUILLO
Theories and models of gradient perception — 35

R. M. MACNAB
Genetics, structure, and assembly of the bacterial flagellum — 77

G. L. HAZELBAUER, R. YAGHMAI, G. G. BURROWS, J. W. BAUMGARTNER, D. P. DUTTON AND D. G. MORGAN
Transducers: transmembrane receptor proteins involved in bacterial chemotaxis — 107

P. MATSUMURA, S. ROMAN, K. VOLZ AND D. McNALLY
Signalling complexes in bacterial chemotaxis — 135

A. DINGWALL, L. SHAPIRO AND B. ELY
Flagella biogenesis in *Caulobacter* — 155

J. P. ARMITAGE, W. A. HAVELKA AND R. E. SOCKETT
Methylation-independent taxis in bacteria — 177

D. R. Zusman, M. J. McBride, W. R. McCleary and K. A. O'Connor
Control of directed motility in *Myxococcus xanthus*   199

D. Oesterhelt and W. Marwan
Signal transduction in *Halobacterium halobium*   219

J. E. Segall
Mutational studies of amoeboid chemotaxis using *Dictyostelium discoideum*   241

P. C. Newell, G. N. Europe-Finner, G. Liu, B. Gammon and C. A. Wood
Chemotaxis of *Dictyostelium:* the signal transduction pathway to actin and myosin   273

J. Van Houten
Chemosensory transduction in *Paramecium*   297

P. C. Wilkinson
Leuococyte chemotaxis: a perspective   323

P. H. Naccache, M. Gaudry, S. Bourgoin and S. R. McColl
Signal transduction in leucocytes: the linkage between receptor and motor   347

K. B. Bacon, N. J. Fincham, D. G. Quinn and R. D. R. Camp
Tissue-derived leucocyte attractant factors in inflammation   379

Index   395

EDITORS' PREFACE

If, as Koestler suggested, politics is the art of the possible and science the art of the soluble, then perhaps the lure of studying chemotaxis is that it is a behavioural system susceptible, in principle, to a complete description. Certainly, ten years ago in a symposium on the same topic (Lackie, J. M. & Wilkinson, P. C., *Biology of the Chemotactic Response*. Cambridge: CUP 1981) it became obvious that it was beginning to be possible to approach a description of chemotaxis in molecular terms. At that time, although a good beginning had been made on prokaryotes, so little was known about signal transduction and the control of the locomotory apparatus in eukaryotes that these topics were not even covered. Although it is clear that we are still a long way from a complete understanding of even the prokaryotic system, the past decade has seen substantial progress, particularly in the field of signal–response transduction. A whole new armoury of techniques have become available and these are being applied gleefully to formerly intractable problems.

Molecular biology is, however, only a set of tools, albeit a powerful, shiny new set, and there has been progress in other areas. There is still a place for careful observation, detailed measurement and mathematical modelling. These have been helped considerably by the advances in computerized motion analysis which allow large numbers of cells to be followed and subtle responses analysed. One of the great attractions of the problem of chemotaxis is that, however much the system is dissected, cloned and sequenced, the results must be related to the whole cell; the problems lie in the biology, even the natural history of the cellular system.

We have laid out the chapters roughly in ascending order of biological complexity, therefore, regretfully, the prokaryotic and eukaryotic systems may seem to be segregated into territories. It is becoming increasingly obvious that this is a false division; similarities are more obvious and conversations across the boundaries more frequent.

The first three chapters deal with the problems of identifying and describing cellular behaviour in general. In Chapter 1, Dunn casts a critical eye over basic assumptions including the long-held belief amongst many workers that bacterial chemotaxis is a misnomer for chemokinesis. This analytical approach is supported by the mathematical modelling of the following two papers where Berg and Tranquillo tackle the problem of describing behaviour in a more mathematical way than do most biologists. Important pointers toward discriminatory experiments emerge from such modelling, as it provides a description of the essential components required for a particular response. The struggle towards numeracy must therefore be made!

Through elegant and patient molecular biology, Macnab and his team of coworkers have cloned and sequenced a large percentage of the 40 or so genes involved in the synthesis of a functional flagellum, and the complex process of controlled synthesis and assembly of the structure can now begin to be described, leading towards a description of the mechanism of rotation. The control of rotational switching of the flagellum in enteric bacteria by chemotactic signals is tackled in the following two chapters. The mechanism of signal transduction in prokaryotes is now understood at the level where site-directed mutagenesis and protein crystallisation can be used to analyse the structural changes involved. This work has revealed that chemosensing is not an isolated cellular activity but part of a large family of environmental sensing systems.

The following chapters deal with the behaviour of four very different bacterial species. The description of motility and taxis in these organisms has not yet reached the level of sophistication found in the enteric bacteria. However, they serve to show both the diversity and homogeneity of behaviour in the prokaryotic world. All but one of the species have proteins that show sequence homology with the methylation-dependent receptors analysed in great detail in enteric bacteria, revealing the early evolutionary history of these receptors. *Caulobacter*, with its motile and non-motile phases, provides an excellent model for the control of synthesis and operation of the behavioural apparatus. Although the mechanisms of gliding and differentiation in *Myxobacterium* are still not well understood, the analysis of behavioural mutants shows similarities to flagellate species. On the other hand, the methylation-independent responses of *Rhodobacter* suggests that, in some environments, a less sophisticated set of responses may still produce a survival advantage. The archaebacterium *Halobacterium halobium* shows the productive combination of molecular biology, biochemistry, biophysics and observation in analysing behaviour, and provides a fascinating bridge to the eukaryotic section of this volume. The requirement for fumarate for switching and the identification of proteins that show homology to the G-proteins of eukaryotes suggests that this supposedly primitive species has a complex and highly developed sensory system. All this suggests that, although methylation-dependent chemotaxis may have provided some of the outside edge, and even the corners, there are many pieces still to be added before the jigsaw of prokaryotic behaviour is finished.

Cellular slime moulds, a favourite group for chemotaxis buffs have yielded substantially to the combined approaches of genetics, biochemistry, and molecular biology. Through the analysis of the defects in behaviour of particular mutants, Segall brings together a diverse set of observations and attempts a synthesis; Newell *et al* describe the second messenger systems operating in control of slime mould behaviour, and both similarities and differences with the signal transduction system of metazoa are revealed.

Paramecium continues to be our best model for the role of membrane excitability in sensing of substances in the environment, and Van Houten relates the responsiveness of *Paramecium* to that of taste and odour chemo-receptors. Memorably, she remarks that *Paramecium* 'resembles a swimming neuron because it displays so many conductances expected of metazoan excitable cells' – and thus provides yet another justification for the study of simple cellular behaviour. Shifting then to metazoan cells, the remaining three chapters deal with mammalian leucocytes. Lymphocytes, long neglected on the grounds that they belonged to immunologists and were uncharted territory for cell biologists, are the main topic of Wilkinson's 'perspective', though he achieves what his title promises in terms of putting leucocyte chemotaxis into its context. The signal–response transduction systems in neutrophils and the second messengers associated with each system are reviewed by Naccache (and there are interesting contrasts with the situation in slime moulds). Finally, Bacon *et al.* take us to the final level of complexity, the behaviour of leucocytes in a pathological process in Man, reminding us that an understanding of basic cell behaviour may eventually be relevant to the rational therapy of human disease.

Different systems have particular advantages – but often bring limitations. We hope that communication between people interested in different aspects of cell biology will be helped by this collection of reviews and that the mathematical modellers, microbiologists, molecular biologists and experimental pathologists will learn by sharing their particular problems and solutions.

We are grateful to Boots PLC, Lilly Research Centre, Roche Products and Yamanouchi (U.K.) for financial support to bring together the speakers at the symposium.

Judith Armitage
June 1990   John Lackie

CONCEPTUAL PROBLEMS WITH KINESIS AND TAXIS

GRAHAM A. DUNN

MRC Muscle and Cell Motility Unit, Biophysics Section, King's College London, 26–29 Drury Lane, London WC2B 5RL

BACKGROUND

The study of kinesis and taxis is concerned with how the external environment influences the motile behaviour of cells and organisms. In the words of Keller *et al.* (1977) 'Responses to environmental stimuli which take the form of directed orientation reactions are called taxes; those which take the form of undirected locomotion are called kineses.' In this view, the concepts of stimulus and response or reaction (I will treat these last two terms as synonymous) are of underlying importance for the distinction between kinesis and taxis. But these innocent-looking terms have caused a great deal of dispute and heated debate. It appears that the difficulties are not so much with terminology but with a failure to agree on the underlying concepts. These concepts are essential tools for both communication and discovery in the field of biological motion and the aim of this paper is to try to discover the root cause of this confusion and to suggest a remedy along the lines proposed in a recent paper by Doucet & Dunn (1990). First, I will give the traditional definitions of kinesis and taxis.

Kinesis

Fraenkel & Gunn (1940) describe kineses as: 'Undirected locomotory reactions, in which the speed of movement or the frequency of turning depend on the intensity of stimulation'.

Keller *et al.* (1977) applied the term to chemical stimuli: 'Chemokinesis. A reaction by which the speed or frequency of locomotion of cells and/or the frequency and magnitude of turning (change of direction) of cells or organisms moving at random is determined by substances in the environment. Chemokinesis is said to be positive if displacement of cells moving at random is increased and negative if displacement is decreased. Two forms of kinesis have been distinguished: ortho-kinesis, a reaction by which the speed or frequency of locomotion is determined by the intensity of the stimulus, klino-kinesis, a reaction by which the frequency or amount of turning per unit time is determined by the intensity of the stimulus.'

Taxis

'The term taxis is used today for directed orientation reactions. Thus, positive and negative photo-taxis mean respectively movement straight towards or straight away from the light. We use the word only for reactions in which the movement is straight towards or away from the source of stimulation.' (Fraenkel & Gunn).

'Chemotaxis. A reaction by which the direction of locomotion of cells or organisms is determined by substances in their environment. If the direction is towards the stimulating substance, chemotaxis is said to be positive, if away from the stimulating substance, the reaction is negative. If the direction of movement is not definitely towards or away from the substance in question, chemotaxis is indifferent or absent. Positive chemotaxis can result in attraction towards the stimulating agent or in retaining the cells in high concentrations of the active substances by avoidance of low concentrations.' (Keller *et al.*, 1977).

ABSTRACTION OF CONCEPTS

A first stage in the abstraction of generally applicable concepts is to isolate those features of biological motion that are held in common by all motile cells, organisms and gametes, etc. In this paper, I will try to avoid reference to a particular cell or organism, except by way of illustration of a general point, since much of the failure to agree on the use of concepts appears to derive from the practice of 'tailoring' them to suit a particular situation.

What is a response?

The most fundamental aspect of biological motion in the context of kinesis and taxis is active translocation: the change in position of a whole organism or cell by means of its own motile machinery. This concept, in turn, depends on the concept of location or position but assigning a definite position to an organism, or especially to a motile cell that continually changes shape and structure as it moves, presents considerable practical and theoretical difficulties to which there is no unique solution. For cells in tissue culture, Brown & Dunn (1989) advocate using microinterferometry to obtain the centre of gravity of the dry mass of the cell as representative of its position. In the case of a moving snake, either the position of the head or of the centre of gravity of the whole animal might be taken as its position. If the concepts of kinesis and taxis are to be robust enough to be unambiguous when applied to any particular situation, then they should not critically depend on the choice of definition of position.

Another problem with the concept of position arises if we wish to include such phenomena as the growth of nerve fibres, blood capillary sprouts, and fungal hyphae within the domain of kinesis and taxis. For such extending protrusions, the word tropism is often borrowed from plant physiology as an alternative to the word taxis. Regardless of terminology, however, the phenomenon of directed movement of the growth cone at the tip of a nerve fibre, for example, is closely related to that of directed movement of a leucocyte. Again, any distinction between taxis and kinesis should not depend critically on the method of assigning a position to the tips of such protrusions.

Another abstraction that is commonly made in connection with kinesis and taxis is that of orientation. In their classic work on kineses and taxes in the lower organisms, Fraenkel & Gunn (1940) consider that the orientation of the long axis of the organism can be a definitive feature for distinguishing kinesis from taxis. This concept is usually unambiguous enough in the case of organisms where the long axis clearly lies in the plane of bilateral symmetry and the direction of translocation bears some definite relation to this axis; but when applied to tissue cells, as with position, there is no unique or 'correct' definition of the long axis and a cell does not inevitably move in a direction parallel with its long axis no matter how this is defined. Dunn & Brown (1990) have suggested that the direction of maximum asymmetry derived from third-order moments may be a better predictor of the direction of movement of a cell. Thus, although the orientation of the long axis might be a useful diagnostic feature for distinguishing between kinesis and taxis, it should be avoided as a basis for the fundamental distinction if the terms are to be general enough to apply to cells as well as lower organisms.

If we consider position and orientation to be primary abstractions, then concepts such as velocity (rate of change in position) and rate of change in orientation might be considered to be secondary abstractions. Further abstractions of relevance to taxis and kinesis, such as acceleration (rate of change in velocity), are based on these. One component of acceleration that is considered to be particularly relevant to the subclassification of kinesis is the rate of change in direction. Unfortunately, this distinction between orthokinesis and klinokinesis is often described in terms of a concept that introduces a new and unnecessary ambiguity: the concept of a turn. Suppose that we want to decide whether a cell is showing a klinokinesis and we therefore want to measure its frequency and magnitude of turning. First, we have to decide what a turn is since cells do not generally make sharp turns. A turn, we might decide, is when the cell has changed its direction of travel by greater than a certain amount within a certain short time. But, how do we measure the magnitude of each turn? We have already defined the turn on the basis of its magnitude. How do we measure the frequency of turning? The maximum possible frequency is decided by how short the time interval is in our definition. The values of magnitude and frequency that we observe are determined as much by our definition as they are by what the cell is doing. This can give rise to entirely the wrong conclusion in extreme cases.

Ideally, then, a robust distinction between kinesis and taxis should be as insensitive as possible to the unavoidable ambiguity inherent in the concept of position and should be independent of all unnecessary ambiguities. Thus I will ignore such properties of an organism as its orientation, and such motile activities as the waving of antennae, and restrict the concept of response to how a geometrical point representing the position of the cell or organism moves in space.

This leads on to the practical issue of what constitutes a response, how do we recognize one? By abstracting the position of a cell or organism, its translocative behaviour can be described entirely by a single vector variable: its velocity. A response is an aspect or feature of the behaviour of this variable (e.g. the magnitude or direction of the velocity) that depends on a particular property of the external environment. Nevertheless, quite sophisticated techniques may be needed to analyse the stochastic properties of this variable and their dependency on external events in order to discover which aspects of behaviour are determined by which properties of the external environment. But this article is a discussion of concepts, and the practical issues will eventually sort themselves out provided that the concepts are robust and unambiguous.

What is a stimulus?

Fraenkel & Gunn (1940) state that 'Anything which can make animal machinery start working is called a *stimulus*,' and add that 'In intact animals the stimuli may be internal ones, such as hormones, or the strains applied to the muscle-spindles, or they may be external, such as changes in intensity of light, or pressure, or in the concentration of some chemical substance.'

This concept of an internal stimulus seems too universal: it would appear that any internal event or change within the organism that influences its motile behaviour can be an internal stimulus. An extreme interpretation might be that it is any event in the complex meshwork of cause and effect linking the remote past to the final action. The organism is a highly interdependent system and any current event is probably influenced by the whole of the internal activity of even a short time earlier.

The concept of an external stimulus seems to be much more manageable: it is any event or change in the external environment that influences the organism's motile behaviour. The external environment is often simpler and can be more easily manipulated than the internal, and so the task of isolating and identifying a single stimulus is not so formidable. Since the study of kinesis and taxis is concerned with how the external environment influences motile behaviour, it would appear that we can disregard the complexities of internal stimuli. Unfortunately, internal stimuli have the unpleasant habit of refusing to be excluded from the traditional description of kinesis and taxis as we shall see later.

Do the terms stimulus and response necessarily describe events?

In common usage and, it would appear, in the definition of a stimulus given by Fraenkel & Gunn, a stimulus is an event – a change in the properties of the environment. Similarly, a reaction or response is also commonly taken to mean a change in the behaviour of a system elicited by a change in the prevailing conditions. This is consistent with the dictionary definition that to 'react' is to 'respond to stimulus, undergo change due to some influence' (Concise O.E.D.). But one problem with restricting these words to a description of events is that, in order to describe comprehensively how behaviour is determined by the external environment, we have also to consider situations in which the environment does not change. It appears from the definitions of kinesis that an organism is showing a kinesis if, for example, its speed is determined by temperature even though the environment is totally uniform, unvarying and the same in all directions. Clearly, one or more of the following is wrong: our concept of stimulus; our concept of kinesis; the idea that a kinesis is inevitably associated with a stimulus.

It is not entirely obvious, in the scheme of Fraenkel & Gunn, whether a traditional stimulus can occur in an unvarying environment or whether it can only occur when the environment changes. In their Table IV they state that the stimulus for a kinesis is a gradient in intensity. While this would seem to imply a stable environment, it is non-uniform and a motile organism could experience a change in its local environment as it moves. It may be that Fraenkel & Gunn consider these self-generated changes to be the effective stimuli. If so, these effective stimuli would appear to be internal since the immediate events causing these changes are internal rather than external.

In the case of a taxis, it is clearer that even the local environment need not necessarily change. Fraenkel & Gunn give the example of a beam of light in Table IV but a better example for this illustration is a stable, uniform magnetic field. Although uniform, such environments are anisotropic – they have different properties in different directions. Few would argue that a magnetotaxis cannot occur in such an unvarying, uniform environment. It could be argued, however, that the stimulus is still a change because the magnetic field, as perceived by the organism, changes in orientation as the organism turns or rotates. But rotation of the organism is obviously not essential for detecting the magnetic field with some conceivable mechanisms and we would not wish to exclude these so arbitrarily. Either the stimulus is the stable magnetic field or magnetotaxis does not require a stimulus.

Whatever we decide, it is still necessary to be able to describe situations in which some property of behaviour depends on some property of the environment regardless of whether change occurs in either. I will assume that a stimulus is a feature or property of the environment that influences the motile behaviour and which may, or may not, vary with time. A response is those aspects of the translocative behaviour that depend on the stimulus. Dependency is the essential concept. Thus an organism moving in a steady, uniform, isotropic environment is showing a kinesis if, say, its speed depends on external temperature. The stimulus is the temperature, even though it does not change, and the response is the speed, even though its mean value may not drift with time. Discovering this dependency almost inevitably requires changing the temperature, or observing similar organisms at two different temperatures, but we should not confuse the concept of stimulus with the method of discovering whether a stimulus exists. This view seems to accord better with the descriptions of kinesis and taxis quoted above.

A disconcerting consequence of this point of view is that any environment is packed full of stimuli for a particular organism and the organism's motile behaviour is packed full of responses. It is impossible to conceive of any earthly motile organism whose behaviour does not depend on many different properties of the environment. Obviously, a cell in tissue culture would not move at all unless each of a myriad of external factors were correctly adjusted. We should therefore describe a kinesis with reference to a particular influential property of the environment and the particular aspect of the behaviour that depends on this property. The case of a taxis is rather different because an environment might conceivably have only a single anisotropic property on which the behaviour depends.

THE DISTINCTION BETWEEN KINESIS AND TAXIS

Does the response differentiate a kinesis from a taxis?

Kinesis and taxis are commonly referred to as responses and so a logical place to start looking for the distinction between them is at the level of the response. To recap, a chemokinesis is a 'reaction by which the speed or frequency of locomotion of cells and/or frequency and magnitude of turning (change of direction) of cells or organisms moving at random is determined by substances in the environment'. Suppose that we place chemokinetic organisms in a confined environment consisting of a chemical gradient. The organisms will initially show a net drift towards the portion of the gradient where their motility is lowest until a steady-state distribution is achieved. But this is also a 'reaction by which the direction of locomotion of cells or organisms is determined by substances in their environment'. It would appear to be both a kinesis and a taxis at the same time and conflicts with the idea that these terms are mutually exclusive.

The traditional view does not exclude the possibility that a gradient of some substances can elicit both a kinesis and a taxis at the same time. Many of the substances that are considered to be chemotactic for neutrophil leucocytes are considered to be also chemokinetic. But such behaviours are seen as a mixture of two responses; the traditional view does not state that a kinesis in a gradient is inevitably also a taxis. Kineses, however, are of interest mainly because they do lead to an uneven distribution of organisms. This inevitably means that some net directional behaviour must occur before this equilibrium distribution is reached. Keller *et al.* and Fraenkel & Gunn were well aware of this problem when formulating their definitions. They did not see it as a major difficulty, however, because kinetic drift is very poorly directed; it is equivalent to a diffusion flux in, say, ink particles undergoing Brownian motion. All that was necessary in order to avoid confusion between kinetic drift and taxis was to say that in a taxis the movement must be 'straight towards' (Fraenkel & Gunn) or 'definitely towards' (Keller *et al.*) some source of directional information in the environment. Keller *et al.* state more specifically that 'The fact that cells or organisms accumulate in response to a chemical gradient is not sufficient proof for directional migration or chemotaxis.' Nevertheless, the definition of taxis now loses its robustness. Several forms of weak directional behaviour that are not kineses might also be excluded from the taxis category depending on how these ambiguous terms are interpreted.

A more serious problem is that a response known traditionally as kinesis-with-memory, sometimes called adaptive kinesis, can lead to a highly directional and sustained pattern of locomotion. All that is necessary for an organism to move up a gradient is for it to compare the current level of stimulus with its memory of some earlier level and to adjust its speed and/or rate of turning according to the difference in the two levels. The behaviour can be very directional if the rate of turning is dependent on the change in signal but the mean velocity vector can also have a highly significant directional component if only the speed is dependent.

It does not seem to be generally realized that a memory is not theoretically essential for this kind of adaptive mechanism. A very similar pattern of behaviour can be achieved if the stimulus that the cell or organism perceives is the time derivative of some scalar quantity in the environment. Suppose that an organism can detect only the rate of change in chemical concentration, dC/dt, in its local environment and suppose that it will only move when dC/dt is equal to zero or very nearly so. In a stable chemical gradient, the stationary organism will perceive that dC/dt is zero and begin to move; but any movement either up or down gradient will immediately cause it to halt. The net result is that the organism will only move in a very directional manner at right angles to the direction of the gradient. In two dimensions this behaviour would lead to a bi-directional pattern of movement and would not be called a taxis but a transverse orientation by Fraenkel & Gunn. Nevertheless, by suitably juggling the rules of the response, the pattern of movement can be made strongly unidirectional. Suppose, for example, that a swimming marine organism turns vigorously until the rate of change in pressure, dP/dt, that it perceives is the maximum that can be achieved for the speed at which it is swimming through the water. Such an organism will turn rapidly until it is swimming vertically downwards (directly up the pressure gradient) when it will stop turning and continue to swim fixedly in this precise direction. Thus the directionality of translocation is not in itself sufficient to distinguish an adapting kinesis from a taxis.

Does the stimulus differentiate between a kinesis and a taxis?

Regardless of the nomenclature that we choose to use, the fact remains that, in the last two examples, we have a highly directional behaviour pattern that arises from a non-directional stimulus. The stimulus, dC/dt, is a scalar quantity that has no directional component; it can, in theory, be detected by a single receptor. Is this, then, the fundamental difference that we are seeking; whether the stimulus is directional or not?

This viewpoint immediately raises a problem. For example, in the case of sensing the direction of a chemical gradient, the stimulus, at the level of a single chemoreceptor, might be the chemical concentration, C. This is also a scalar quantity and is obviously non-directional. But there is directional information in the local environment. Two or more receptors on different parts of the organism could compare signals and obtain at least some directional information. Thus whether or not the stimulus is considered to be directional resolves to a question of scale. If the whole local environment, rather than just the immediate environment of a receptor, is considered to contain the stimulus, then a chemical gradient might constitute a directional stimulus and chemotaxis can be seen as a valid form of taxis on this basis.

On this view, then, the stimulus is what the organism as a whole perceives at any instant and either immediately or eventually acts upon. The information detected by a single receptor is not necessarily the stimulus and we should call this something else: perhaps a signal. If the information received by the whole receptor system is not directional, then the response cannot be a taxis. The adaptive kinesis is excluded in this way from being classed as a taxis regardless of how directional the resulting behaviour is. This is because the whole receptor system is still only detecting non-directional information.

Two or more receptors comparing non-directional signals are not the only way of extracting directional information from the local environment. A single receptor detecting a non-directional signal can perform the same function if it can be waved about. A sophisticated organism could use a memory of the signals corresponding to successive receptor positions to build up a picture of the directional information in its immediate environment. Much simpler systems without a memory could also be effective as I will explain later.

On the basis of whether or not the stimulus is directional, it would seem that all such mechanisms should be classed as taxes. Indeed, Fraenkel & Gunn (Table IV) state that a taxis can be achieved 'by comparison of intensities of stimulation which are successive in time'. We have now had to expand the time-scale as well as the distance-scale in order to encompass the stimulus but this should not worry us too much because any system for detecting the direction of a local chemical gradient requires time, especially if the average concentration is very low.

We must now face the problem of what is the essential difference between this sort of tactic mechanism and an adaptive kinesis? In one case, the organism, which may be stationary, waves a receptor about in order to extract the directional information from its local environment. In the other case, the receptor may be fixed with respect to the organism and the organism must move in order to detect the directional information contained in the environment as a whole. The essential difference seems to be that the kinetic organism needs to translocate in order to perceive the directional information in the environment. This problem of discovering whether an organism needs to translocate is going to be very difficult to answer in those cases where the organism always does translocate! The distinction also has the undesirable feature that it is highly sensitive to how we define position. In order to know whether a cell or a snake is showing a kinesis or a taxis on this basis, we would have to decide whether the protrusion of a process or a movement of the head, respectively, constitutes a translocation or merely a local change in the position of a receptor.

But is this distinction all that fundamental? Consider the very hypothetical case of two cells each with a single dC/dt receptor and each placed in a chemical gradient. In one cell, the receptor wanders around at random in its surface and, if the signal it receives is zero or positive, the cell sets off moving in the same direction as the receptor. If the signal is negative, the cell either stops or remains stationary. The receptor on the other cell is fixed to its surface but the cell nevertheless obeys the same rules, becoming stationary if the signal is negative. Both cells will behave in a similar manner and only move in directions which take them into increasing or unchanging concentrations. Yet only one cell can detect a directional stimulus: only one cell can obtain directional information from its immediate environment without moving. The interesting thing is that these cells will also behave in a very similar way in a kinesis test. If both are placed in uniform environments in which C is changing, they will both remain stationary as C decreases and only move when C is constant or increasing.

If this is the basic distinction between kinesis and taxis, it seems highly unsatisfactory that it should be so trivial. For these two cells, the signals received by the receptors are identical; the rules of responding to the signals are very similar; and the resulting patterns of behaviour are practically indistinguishable. Any satisfactory scheme would either class them both as kinetic or both as tactic. This problem cannot be dismissed as being merely hypothetical since differential receptors are probably more commonplace in the biological world than absolute receptors.

Does the nature of the mechanism distinguish kinesis from taxis?

One possible answer might be that neither of the two hypothetical cells is showing a taxis. It could be argued that a tactic mechanism must involve two or more receptors that compare signals and that using 'tricks' such as a dC/dt receptor in order to gain entry to the taxis club is an unacceptable form of cheating. But the number of receptors is not a valid criterion. Geotaxis, magnetotaxis, phototaxis and related phenomena theoretically need only a single receptor that detects a vector signal. A single statocyst, for example, can detect the direction of gravity without comparing signals with another receptor. We have seen that the time taken to detect directional information also cannot be a satisfactory criterion. It is doubtful whether the whole range of possible detection mechanisms can be divided naturally into two classes by any criterion. Even if it could, the distinction between kinesis and taxis would then hang on rather arbitrary details of the detection mechanism and the original definitions of kinesis and taxis certainly did not prepare us for this conclusion.

AN ALTERNATIVE VIEW

The root of the problem seems to lie with the concepts of stimulus and response. We did not entirely eradicate the problem of the self-generated stimulus by deciding that a stimulus need not necessarily be a change in the external environment. The term stimulus refers only to the information that is acted upon by the organism: a stimulus should obviously not be so called if it fails to elicit a response because then it does not stimulate.

But herein lies the problem. In the case of a potential stimulus such as temperature or chemical concentration, only the external environment in the immediate vicinity of the organism can have an influence. And yet this local environment is determined by the movement of the organism as well as by the structure of the global environment. The organism (co)determines its own stimulus. In a non-uniform world, this self-generated stimulus contains information about the organism's own behaviour as well as information about the external world. For an organism with a single

dC/dt receptor, its local environment when moving up a chemical gradient can be indistinguishable from its local environment when stationary and subjected to an increasing chemical concentration. The difference between the two situations lies not in the information content of the stimulus or in the immediate response of the organism but in the type of information contained in the environment as a whole.

Doucet & Dunn (1990) propose the view that what is important in the distinction between kinesis and taxis is whether directional information in the environment is transferred to the pattern of behaviour. The details of how an organism extracts the directional information from the environment, if it does, are not important for making the distinction even though they are of central importance to understanding the mechanism. Also, it is not the details of the immediate response of the organism, for example whether its speed is dependent on the stimulus, but the whole pattern of behaviour that reveals whether its direction of movement depends on external information.

And so the proposed solution is to expand the time and distance scales even more and take a global view of the behaviour. Some property of the whole accessible environment, irrespective of the organism's movement, is the stimulus field. Some dependent property of the whole pattern of behaviour is the response field. A taxis occurs if directional information passes from the stimulus field to the response field and a kinesis occurs if non-directional information passes from the stimulus field to the response field. An identical mechanism can yield either a taxis or a kinesis depending on whether the external environment contains directional information.

However, the mechanism by which information is transferred cannot be entirely ignored. This is because a scalar field, such as a non-uniform field of chemical concentration, C, contains directional information – the gradient of C denoted by the vector field grad C. Even if only the non-directional scalar information is transferred from the stimulus field to the response field, the response field can contain the same non-uniformities as the stimulus field and hence it can contain the same directional information as the stimulus field. Often the only way of deciding this question of what sort of information is being transferred is by experiment. But if the gradient is linear, i.e. if grad C is uniform, then any non-uniformity of the response field indicates that at least some scalar, positional information has been transferred. Conversely, if the response field is uniform but contains directional information, then pure directional information has been transferred. It is only when we wish to know whether directional information is being transferred along with positional information that further experiment is required. For this, it is necessary to study how behaviour depends on the scalar quantity in uniform environments and hence to predict, using theory or modelling, the response field that would result from a transfer of scalar information from a non-uniform stimulus field.

Any discrepancy between prediction and experiment would then indicate that directional information has also been transferred.

As discussed previously, a traditional adapting kinesis mechanism and a two-receptor taxis mechanism can lead to very similar behaviour patterns in a gradient. Viewed in the light of information transfer, this is because the information available to the two types of mechanism can be very similar. Consider two highly hypothetical organisms in a stable, linear gradient. One compares the absolute concentration signals from two receptors, one on its front and one at its rear, separated by a distance d. The information, C_1–C_2, that it obtains can be shown to be equal to $d.\cos(\Theta).|\mathrm{grad}(C)|$- where Θ is the angle between its direction of travel and the direction of the gradient. The other has a single receptor that detects dC/dt. The information that it receives is equal to $s.\cos(\Theta).|\mathrm{grad}(C)|$ where s is its speed of travel. Neither organism receives any information about the absolute C in its local environment. In a linear gradient, neither can 'know' where it is in the gradient because $|\mathrm{grad}(C)|$ is constant throughout the environment. If d and s are also both constant, then the signal depends only on $\cos(\Theta)$ in both cases. The dependent properties of the behaviour pattern could be identical in both cases and, since only directional information is transferred, both organisms are showing a taxis.

In summary, Doucet and I advocate that the distinction between a kinesis and a taxis depends on whether positional or directional information respectively is transferred from the stimulus field to the response field. Thus it does not depend on the stimulus, or on the response or on the mechanism but on a global view of the information contained in the environment and in the whole pattern of behaviour. For example, an organism whose response is to vary its speed according to a dC/dt stimulus can show either a kinesis or a taxis depending on whether the environment is uniform or not. Also, this distinction does not depend critically on how position is defined. An analysis of the pattern of behaviour will usually reveal whether or not directional information has been transferred regardless of the exact method of assigning position. Transfer of directional information might also be detected by examining the orientation of organisms or the axes of maximum asymmetry of cells. Whether such a transfer is called a taxis or merely evidence for a taxis depends on whether we wish to restrict this term to a description of translocative behaviour.

REFERENCES

Brown, A. F. & Dunn, G. A. (1989). Microinterferometry of the movement of dry matter in fibroblasts. *Journal of Cell Science*, **92**, 379–89.

Doucet, P. G. & Dunn, G. A. (1990). Distinction between kinesis and taxis in terms of system theory. *Lecture Notes in Biomathematics* (in press).

Dunn, G. A. & Brown, A. F. (1990). Quantifying cellular shape using moment invariants. *Lecture Notes in Biomathematics* (in press).

Fraenkel, G. S. & Gunn, D. L. (1940). *The Orientation of Animals: kineses, taxes and compass reactions*. Oxford University Press (Dover edition 1961).

Keller, H. U., Wilkinson, P. C., Abercrombie, M., Becker, E. L., Hirsch, J. G., Miller, M. E., Ramsey, W. Scott & Zigmond, Sally H. (1977). A proposal for the definition of terms related to the locomotion of leucocytes and other cells. *Clinical and Experimental Immunology*, **27**, 377–80.

STRATEGIES FOR CHEMOTAXIS

M. J. SCHNITZER*, S. M. BLOCK†, H. C. BERG†, E. M. PURCELL*

Departments of Physics*, and Cellular and Developmental Biology†, Harvard University, Cambridge MA 02138, and The Rowland Institute for Science†, Cambridge MA 02142, USA

INTRODUCTION

The problem of migration of cell or animal populations has often been formulated in terms of a differential equation containing terms for both diffusion and drift. Controversy has arisen over the correct form of this equation when the diffusion coefficient varies over a region of space. Here we show, from a model for a one-dimensional random walk, that different microscopic mechanisms for variation in the diffusion coefficient require distinct differential equations. From these derivations we argue, in particular, that if behaviour is determined locally (i.e. does not depend on earlier events) and speed is constant, organisms cannot actively accumulate. They do so only where speeds are low. However, spatial variations in diffusion coefficent will affect the way in which cells approach equilibrium. One has to be precise about the conditions imposed by a given experimental set-up.

Earlier work in which these microscopic mechanisms were not properly distinguished includes that of Patlak (1953), Keller & Segel (1971), and Lapidus (1980, 1981). More recent work in which these distinctions have been recognized, explicitly or implicitly, includes that of Futrelle (1982) and Rivero *et al*. (1989).

To be certain of our ground, we modelled the behaviour by Montecarlo simulation. In the interest of understanding wild-type *Escherichia coli* and mutants defective in adaptation (*cheR cheB* mutants), we included mechanisms in which turning frequency depends on a measurement of concentration made over the recent past, or in a comparison of two such measurements made sequentially. If speed is constant, we find that cells will drift up gradients of attractants only if they are able to make temporal comparisons.

Students of animal behaviour use a different terminology (cf. Diehn *et al.*, 1977). A strategy in which speed is constant but turning frequency depends on the local intensity of a stimulus is called 'klinokinesis without adaptation'. In this case, the equilibrium distribution is uniform. A strategy in which speed is constant but in which turning frequency depends on temporal comparisons of stimulus intensity is called 'klinokinesis with adaptation'. In this case, if turns are suppressed when organisms move up a gradient, they will accumulate near the top of that gradient. A strategy in which turning frequency is constant but in which speed depends on the local intensity of a stimulus is called 'orthokinesis'. Here, organisms tend to accumulate where speeds are low.

Clearly, the term 'bacterial chemotaxis' is a misnomer. The phrase was coined by Pfeffer at a time when he believed that bacteria could steer directly toward the source of a chemical attractant (Pfeffer, 1884; reviewed by Berg, 1975). Molecular biologists have been more interested in understanding the genetics and biochemistry of chemotaxis (i.e. of chemoklinokinesis with adaptation) than in being scrupulous about nomenclature. So the name has stuck. It is commonly used to refer to any directed movement of bacteria towards, or away from, chemicals, regardless of the underlying mechanism.

ANALYTICAL TREATMENT

We were led to a protocol for deriving diffusion equations by the following thought experiment. Fill a long, capped pipe with an ideal gas at a low pressure, so that the mean-free path of a molecule is comparable to the length of the pipe. Keep the pipe at constant temperature, so that the mean speed of a molecule is everywhere constant. Now, add a number of solid objects to the pipe, more at one end than at the other. Collisions with these obstructions will increase the turning frequency of the gas molecules. Will this lead to an increase in the mean number of molecules per unit volume of free space at one end of the pipe as compared to the other? Any such accumulation would increase the local pressure of the gas. If so, we ought to be able to harness the pressure difference between the two ends of the pipe in order to do external work and, thereby, achieve perpetual motion. So, naïvely, we would not expect molecules to accumulate in regions of enhanced turning frequency.

Consider the following one-dimensional analogue. Let an ensemble of non-interacting particles move parallel to the axis of a long pipe. Distribute a series of semi-permeable barriers along that axis. Half of the surface of each barrier is covered with openings, so that whenever a particle reaches a barrier, it has an equal probability of bouncing back or continuing in the original direction. In either event (by construction) it adopts the speed characteristic of the local region, as defined in Fig. 1. In addition, we assume that the concentration of particles, C, in particles per unit volume, changes slowly enough over space that we can take it to be constant over a region of length δ. In other words, $\delta(\partial C/\partial x) \ll C$. We also assume that, when δ and τ vary over space, they do so slowly enough that we can take each of these quantities to be constant over a region of length δ. In order words, $\delta(\partial \delta/\partial x) \ll \delta$ and $\delta(\partial \tau/\partial x) \ll \tau$. These assumptions are made so that we can neglect higher-order terms in our derivations.

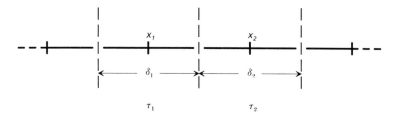

Fig. 1. A series of semi-permeable barriers (shown as vertical dashed lines) distributed along the x-axis, dividing the space into region 1, region 2, etc. The distance between the barriers adjacent to the point x_1 is δ_1 and the travel time between these barriers (in either direction) is τ_1.

Now, returning to Fig. 1, what is the particle flux, J, across the barrier separating region 1 and region 2? Half of the particles in region 1 are moving toward region 2, and half of the particles in region 2 are moving toward region 1, at velocities of δ_1/τ_1 and δ_2/τ_2, respectively. Half of the particles in each group are destined to pass through the barrier and half to be reflected. Thus, the net number of particles per unit area and unit time that cross the barrier from left to right (in the positive x-direction) is:

$$J = -(1/4)[C(x_2)(\delta_2/\tau_2) - C(x_1)(\delta_1/\tau_1)]. \qquad (1)$$

The simplest case is one in which the distance between barriers, δ, and the travel time between barriers, τ, are constant. In this case, $\delta_1 = \delta_2 = \delta$, $\tau_1 = \tau_2 = \tau$, and

$$J = -(\delta/4\tau)[C(x_2) - C(x_1)], \qquad (2)$$

which can be written

$$J = -(\delta^2/4\tau)[C(x_2) - C(x_1)]/\delta. \qquad (3)$$

Since C changes slowly over space, $[C(x_2) - C(x_1)]/\delta \approx \partial C/\partial x$, and we arrive at Fick's First Equation:

$$J = -D(\partial C/\partial x), \qquad (4)$$

with the diffusion coefficient $D = \delta^2/4\tau$. [*Note:* in a derivation in which particles at region x_n jump every τ seconds with probability 1/2 to region x_{n-1} and with probability 1/2 to region x_{n+1}, $D = \delta^2/2\tau$ (cf. Berg, 1983). The difference of a factor of 2 arises because in the barrier model, half of the particles in region x_n are destined to remain in that region after an interval τ. But this does not change any of the basic physics.] Note also that if particles had been subjected to an external force, eqns 1–4 would have contained an additional term of the form Cv_{drift}, where v_{drift}

is the drift velocity generated by that force. The equations, as they stand, refer only to the flux due to random motility.

To see what happens at equilibrium, we cap the pipe with reflective lids and wait for the particles to move around until $J=0$. Then by eqn 4, $\partial C/\partial x=0$. This implies that at equilibrium, C is uniform. There is no accumulation.

Let us generalize this derivation to allow for situations in which D changes over space. We can effect this change in four ways. *Case 1*: Vary the barrier interval, δ, and the travel time, τ, but keep the speed, δ/τ, constant. *Case 2*: Keep the barrier interval, δ, constant, but vary the travel time, τ. (The speed is not constant.) *Case 3*: Vary the barrier interval, δ, but keep the travel time, τ, constant. (The speed is not constant.) *Case 4*: Vary the barrier interval, δ, the travel time, τ, and the speed, δ/τ.

Case 1 (δ/τ constant)

We reach eqn 1, as before. Since the ratio δ/τ is constant, eqn 1 factorises to give eqn 2. With the assumption that δ varies slowly over space, eqns 3 and 4 follow, with $D=\delta^2(x)/4\tau(x)$. Note that D varies with position. When $J=0$, $\partial C/\partial x=0$, as before, and the equilibrium distribution is uniform. It does not matter how we arrange the barriers: the equilibrium distribution is uniform, provided that the speed is constant. This is the conclusion that we reached earlier in our thought experiment. But here we have said nothing about ideal gases: the proposition holds equally well for self-propelled objects, such as bacteria.

Case 2 (δ constant)

Eqn 1 now factors to give

$$J = -(\delta/4)[C(x_2)/\tau_2 - C(x_1)/\tau_1]. \tag{5}$$

To distinguish contributions to J due to spatial variations in C and τ, we add and subtract (within the brackets) $C(x_2)/\tau_1$ and group terms:

$$J = -(\delta/4)\{C(x_2)[1/\tau_2 - 1/\tau_1] + [C(x_2) - C(x_1)]/\tau_1\}. \tag{6}$$

This can be written

$$J = -(\delta^2/4)\{C(x_2)(1/\tau_2 - 1/\tau_1)/\delta + (1/\tau_1)[C(x_2) - C(x_1)]/\delta\}. \tag{7}$$

By our earlier assumptions, $(1/\tau_2 - 1/\tau_1)/\delta \approx \partial(1/\tau)/\partial x$, and $[C(x_2) - C(x_1)]/\delta \approx \partial C/\partial x$. Since C and t are slowing varying:

$$J = -(\delta^2/4)\{C[\partial(1/\tau)/\partial x] + (1/\tau)(\partial C/\partial x)\}. \tag{8}$$

But this is just

$$J = -D(\partial C/\partial x) - C(\partial D/\partial x) = -\partial(DC)/\partial x, \tag{9}$$

with $D=\delta^2/4\tau(x)$. Now, when $J=0$, DC is constant, and C is inversely proportional to D. The particles accumulate where the travel time is large, i.e. where the speed, δ/τ, is low.

Case 3 (τ constant)
Eqn 1 now factors to give

$$J = -(1/4\tau)[C(x_2)\delta_2 - C(x_1)\delta_1]. \tag{10}$$

By adding and substracting (within the brackets) $C(x_2)\delta_1$ and grouping terms we get

$$J = -(1/4\tau)\{C(x_2)(\delta_2 - \delta_1) + [C(x_2) - C(x_1)]\delta_1\}. \tag{11}$$

Proceeding as before,

$$J = -(\delta/4\tau)\{C(x_2)(\delta_2 - \delta_1)/\delta + \delta_1[C(x_2) - C(x_1)]/\delta\}, \tag{12}$$

$$\approx -(\delta/4\tau)[C(\partial\delta/\partial x) + \delta(\partial C/\partial x)], \tag{13}$$

or, given $D = \delta^2(x)/4t$,

$$J = -D(\partial C/\partial x) - C(\partial D/\partial x)/2. \tag{14}$$

Now, when $J = 0$, $D^{1/2}C$ is constant, and C is inversely proportional to $D^{1/2}$.

Case 4 (δ, τ, δ/τ all varying)
By adding and subtracting terms within the brackets of eqn 1 and approximating derivatives, as before, we obtain,

$$J = -(\delta/4)\{C\delta[\partial(1/\tau)/\partial x] + (C/\tau)(\partial\delta/\partial x) + (\delta/\tau)(\partial C/\partial x)\}. \tag{15}$$

This equation cannot be formulated in terms of a diffusion coefficient, but one can show that the equilibrium particle density varies inversely with the speed. In fact, in all four cases, the flux equation can be rewritten as

$$J = -(\delta/4)[v(\partial C/\partial x) + C(\partial v/\partial x)], \tag{16}$$

where v is the speed, δ/τ. If v is constant, the equilibrium particle density is uniform. If v varies, the equilibrium particle density is inversely proportional to v. If, in this situation, one starts with a uniform distribution, then particles will drift until their concentration becomes inversely proportional to v. By eqn 16, the drift velocity is $-(\delta/4)(\partial v/\partial x)$; if v increases with x, the particles drift in the $-x$ direction. For cases 2 or 3 (eqns 9 or 14) this drift velocity is $-\partial D/\partial x$ or $-(\partial D/\partial x)/2$, respectively.

Eqn 4 has been adopted by a number of workers in the field of bacterial chemotaxis. Lapidus (1981) derived eqn 9. Futrelle (1982) derived eqn 14 and noted the different equilibrium distributions expected from eqns 4, 9, and 14.

It is known that the way in which *E. coli* modulates its behaviour is by extending runs (relatively straight segments of its track) when the direction of travel is favourable; changes in swimming speed are small (Berg & Brown, 1972). Therefore, the present analysis implies that *E. coli* cannot do chemotaxis solely on the basis of local cues: it must monitor concentration over a finite interval of time. This also is required by counting statistics: a cell cannot measure the concentration of chemicals in its environment without taking a reasonable period of time. The standard deviation in a count is proportional to the square-root of the count, and the counting rate is limited by diffusion of chemicals in the vicinity of the cell (Berg & Purcell, 1977). On the other hand, tumbles are not instantaneous, so, if a cell tumbles frequently enough, its average speed will fall. Eqn 16 implies that, in the absence of other stimuli, such cells will tend to accumulate where the average speed is small. For wild-type cells, this is a minor effect. It will not be considered further here.

OTHER THEORIES

It is worth noting where we disagree with earlier work. Patlak (1953) presents a generalized formulation of the random walk. In addition to allowing for the possibility of a preferred direction, he assumes that there are distributions of possible values for δ, τ, and v, all of which depend on position and time. This leads to an elaborate diffusion equation. In examining the limit in which there is no preferred direction and no correlation between runs, Patlak arrives at a simpler equation, $\partial C/\partial t = \nabla^2(DC)$, which is equivalent to $J = -\nabla(DC)$, the three-dimensional analogue of our eqn 9. Unfortunately, in making this simplification, he inadvertently holds the mean run length constant (by setting the deviation from the mean run length equal to zero, p. 329). As we have seen (*Case 2*) the equation $J = -\partial(DC)/\partial x$ is not valid in cases in which the mean run length varies while the speed remains constant. Patlak and those who rely heavily on his work (e.g. Okubo, 1980; Doucet & Wilschut, 1987; Turchin, 1989) believe incorrectly that simply altering the frequency of turning as a function of position without changing speed can result in accumulation. One encounters misleading phrases such as 'the well-known phenomenon of trapping ... by enhancement of the turning frequency' (Alt, 1980) or the 'flypaper effect' (Stock & Stock, 1987).

Keller & Segel (1971) derive an equation in which the flux due to random motility is given by Fick's Equation (with $D = \mu$, which they call the mobility) and the flux due to chemotaxis is given by a drift velocity times the concentration (with drift velocity equal to the product of a chemotaxis coefficient, χ, and the slope of the gradient of the chemoattractant). In their derivation, the step time varies with the concentration of the attractant but the step length is held constant, so they have, in fact, assumed that the speed changes as a function of position (*Case 2*). As a consequence, the drift velocity is the gradient of the diffusion coefficient, in disguise. Sight is lost of this in later applications, where the mobility and chemotactic coefficients are treated as phenomenological parameters (e.g. Nossal, 1980). Segel (1984) considers it a matter of taste whether or not one regards $-(\partial D/\partial x)$ as an effective drift, 'indeed not brought about by intrinsic directional preferences at point x, but rather due to spatial differences in the vigor of random motion'. Our point is that one must be precise in defining what is meant by the word 'vigor'. The consequences of varying δ or τ, or both, are not the same (*Cases 2–4*), and if δ/τ happens to be constant, the effective drift vanishes.

Lapidus (1981) derives a diffusion equation from a random walk with a constant step length and a stepping probability per unit time that varies spatially. Within the limit that steps in either direction are equally probable, his equation reduces to $\partial C/\partial t = \partial^2(DC)/\partial x^2$, which is equivalent to our eqn 9. Lapidus concludes that 'it is clear that cells aggregate at those places where D is small'. However, if the step length is held constant while the stepping probability per unit time varies spatially, then the cell's speed will also vary spatially. Lapidus's mistake is assuming that this particular flux equation is generally applicable; aggregation will not occur when the speed is constant.

In earlier work, Lapidus (1980) assumes that the diffusion coefficient is a function only of the local concentration of a chemical attractant and also that Fick's Equation (eqn 4) is valid. He finds numerically that cells, initially distributed far from equilibrium, transiently shift towards regions in which the diffusion coefficient is large (i.e. move up the gradient of a substance that suppresses tumbles). However, this 'pseudochemotaxis', as he calls it, fades with time, and the distribution becomes uniform. This analysis is correct. But once again, it is not generally applicable. It applies only when the speed is constant.

The moral of all this is that different assumptions about the microscopic behaviour of particles or cells require distinct diffusion equations. One cannot claim, for example, that $J = -\partial(DC)/\partial x$ and then assert that D varies over space solely because of differences in turning frequency. Nor can one claim that $J = -D(\partial C/\partial x)$, and then assert that D varies because of differences in speed.

Recent workers have been more careful. Rivero *et al.* (1989) derive a flux equation (their eqn 14) in which the dependence on turning frequency and speed are made explicit. In the absence of chemotaxis and at constant speed, their equation reduces to Fick's Equation, as it should.

COMPUTER EXPERIMENTS

Some of the assumptions on which our derivations are based, notably that all parameters change slowly over space, might be restrictive. Are there difficulties that arise from neglected higher-order terms? In the interest of uncovering any such problems and extending the analysis to situations that are difficult to solve analytically, e.g. to bacterial chemotaxis, we developed a series of Montecarlo simulations.

Equilibrium distributions

In the first simulation, a probability per step that a tumble will occur is assigned to every point in a one-dimensional lattice, either with reflecting or periodic boundary conditions. When the cell arrives at a lattice point, a random number is selected, uniformly distributed between 0 and 1. If the value is less than the tumble probability, the next step is taken in the opposite direction; otherwise, it is taken in the same direction. (Tumbles could just as well have been defined so that the cell sets off with equal probability in either direction, rather than reversing, but this would merely double the mean run length without changing any essential feature.) Before the computer writes anything down, the cell is allowed to walk along the lattice for a number of steps which is large compared to that required for an ensemble of such cells to approach the equilibrium distribution. After this initialization, the program is allowed to run for a much larger number of steps, while the computer records the number of visits to each lattice point. We assume that a collection of non-interacting cells will distribute themselves at equilibrium with concentrations proportional to these visitation numbers (i.e. to the fraction of time that the cell spends in a given region of space).

Case 1 (speed constant)

After many trials with lattices of different sizes, with both reflecting and periodic boundary conditions, and with different probabilities of tumbling, we found no evidence that cells moving at constant speed accumulate in regions of higher tumble probability. An example with reflecting boundary conditions is shown in Fig. 2(a). In the analytical derivation for this case, the tumble probability is 1/2. Here, it can be much smaller, so that a cell can continue for many lattice points before being reflected. This allows us to probe for changes in behaviour over distances which are short compared to the mean run length. If the tumble probability is approximately constant, runs are distributed exponentially, with a mean length equal to the reciprocal of the tumble probability per unit length, as is the case for *E. coli* (Berg & Brown, 1972).

Cases 2–3 (variable speed)

The essential information provided by the computer simulations just described is that changes in tumble probability have no effect upon the number of visits to each lattice point (the total amount of time per unit length spent in a given region of space). All that the computer does is to select a random number when the cell arrives at a given lattice point and to record the visit. The same process would be carried out in a simulation in which the lattice spacing or the step time varied. For example, if in one region the lattice spacing were doubled and the step time were held constant (i.e. if the speed were doubled), the number of visits to the lattice points would remain the same, but the number of visits per unit length would halve. Thus, the equilibrium particle density would halve. If the lattice spacing were held constant and the step time were halved (again, if the speed were doubled) the time spent in a given region would halve. Thus, the equilibrium particle density would halve. A cell that changes its speed as a function of position will spend a total amount of time in a given region of space that is inversely proportional to the local speed. Thus, we arrive at the distribution predicted by eqn 16 (and by inference eqns 9 and 14).

Measurements extending over time

In these simulations, we follow the strategy used in the first simulation but assign the tumble probability for a given lattice point in a manner that depends upon the cell's past history, in particular, upon the concentrations of an attractant that the cell has recently measured. The weightings are approximations to those determined from measurements of responses of tethered cells to impulsive stimuli (cf. Segall *et al.*, 1986). We use reflecting boundary conditions. To simulate the behaviour of wild-type cells, we assign a weighting factor $1/N$ to the concentrations sensed at the previous N lattice points, and a weighting factor $-1/M$ to the concentrations sensed at the M lattice points before that, and take the sum. This is equivalent to taking the difference between two sequential averages, and approximates to a derivative operator with memory. If the cell has been moving up the gradient, this sum is positive. If so, we reduce the probability of a tumble from a baseline value by an amount (gain × sum). The gain is picked so that a cell moving steadily up the gradient still has a finite probability of tumbling. If the cell has been moving down the gradient, the sum is negative. In this event, the tumble probability is set to the baseline value. Thus, runs are lengthened as a consequence of favourable measurements but not shortened as a consequence of unfavourable ones (cf. Brown & Berg, 1974). In simulating the behaviour of cells that do not adapt (e.g. *che*R *che*B mutants, cells that cannot methylate or demethylate their membrane transducers; cf. Segall *et al.*, 1986) we assign a weighting factor $1/N$ to the concentrations sensed at the previous N lattice points and take the sum. As before, we reduce the probability of a tumble from the baseline value by an amount (gain × sum), where the gain has a smaller value

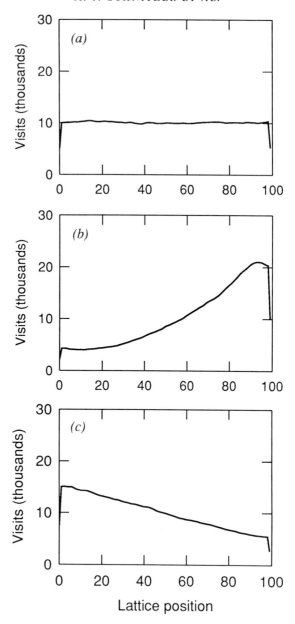

than before. Once again, the gain is chosen so that cells near the top of the gradient have a finite probability of tumbling. Thus, tumbles are suppressed on the basis of measurements made over the previous N lattice points, but temporal comparisons are not made. The results of these simulations are shown in Figs. 2(b) and 2(c). Wild-type cells drift up the gradient, Fig. 2(b), while cells that fail to make temporal comparisons drift down, Fig. 2(c).

Approach to equilibrium

These simulations are done in the same way, except that we start cells at the centre of a lattice and record where they end up after a number of steps, n. The specifications for setting the tumble probabilities are the same as before. Fig. 3 shows a control experiment, in which the tumble probabilities are constant. Figs 4, 5, and 6 refer to the same experiments as Figs 2(a), 2(b), and 2(c), respectively. In Fig. 3 the cells spread symmetrically (the distribution is a Gaussian). In Fig. 4 they spread more rapidly in the direction in which tumbles are suppressed (up the gradient), and then they relax to a uniform distribution. In Fig. 5 the cells drift up the gradient, and in Fig. 6 they drift down.

OTHER COMPUTER EXPERIMENTS

Other simulations have been run in which organisms move at constant speed and modulate their turning frequencies. Some workers reach conclusions that appear similar to ours and others do not (Rohlf & Davenport, 1969; Van Houten & Van Houten 1982; Doucet & Drost, 1985; Doucet & Wilschut, 1987; and Dusenbery, 1989). One possible source of

Fig. 2. Equilibrium distributions for cells moving at constant speed. (a) Tumble probabilities defined locally (pseudochemotaxis). The tumble probability per step varied linearly from 0.3 near the left reflecting boundary (position 0) to 0.03 near the right reflecting boundary (position 99). The tumble probabilities at the boundaries were set equal to 1. At lattice point 50 the tumble probability was 0.135, corresponding to a mean run length of 7.4 steps. One million steps were logged. A similar distribution was obtained from a computation in which the tumble probability was constant (0.135 everywhere; data not shown). A portable pseudorandom number generator was used that is known to be free from sampling artifacts (Press *et al.*, 1988). A different seed was used for each figure shown in this paper. (b) Tumble probabilities defined on the basis of the difference between two sequential averages (real chemotaxis). The number of lattice points used for the two averages was $N = 7$ and $M = 21$. The baseline probabability was 0.135. The gain was chosen so that a cell moving steadily up the gradient (toward the right) had a tumble probability 0.03. See the text. One million steps were logged. (c) Tumble probabilities defined on the basis of a single average (no adaptation). The same gradient was sensed as in the previous computation. The number of lattice points used for the average was $N = 7$. The baseline probability was 0.135. The gain was chosen so that a cell moving near the right reflecting boundary had a tumble probability 0.03. Note that the tumble probability depends on position; for a cell moving near the left boundary, it was 0.111. See the text. One million steps were logged.

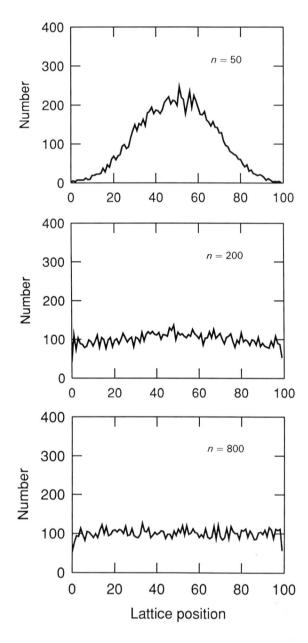

Fig. 3. Approach to equilibrium at constant tumble probability (simple diffusion). Ten thousand cells were released at lattice point 50 and allowed to step for $n = 50$ times (top), 200 times (middle), or 800 times (bottom). The number of cells ending up at each lattice point was recorded. The tumble probability was 0.135 at all points. Note the symmetrical spreading.

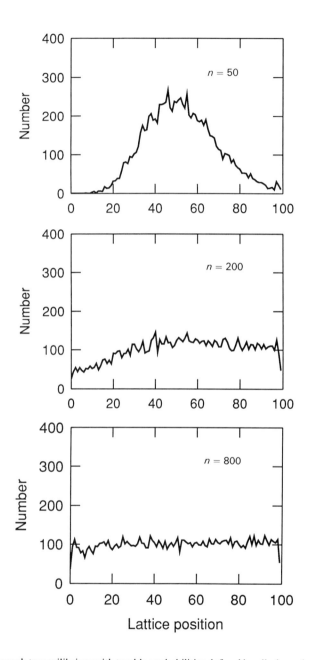

Fig. 4. Approach to equilibrium with tumble probabilities defined locally (pseudochemotaxis). The computation was carried out as in Fig. 3 with the tumble probabilities set as in Fig. 2(a). Note the feint to the right (in the direction of lower tumble probability).

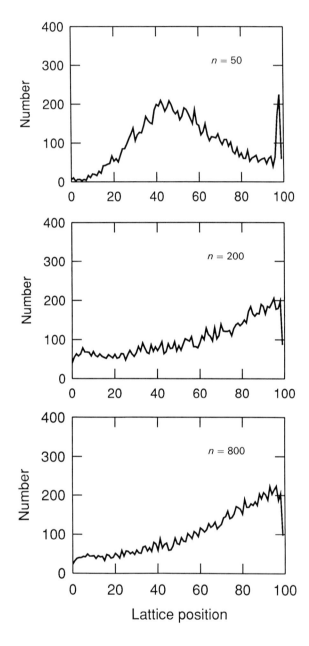

Fig. 5. Approach to equilibrium with tumble probabilities defined on the basis of the difference between two sequential averages (real chemotaxis). The computation was carried out as in Fig. 3 with the tumble probabilities set as in Fig. 2(b). Note the progressive shift to the right (up the gradient of an attractant).

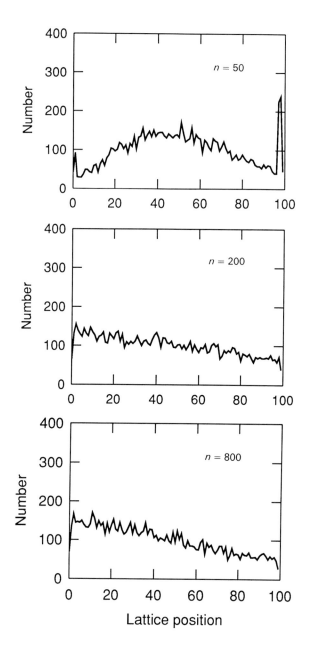

Fig. 6. Approach to equilibrium with tumble probabilities defined on the basis of a single average (no adaptation). The computation was carried out as in Fig. 3 with the tumble probabilities set as in Fig. 2(c). Note the progressive shift to the left (down the gradient of an attractant).

disagreement is made particularly clear by Figs 4 and 2(a), namely, that, if one follows the displacements of cells for a limited period of time, the conclusions reached might well be different than if one were to determine equilibrium distributions. This is why Lapidus (1980) coined the term 'pseudochemotaxis'. Note that the transient movement in Fig. 4 is up the gradient, i.e. in the direction in which tumbles are suppressed. If one adds sources and sinks and computes fluxes, the differences can appear even more dramatic. For example, if, in the experiment of Fig. 4, the boundaries at positions 0 and 99 are made absorbing, the fraction of cells arriving at position 99 is substantially larger than that arriving at position 0 (by a factor of 2.7). But this merely reflects the fact the the diffusion coefficient is substantially larger over the right half of the figure than over the left. It does not tell us what the equilibrium distribution might be.

DISCUSSION

Patterns of accumulation in bacteria resulting from changes in swimming speed are well known. Vivid descriptions have been given by Metzner (1920; reviewed by Berg, 1975), who watched cells of *Spirillum volutans* as they responded to chemicals placed near the edge of a coverslip. In some cases, the cells became trapped, with their flagellar bundles spinning in a head–head or tail–tail configuration (i.e. working against one another); in other cases the cells shuttled back and forth rhythmically a body length or less, or stopped moving altogether. Clayton (1957) analysed such patterns quantitatively, both in theory and experiment, using the phototactic organism *Rhodospirillum rubrum*. A distinction between effects arising from modulation of turning frequency, as opposed to those arising from modulation of swimming speed, was drawn by Gunn *et al.* (1937) and Fraenkel & Gunn (1940), who assigned to them the terms 'klinokinesis' and 'orthokinesis', respectively. The definition of klinokinesis was based on the work of Ullyott (1936), who studied the photoresponses of a flatworm, *Dendrocoelum lacteum*; however, this work did not stand the test of time (cf. Gunn, 1975). Fraenkel & Gunn argued (correctly) that organisms moving at constant speed could not accumulate (alter the shape of a uniform distribution) by modulating their turning frequencies, unless they could adapt (compare the stimulus in the present with that in the past). This conviction was undermined by the analysis of Patlak (1953), which we have found to be in error. However, as stressed in the previous section, variations in diffusion coefficient will affect the way in which cells approach equilibrium. One of the best ways to determine whether cells actively accumulate is to start with a uniform distribution and then to ask whether that distribution changes with time. This can be done experimentally with the layered-gradient assay (Dahlquist *et al.*, 1972; Weis & Koshland, 1988).

The results of Fig. 2 can be rationalized as follows. Imagine yourself at an arbitrary lattice point watching cells arriving from the left or the right. In (a) the probability that a cell will tumble does not depend on whether it has arrived from the left or the right, so the same fraction of cells of either type will cross the boundary. At equilibrium, when the net flux is zero, an equal number of cells must arrive from the left or the right. Therefore, their distribution must be uniform. In (b) the probability of a tumble is smaller when a cell arrives from the left than when it arrives from the right, because cells have been taking derivatives with respect to time, and some have been going up the gradient, while others have been going down. More of the cells that arrive from the left will cross the barrier than those that arrive from the right. Therefore, at equilibrium, there must be fewer cells on the left than on the right. In (c), the situation is reversed, because the probability that a cell will tumble is larger if it has arrived from the left, where the concentration of the tumble-suppressing substance is smaller, than if it has arrived from the right. Therefore, at equilibrium, there must be more cells on the left than on the right.

We began this work with the hope of learning whether the weighting function used by *E. coli* for making temporal comparisons (called the impulse response) reflected an optimum design (cf. Block *et al.*, 1982; Segall *et al.*, 1986; Berg, 1988). *En route*, we became aware that there was a great deal of confusion about the success of more rudimentary strategies. These became the main subject of the present paper. In the linear, noiseless gradients of the sort considered in Figs 2(b) and 5, similar drift rates are obtained for weighting functions of the same total width, regardless of the relative size of the positive and negative lobes (assuming that the areas of the two lobes are the same, i.e. that the cells adapt). We know from an analysis of how cells count molecules that in the real world the lobes should have a width of the order of 1 second (Berg & Purcell, 1977). The first lobe cannot be much longer than this, because then several tumbles could occur before a decision were reached, and the results would not be relevant. In addition, the baseline tumble probability cannot be much lower than it is, because rotational Brownian movement carries a cell off course by as much as 90° within 10 seconds. In wild-type cells, the widths of the two lobes are about 1 and 3 seconds, respectively. Is the ratio of these values a matter of physics or biochemistry? One way to find out would be to extend the simulations by working in three dimensions, making the gradients lumpy rather than smooth, adding the fluctuations expected in counting molecules, and including the meandering due to rotational Brownian movement. Drift rates computed in this manner for cells using the wild-type impulse response in a smooth, noiseless gradient agree well with drift rates measured experimentally (Berg, 1988). A broader range of parameters remains to be explored.

CONCLUSIONS

We re-examined the problem of migration of motile organisms in spatial gradients of chemical attractants. We showed analytically and by Montecarlo simulation that organisms whose turning frequencies (tumble probabilities) depend solely on the local concentration of an attractant, but whose speeds remain constant, do not accumulate at the top of such a gradient: once uniformly distributed, they remain uniformly distributed. On the other hand, organisms whose swimming speeds depend on the local concentration of an attractant do accumulate in regions where the speeds are low. This analysis resolves a long-standing controversy in the literature that arose because different microscopic mechanisms for generating variations in the diffusion coefficient (a macroscopic parameter) were not properly distinguished. These mechanisms lead to distinct diffusion equations. We extended the Montecarlo simulation to non-local strategies and found that cells that respond (by suppressing tumbles) to concentrations of an attractant sensed over the recent past, but do not make temporal comparisons, drift down rather than up the gradient. Cells that compare concentrations sensed over the recent past with those sensed earlier are able to drift up the gradient. This is the strategy used by *E. coli* for chemotaxis.

ACKNOWLEDGEMENTS

We thank David Blair for helpful discussions. This work was supported by the Rowland Institute for Science.

REFERENCES

Alt, W. (1980). Biased random walk models for chemotaxis and related diffusion approximations. *Journal of Mathematical Biology*, **9**, 147–77.
Berg, H. C. (1975). Chemotaxis in bacteria. *Annual Review of Biophysics and Bioengineering*, **4**, 119–36.
Berg, H. C. (1983). *Random Walks in Biology*, p. 18. Princeton: Princeton University Press.
Berg, H. C. (1988). A physicist looks at bacterial chemotaxis. *Cold Spring Harbor Symposia on Quantitative Biology*, **53**, 1–9.
Berg, H. C. & Brown, D. A. (1972). Chemotaxis in *Escherichia coli* analysed by three-dimensional tracking. *Nature, London*, **239**, 500–4.
Berg, H. C. & Purcell, E. M. (1977). Physics of chemoreception. *Biophysical Journal*, **20**, 193–219.
Block, S. M., Segall, J. E. & Berg, H. C. (1982). Impulse responses in bacterial chemotaxis. *Cell*, **31**, 215–26.
Brown, D. A. & Berg, H. C. (1974). Temporal stimulation of chemotaxis in *Escherichia coli. Proceedings of the National Academy of Sciences, USA*, **71**, 1388–92.
Clayton, R. K. (1957). Patterns of accumulation resulting from taxes and changes in motility of micro-organisms. *Archiv für Mikrobiologie*, **27**, 311–19.
Dahlquist, F. W., Lovely, P. & Koshland, D. E., Jr (1972). Quantitative analysis of bacterial migration in chemotaxis. *Nature New Biology*, **236**, 120–3.

Diehn, B., Feinlieb, M., Haupt, W., Hildebrand, E., Lenci, F. & Nultsch, W. (1977). Terminology of behavioral responses of motile microorganisms. *Photochemistry and Photobiology*, **26**, 559–60.
Doucet, P. G. & Drost, N. J. (1985). Theoretical studies on animal orientation: II. Directional displacement in kineses. *Journal of Theoretical Biology*, **117**, 337–61.
Doucet, P. G. & Wilschut, A. N. (1987). Theoretical studies on animal orientation. III. A model for kinesis. *Journal of Theoretical Biology*, **127**, 111–25.
Dusenbery, D. B. (1989). Efficiency and the role of adaptation in klinokinesis. *Journal of Theoretical Biology*, **136**, 281–93.
Fraenkel, G. S. & Gunn, D. L. (1940). *The Orientation of Animals: Kineses, Taxes and Compass Reactions*. Oxford: Clarendon Press. Reprinted with additional notes in 1961, New York: Dover Publications.
Futrelle, R. P. (1982). Dictyostelium chemotactic response to spatial and temporal gradients. Theories of the limits of chemotactic sensitivity and of pseudochemotaxis. *Journal of Cellular Biochemistry*, **18**, 197–212.
Gunn, D. L. (1975). The meaning of the term 'klinokinesis'. *Animal Behaviour*, **23**, 409–12.
Gunn, D. L., Kennedy, J. S. & Pielou, D. P. (1937). Classification of taxes and kineses. *Nature, London*, **140**, 1064.
Keller, E. F. & Segel, L. A. (1971). Model for chemotaxis. *Journal of Theoretical Biology*, **30**, 225–34.
Lapidus, I. R. (1980). 'Pseudochemotaxis' by micro-organisms in an attractant gradient. *Journal of Theoretical Biology*, **86**, 91–103.
Lapidus, I. R. (1981). Chemosensory responses of microorganisms: asymmetric diffusion as the limit of a biased random walk. *Journal of Theoretical Biology*, **92**, 345–58.
Metzner, P. (1920). Die Bewegung und Reizbeantwortung der bipolar begeißelten Spirillen. *Jahrbücher für wissenschaftliche Botanik*, **59**, 325–412.
Nossal, R. (1980). Mathematical theories of topotaxis. *Lecture Notes in Biomathematics*, **38**, 410–39.
Okubo, A. (1980). *Diffusion and Ecological Problems: Mathematical Models*, Chpt. 5. Berlin, Heidelberg, New York: Springer-Verlag.
Patlak, C. S. (1953). Random walk with persistence and external bias. *Bulletin of Mathematical Biophysics*, **15**, 311–38.
Pfeffer, W. 1884. Locomotorische Richtungsbewegungen durch chemische Reize. *Untersuchungen aus dem botanischen Institut in Tübingen*, **1**, 363–482.
Press, W. H., Flannery, B. P., Teukolsky, S. A & Vetterling, W. T. (1988). *Numerical Recipes in C*, p. 210. Cambridge: Cambridge University Press.
Rivero, M. A., Tranquillo, R. T., Buettner, H. M. & Lauffenburger, D. A. (1989). *Chemical Engineering Science*, **44**, 2881–97.
Rohlf, F. J. & Davenport, D. (1969). Simulation of simple models of animal behavior with a digital computer. *Journal of Theoretical Biology*, **23**, 400–24.
Segall, J. E., Block, S. M. & Berg, H. C. (1986). Temporal comparisons in bacterial chemotaxis. *Proceedings of the National Academy of Sciences, USA*, **83**, 8987–91.
Segel, L. A. (1984). Taxes in cellular ecology. *Lecture Notes in Biomathematics* **54**, 407–24.
Stock J. and Stock, A. (1987). What is the role of receptor methylation in bacterial chemotaxis? *Trends in Biochemical Sciences*, **12**, 371–5.
Turchin, P. (1989). Beyond simple diffusion: models of not-so-simple movement of animals and cells. *Comments on Theoretical Biology*, **1**, 65–83.
Ullyott, P. (1936). The behaviour of *Dendrocoelum lacteum* II. Responses in non-directional gradients. *Journal of Experimental Biology*, **13**, 265–78.

Van Houten, J. & Van Houten, J. (1982). Computer simulations of *Paramecium* chemokinesis behavior. *Journal of Theoretical Biology*, **98**, 453–68.

Weis, R. M. & Koshland, D. E., Jr (1988). Reversible receptor methylation is essential for normal chemotaxis of *Escherichia coli* in gradients of aspartic acid. *Proceedings of the National Academy of Sciences, USA*, **85**, 83–87.

THEORIES AND MODELS OF GRADIENT PERCEPTION

ROBERT T. TRANQUILLO

Department of Chemical Engineering and Materials Science, University of Minnesota, Minneapolis MN 55455, USA

Nothing in life is certain except death and taxes.   *Benjamin Franklin*

Nothing in life is certain except death and *taxis*.   *Lee A. Segel*

Everything should be made as simple as possible, but not simpler.   *Albert Einstein*

The goal of understanding the mechanisms by which blood and tissue cells bias their direction of migration towards higher concentration of chemoattractants, that is, exhibit chemotaxis, has engaged many experimentalists and even a fair number of theoreticians in recent years. Motivation stems from both the crucial medical importance of chemotaxis in homeostasis and pathophysiology (generally accepted but not as yet decisively proven) and the intrinsic scientific fascination evoked by watching cells meander up chemoattractant gradients in the laboratory. A major thrust towards this goal has been provided by the development of sophisticated instruments and specialized probes, allowing many molecular details of the three basic components of chemotaxis, namely chemoattractant-receptor binding and dynamics, signal transduction, and the activation of motility, to be elucidated, in many cases in a quantitative fashion. Despite this explosion of molecular information, however, a basic understanding of how the cell translates the information contained in a chemoattractant gradient into directed movement, i.e. spatially and temporally integrates the receptor signals generated over its dimension into co-ordinated, yet alone directed, locomotion, is still lacking, although a number of plausible thinking models have been refined over time. The relative simplicity of these thinking models stands, however, in stark contrast to the emerging molecular complexity, and raises the question of how much of the complexity must our thinking models accommodate to enable that basic understanding? Closely related to the pursuit of this goal is identifying the source of inherent randomness in chemotaxis – why do cells meander rather than migrate directly up a gradient, occasionally even transiently migrating down the gradient? Mathematical models offer the possibility of dealing with both the complexity and randomness, and the main subject of this paper is to present a new model framework which does so.

First, the groundwork is provided. The polymorphonuclear neutrophil leucocyte (neutrophil) generally recognized as a model for chemotactic blood and tissue cells, is treated exclusively throughout. In Section 1, a brief summary of pertinent experimental data and observations and a discussion of salient mechanistic concepts for developing and validating any mathematical model of neutrophil chemotaxis is given. This is followed in Section 2 by a review of the theoretical concepts which have been proposed to explain how stochastic receptor-sensing of chemoattractant influences the direction of movement. It will become evident that the random directional component of chemosensory movement is exploited in mathematical modelling to elucidate the salient biochemical and biophysical mechanisms for which account must be made. The first generation models reviewed in Section 3 all share a common premise: that the direction of movement is determined by a spatial or temporal difference in receptor occupancy, which fluctuates in time because receptor binding is inherently a stochastic process. After discussing the limitations of these models, a second generation model formulated to overcome these limitations is reviewed in more detail in Section 4. It is again based on fluctuations in chemoattractant receptor binding, but, in contrast to the simpler premise above, the direction of movement is regulated by the spatio-temporal pattern of a chemoattractant receptor signal and integrated signal–response coupling. After discussing the useful results and ongoing extensions of the second generation model, a new third generation model is described and some preliminary results are discussed in Section 5. This third generation model removes the inherent limitation of the second generation models which do not predict extension and retraction of lamellipodia, phenomena which seem to be fundamental to an ultimate understanding of neutrophil chemotaxis.

1. ESSENTIAL DATA AND OBSERVATIONS FOR MODELLING NEUTROPHIL CHEMOTAXIS

The intention here is to summarize only some of the basic facts and common speculation suggested by the facts. The review by Devreotes & Zigmond (1988) is quite comprehensive in these respects. See also Wilkinson (this volume). To economize on space (and with apologies to many authors) reference is made mainly to this and other reviews, as appropriate.

Neutrophils are amoeboid cells which translocate by locomotion, the continuous extension of pseudopods which allow transmission of motile force to the environment. Pseudopod extension, usually in the form of flat lamellar projections termed lamellipodia, is one of a series of complex biochemical and biophysical responses elicited when chemoattractant binds to specific surface receptors, others being the development of behavioural and morphological polarity, enhanced adhesion, pinocytosis, endocytosis, exocytosis, and locomotion. Chemotaxis is the direct consequence of preferential pseudopod extension towards higher chemoattractant concentrations, as modulated by the cell polarity. It will become evident that all these responses can be integrated under the chemotaxis response to spatial chemoattractant gradients. However, many illuminating observations come from the simpler situation of subjecting neutrophils to spatially uniform chemoattractant concentrations in various modes. In what follows, it is assumed for simplicity that chemoattractant molecules binding to receptors are fluid-phase rather than substratum-bound, although the effects of the latter on chemoattractant-stimulated migration may be significant.

Uniform chemoattractant concentrations

In the absence of chemoattractants (and any other non-chemotactic stimulatory factors), a neutrophil is spherical and apparently quite homogeneous with respect to spatial distributions of molecular components. When subject to a step increase in chemoattractant concentration to some constant value in the range of the equilibrium dissociation constant for the chemoattractant–receptor complex, K_d, lamellipodia form randomly within seconds over the entire cell surface. After several minutes, the cell develops an anterior–posterior morphology as lamellipodia become localized to the anterior, or leading edge. This occurs both for neutrophils in suspension and on a surface. If the surface permits a suitable adhesive interaction, the cell commences locomotion, which continues for hours if the chemoattractant concentration is maintained (on two-dimensional substrata, adhesion of lamellipodia is a prerequisite; in three-dimensional fibrillar matrices, it is not). The polarity is behavioural as well as morphological, with new lamellipodia forming most frequently around the leading edge and only rarely at the trailing edge (i.e. the uropod). This is obviously essential for efficient locomotion; at chemoattractant concentrations much greater than K_d lamellipod formation is still frequent but there is little polarity and, consequently, locomotion is quite inefficient.

Neutrophils migrating in a uniform chemoattractant concentration are observed to translocate in the same general direction on a short time-scale (i.e. minutes), a phenomenon termed 'directional persistence'. On a slightly longer time-scale cells exhibit directional changes during translocation, which can vary from a continuous turn (i.e. the cell maintains its polar morphology during the directional change) to an abrupt redirection (i.e. the directional change is due to a morphological repolarization). On a long time-scale (i.e. hours), the cumulative result of many directional changes is a cell path with a meandering appearance that can be well described as a random walk. For neutrophils exhibiting random motility, the mean speed is known to depend on the chemoattractant concentration (orthokinesis), with typical values being 10–30 µm/min on a protein-coated surface (the typical speed of an extending pseudopod can be significantly greater). That neutrophil exhibit a concentration dependance of the random turning behaviour (klinokinesis) has not been definitively established because, only recently, has turning behaviour been characterized with objective indices (see Section 4).

Discontinuous temporal chemoattractant gradients

If the neutrophil is subject to a second step increase in chemoattractant concentration (e.g. from 0.01 K_d to 0.1 K_d), it transiently ceases locomotion and loses polarity as lamellipodia again form rapidly over the cell surface (the time-course and spatial extent of lamellipod formation depends on the two chemoattractant concentrations). Again, after a period of minutes, the neutrophil regains polarity and exhibits locomotion in a way almost indistinguishable from that in the first chemoattractant concentration, a phenomenon referred to as behavioural 'adaptation'. A working definition of adaptation is that it is a reversible time-dependent adjustment of sensitivity to the current chemoattractant concentration such that the response evoked by the change to the current chemoattractant concentration ceases (here, the response being nonlocalized lamellipodial activity). This, effectively, resets the sensitivity baseline so that the response can again be elicited when the cell encounters a higher chemoattractant concentration. Another important observation is the behavioural response to a step decrease in chemoattractant concentration. In this case, the neutrophil transiently ceases locomotion and loses polarity as lamellipodia are rapidly withdrawn from the leading edge and blebs form over the cell surface. However, once again after a period of minutes, adaptation occurs and a polarized cell exhibiting locomotion is observed. Observations of adaptation to temporal stepwise increases and decreases in chemoattractant concentration inspired the temporal mechanism of neutrophil chemotaxis, where lamellipodial activity is being determined by differential changes in chemoattractant concentration as the cell moves in a spatial chemoattractant gradient, and support the related pseudospatial mechanism subsequently proposed (see 'Thinking models' below).

Continuous temporal chemoattractant gradients

Behavioural adaptation is closely related to biochemical adaptation, the reversible time-dependent adjustment of concentrations of receptor-mediated second messengers and of motility-associated proteins. Using complementary fluorescence and flow cytometric methods, Sklar and co-workers (Sklar & Omann, 1990) have investigated a number of chemoattractant-mediated responses to continuous temporal gradients by infusing chemoattractant into a cell suspension at a prescribed rate. Intracellular Ca^{2+} changes, O_2^- production, and F-actin generation all occur over a longer time-course with smaller magnitude when the temporal gradient is decreased, i.e. the rate of adaptation approaches the rate of receptor-mediated stimulation and the ability of the cell to sense the temporal gradient diminishes. The same is observed for right-angle light scatter, which correlates with lamellipod formation. This again implies a critical relationship between the time-scales of adaptation and of chemoattractant concentration change which determines the ability of neutrophils to sense and respond to concentration changes when migrating in a spatial chemoattractant gradient.

Continuous spatial chemoattractant gradients

Chemotaxis is usually defined in terms of biased turning towards higher chemoattractant concentrations in a steady-state spatial chemoattractant gradient. Neutrophil chemotaxis has been almost exclusively characterized, to date, using chambers designed to yield quasi-steady spatial gradients during a certain window of time (i.e. the time during which the chemoattractant concentration change with time at a particular point is negligible compared to the change in space at that point). This circumvents the question of how a significant simultaneous temporal gradient influences the cell behaviour in a spatial gradient, but recent experiments have demonstrated that neutrophils orient similarly in spatial gradients on which a positive, zero, or negative temporal gradient is superimposed, implying that a spatial gradient is sufficient to elicit chemotaxis (Lauffenburger *et al.*, 1987). Orientation is defined by the alignment of the cell polarity axis and, hence, the direction of movement with the direction of the chemoattractant gradient (see Dunn, this volume). Orientation is a consequence of the cell's biased turning behaviour and its measurement avoids the subjective definition of a turn in what is largely a continuous cell path. The best quantitative data on neutrophil chemotaxis is based on Zigmond's visual assay, where orientation is measured as a function of the chemoattractant absolute concentration and gradient steepness, which fully characterize the quasi-steady linear spatial gradient formed in the assay. Some of Zigmond's data for rabbit neutrophils orientating in gradients of N-formylated chemotactic peptides, a class of chemoattractants which are analogues of bacterial metabolites, are reproduced here. Fig. 1(*a*) shows the percentage of polarized cells with orientation towards higher chemoattractant concentration ('correct' orientation, O_C) as a function of chemoattractant concentration for a fixed gradient of fixed steepness $\varepsilon = 1.6\%$. The parameter ε, ($\Delta/C \cdot 100$), is the fractional concentration difference across the effective cell diameter (taken here as 10 μm), where Δ is the concentration difference and C is the chemoattractant concentration at the cell midpoint. Fig. 1(*b*) shows

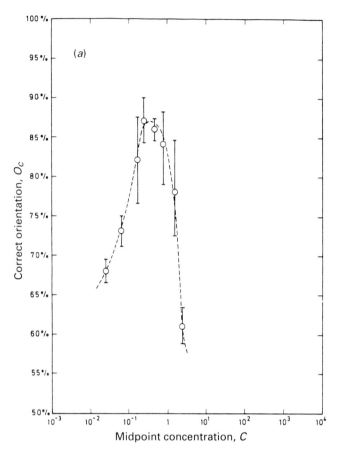

Fig. 1. Plots of experimental data from the visual orientation assay of Zigmond showing the percentage of cells with 'correct' orientation, O_C, towards higher chemoattractant concentration vs (a) C for a fixed ε (= 1.6%), and (b) ε for a fixed C (= K_d) for rabbit neutrophils responding to (a) FNLLP gradients (Zigmond, 1981), and (b) FMMM gradients (Zigmond, 1977). C is the concentration at the cell midpoint (scaled to the K_d value) and ε is the steepness of the gradient across the cell (see text for definition).

the dependence of O_C as a function of ε for a fixed $C = K_d$. Of note is that the neutrophil exhibits significant orientation bias ($O_C = 75\%$) for $\varepsilon \approx 1\%$.

Time-lapse microscopic examination of the cells shows that chemotactic movement is not fundamentally different from random motility. The cells exhibit the same polar morphology in both cases, the main difference being the preferential extension of lamellipodia from the leading edge in the direction of higher chemoattractant concentration. This suggests that the biased turning in chemotaxis derives mainly from a difference in the spatiotemporal pattern of receptor signals which activate the motile mechanisms

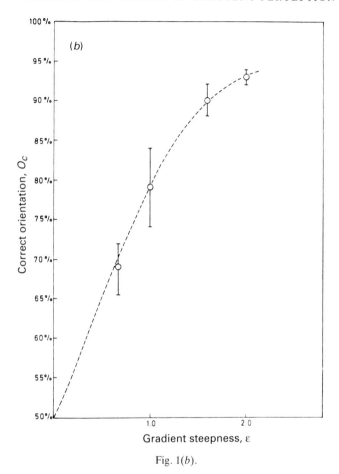

Fig. 1(b).

and not from an alteration in the intrinsic motile mechanisms. Attention is now briefly turned to receptor signal generation and motility activation.

Receptor binding and dynamics

Although the chemotactic peptide receptors appear monovalent, they might not be homogeneous, and they certainly do more than just reversibly bind chemoattractant molecules on the cell surface (Sklar & Omann, 1990). Obviously, it is essential to understand the mechanisms which determine the spatial distribution of receptors capable of transduction in order to understand neutrophil chemotaxis. The emerging picture is that upon binding a chemoattractant molecule, the mobile receptor–chemoattractant complex has at least three possible fates. It can simply dissociate the chemoattractant molecule, it can encounter a G-protein to form a ternary complex from which transduction proceeds (see below), or it can become cytoskeletally associated to a virtually non-dissociating receptor–chemoattractant complex, segregated from further G-protein activation and targeted for receptor-mediated endocytosis (internalization). The lifetime of the ternary complex, or 'signalling' form of the receptor, is very brief (seconds) compared to the receptor/chemoattractant complex (minutes). Since internalization occurs largely at the uropod, and free chemotactic peptide receptors appear to be at higher density on the cell anterior, the implication is that the free receptors on the surface represent a dynamic state involving (Singer & Kupfer, 1986): (1) binding and signalling at the anterior before being transported via the cytoskeleton to the uropod, (2) internalization and processing through the endocytic pathway, and (3) reinsertion in conjunction with membrane recycling at the anterior 10/20 min post-internalization. Down-regulation, the decrease in the number of receptors exhibiting saturable binding on the cell surface with increasing chemoattractant concentration as receptors are internalized faster than they can be processed and reinserted, is consistent with this picture. The non-stimulated rabbit neutrophil expresses 50 000 chemotactic peptide receptors which bind formyl-Norleucyl-Leucyl-Phenylalanine (FNLLP), which serves as the case study in this paper, decreasing with FNLLP concentration to 6000 receptors at concentrations which saturate binding. Notice that down-regulation provides for some global adaptation, since the total number of bound receptors on the neutrophil does not then vary as much with concentration, and probably some spatially local adaptation, since the cytoskeleton-induced segregation will act to reduce the number of transducing receptors where the bound receptor density is high. There is no firm evidence yet that the chemotactic peptide receptor becomes covalently modified upon binding, an adaptation mechanism common in other chemoattractant receptor systems such as the cAMP receptor on *Dictyostelium discoideum* amoebae, which is the other extensively studied chemotactic eukaryotic cell.

Receptor signal transduction and motility activation

As indicated above, a receptor–chemoattractant complex can associate with a G-protein, which upon binding GTP and dissociation of the activated G-protein initiates transduction via a complex biochemical cascade (Omann *et al.*, 1987; Sklar & Omann, 1990; Newell *et al.*, this volume, Naccache, this volume). The activated G-protein initiates the transduction cascade by dissociating into subunits, one of which activates phospholipase C. Through what appears to be a ubiquitous pathway, membrane-associated PIP_2 is then hydrolysed into IP_3 and DAG, leading to an increase in intracellular Ca^{2+} and other second messengers, and activation of protein kinase

C and other kinases with stimulatory protein targets. Opposed to these stimulatory pathways are inhibitory pathways which are initiated by an activated G-protein subunit which activates adenylate cyclase, leading to activation of cAMP-dependent kinases with inhibitory protein targets. A critical issue is the spatial localization of the second messengers associated with these pathways as it determines whether stimulation and activation are local to the initiating receptor or global for the cell. This requires knowledge of the transport (e.g. diffusion, convection) and kinetic (e.g. formation, degradation, sequestration) mechanisms associated with each messenger. Adaptation may also exist at the G-protein level as prior exposure to chemoattractant reduces the basal GTPase activity upon subsequent chemoattractant exposure in a heterologous fashion. There are many examples of tremendous 'magnitude amplification' in these pathways, i.e. the formation of one species leading to the formation of tens to millions of others, suggesting that no transduction species is present in fewer numbers than the chemoattractant receptors. For example, a single bound receptor is estimated to activate up to 5–10 G-proteins/min and thus may activate 10–20 G-proteins during the mean time of two minutes before it dissociates the chemoattractant molecule or becomes cytoskeletally segregated. One activated G-protein may lead to the assembly of 150 000 G-actin monomers into filamentous F-actin during its mean 15 s signalling lifetime. It has been estimated that there are 600 000 G-proteins which serve as a common pool for all of the neutrophil chemoattractant receptors (listed in Section 2), so that, except for the case of many chemoattractants being present simultaneously at saturating concentrations, the number of G-proteins far exceeds the number of bound receptors.

Lamellipod formation correlates temporally and spatially with rapid and extensive assembly of G-actin into F-actin, although the second messengers controlling assembly are not yet elucidated (Cassimeris & Zigmond, 1990; Naccache, this volume). There is evidence that the concentrations of free G-actin and F-actin barbed-ends (to which G-actin preferentially associates) both increase upon chemoattractant binding. Although F-actin is concentrated in lamellipodia, along with various actin binding proteins, it is present in the cortex underlying the plasma membrane in the cell body and uropod as well. Conventional myosin II becomes localized in the uropod and its affinity for actin is enhanced upon chemoattractant binding via activation of a myosin kinase.

The connection between the formation of an actin network and the development of motility forces which are co-ordinated into locomotion via lamellipodial activity is under intense theoretical as well as experimental investigation. Most theories involve the osmotic forces that can develop in a fibrillar F-actin network (essentially a gel) and the contraction forces that can occur if distributed actomyosin complexes are present (see Section 5), and there is experimental evidence for such complexes in cells and in reconstituted networks. Myosin II is not necessary for locomotion of *Dictyostelium* amoebae, but myosin I is observed in the leading edge of migrating amoebae, salvaging a likely role for actomyosin contraction forces in amoeboid locomotion, although neither myosin I nor an analogue has yet been reported for the neutrophil.

Thinking models

A number of thinking models have been proposed in recent years to account for the accumulating facts and observations (and speculation), with the goal of understanding the mechanisms by which neutrophils spatially and temporally integrate the receptor signals generated over their dimension into co-ordinated locomotion that is biased up a chemoattractant gradient (Lackie, 1986). The early proposition that the neutrophil uses a spatial mechanism, i.e. that the direction of movement in a gradient is determined by a receptor-measured concentration difference associated with two positions on the cell at one point in time, rather than a temporal mechanism, i.e. associated with two points in time at one position (Zigmond, 1982), served to stimulate theoretical analysis (see Section 2) as well as more experimental work. However, the obvious occurrence of adaptation and the clearly local but globally co-ordinated nature of lamellipodial activity has stimulated further modelling. The pseudospatial mechanism accounts for both: an extending lamellipodium continues to extend or retracts depending on whether the chemoattractant concentration sensed by the receptors on the lamellipodium increases or decreases, and the direction of locomotion is determined by the result of competition between lamellipodia occurring simultaneously around the cell (an influence of cell polarity on the local potential for lamellipodium formation is necessary to make this mechanism consistent with observations). Other modelling has been motivated by the observation that O_C is proportional to the predicted difference in the number of bound receptors across a cell, assuming a uniformly-distributed population of homogeneous monovalent receptors that only reversibly bind chemoattractant molecules (Fig. 2). Making mechanistic inferences from this correlation is dubious since at least the first assumption about the receptors appears wrong (see Section 3); none the less, that orientation data, along with forthcoming data correlating the spatio-temporal pattern of lamellipodia activity with the direction of locomotion, mandate a mechanistic explanation.

In the chronology of models beginning with Section 3, a trend towards accounting for the biomechanics of lamellipod formation and cell movement in addition to the kinetics of receptor signalling (consistent with the pseudospatial mechanism) will become evident. This reflects current thinking that the time-scale of directional changes (minutes) cannot be explained simply in terms of receptor binding and transduction events (seconds) and the

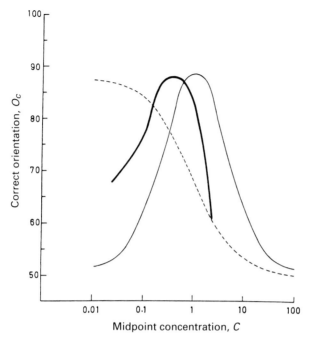

Fig. 2. A comparison of candidate orientation signals with the experimental orientation data (thick solid line) for rabbit neutrophils responding to FNLLP gradients (same data as Fig. 1(a)). The thin solid line shows the predicted differential in *absolute* receptor occupancy for a cell in the gradient and the dashed line the predicted differential in *fractional* receptor occupancy (both scaled to make maximum values the same as the data). The predictions assume a uniformly distributed population of homogeneous monovalent receptors that only reversibly bind chemoattractant molecules.

desirability of having a dynamical model which accounts for the continuous nature of cell movement in random motility and chemotaxis (hours).

2. THEORETICAL CONCEPTS IN RECEPTOR-SENSING

Receptor-sensing for the purpose of initiating and regulating specific cell responses and functions appears to be ubiquitous. In particular, neutrophils use information in the form of signals generated by chemoattractant receptor binding to regulate their speed and direction of movement. A neutrophil expresses several populations of 10^3–10^5 receptors for different chemoattractants distributed over the cell surface, including N-formyl peptides, C5a, PAF, TNF, and LTB_4, each of which binds its complementary chemoattractant molecules in a highly specific fashion. Thus, each receptor population essentially acts as a sensor to measure the extracellular concentration of a chemoattractant, and, as with any sensor, will have an associated measurement error (Berg & Purcell, 1977). 'Receptor noise' will be seen to be crucial in understanding the directional characteristics of cell movement in a chemoattractant gradient, where gradient will mean a steady-state, one-dimensional, linear concentration gradient hereafter.

It is appropriate to consider first sources of stochastic phenomena that exist and might contribute to the random directional component of cell movement. These include thermal fluctuations in the chemoattractant concentration local to the cell, fluctuations in the number of receptor binding events, and random intracellular responses by the cell to a receptor binding event. Using the formalism of the cell as a sensory system, with the signal being the local chemoattractant concentration, the stochastic nature of chemosensory movement might arise from true signal noise, noise in the signal measurement process, and/or a noisy transformation of the measured signal.

The following analysis suggests that, for neutrophils, noise associated with fluctuations in receptor binding swamps noise associated with fluctuations in the local chemoattractant concentration. Consider a hypothetical sampling volume (volume V) associated with a monovalent receptor that only reversibly binds chemoattractant molecules (i.e. no affinity conversion, ternary complex formation, internalization, etc) present at true uniform concentration C_∞ (*true* refers to the macroscopic concentration associated with averaging over an infinite volume). At any instant, the number of chemoattractant molecules that the sampling volume contains will deviate from the expected number (VC_∞) due to thermal fluctuations (i.e. molecules continuously enter and leave the sampling volume in a random fashion due to Brownian motion). Assuming the sampling volume is spherical with radius R determined by the root-mean-square distance that a chemoattractant molecule will diffuse during the mean time that a receptor is bound (to ensure statistical independence of binding events), then $R \propto (D_{CA}/k_r)^{1/2}$, where D_{CA} is the chemoattractant diffusion coefficient in the extracellular medium and k_r is the reverse (dissociation) rate constant for the chemoattractant–receptor complex. For FNLLP diffusing in a typical aqueous or dilute gel medium, $D_{CA} = 10^{-5}$ cm^2/s. Assuming for this order-of-magnitude calculation that the receptor for FNLLP exists in a single affinity state with $k_r = 0.4$ min^{-1} and $K_d = 2 \times 10^{-8}$ M (Sullivan & Zigmond, 1980) ($K_d = k_r/k_f$, where k_f is the forward (binding) rate constant). Choosing $C_\infty = K_d$, since random motility and chemotaxis are optimal near this chemoattractant concentration, the relative noise due to thermal fluctuations is $\approx 10^{-6}\%$, defined as $\sigma/C_\infty \cdot 100$ where σ is the standard deviation for the concentration in the sampling volume relative to C_∞, based on the standard result (see Tranquillo (1990) for all mathematical details relating to the equations stated in Sections 2–4):

$$\left(\frac{\sigma}{C_\infty}\right)^2 = \frac{1}{C_\infty V N_{Av}} \qquad (1)$$

Consider now the relative noise due to fluctuations in receptor binding for a population of total N_T receptors binding chemoattractant molecules present at true uniform concentration C_∞ (i.e. neglecting thermal fluctuations). To keep the calculation simple and consistent with that leading to eqn (1), we again assume a homogeneous population of monovalent receptors that only reversibly bind chemoattractant molecules. In this case, the binding of a receptor is simply a reversible bimolecular reaction and, like any chemical reaction, inherently a stochastic process: there will be a distribution of times a receptor remains bound around some mean time and likewise for the time a receptor remains free. The consequence of these times being random variables along with N_T being a finite number for the cell is that the instantaneous fraction of bound receptors, I, for the population of N_T receptors will deviate in a random fashion at almost every instant in time from the mean, P, associated with an ensemble average ($N_T \to \infty$). Since C_∞ is constant and binding is assumed to have attained equilibrium, the equilibrium binding relationship (i.e. in the mean) for such a receptor population is simply $P = C_\infty/(K_d + C_\infty)$. If the cell had the information P then it could 'measure' C_∞ by the equivalent relationship $C_\infty = K_d \cdot P/(1-P)$. However, since the cell only has the information I, which deviates from P, the cell's concentration measurement, C, necessarily deviates from C_∞. Choosing $N_T = 12\,500$, which is one-half (see below) the measured value of the total number of saturable receptors at $C_\infty = K_d$ (Sullivan & Zigmond, 1980), the relative noise due to fluctuations in the fraction of bound receptors is $\approx 2\%$ based on the result:

$$\left(\frac{\sigma_C}{C_\infty}\right)^2 = \frac{P}{N_T(1-P)^3} \qquad (2)$$

This shows that the noise associated with true fluctuations in the local chemoattractant concentration ($\approx 10^{-6}\%$) is insignificant compared to perceived fluctuations in the local chemoattractant concentration associated with true fluctuations in receptor binding ($\approx 2\%$).

For a given true chemoattractant concentration history (omitting the subscript '∞'), $C(t)$, the receptor parameters, k_f, k_r, and N_T completely define the equations which describe the time evolution of P and I:

$$\frac{dP}{dt} = k_f C(1-P) - k_r P \qquad (3)$$

$$dI = [(k_f C(1-P) - k_r P) - (k_f C + k_r)(I - P)]dt + \frac{1}{N_T^{1/2}}[k_f C(1-P) + k_r P]^{1/2}dW \qquad (4)$$

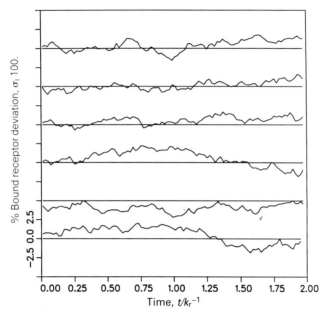

Fig. 3. Simulations of receptor binding fluctuations for a receptor population (see text for characteristics). The percentage deviation of I from P ($\sigma_I.100$) is plotted vs time (scaled to the mean time a receptor remains bound, k_r^{-1}). The scale on the ordinate applies directly to the simulation at the bottom, and to the other simulations upon translation down to $\sigma_I = 0$.

Unlike the familiar kinetic binding equation describing P (a deterministic differential equation which applies in the limit $N_T \to \infty$), a stochastic differential equation which applies in the limit of large but finite N_T describes I (Gardiner, 1985). A 'noise term' appears in this equation, which involves an increment of the Wiener process, dW, a random variable with zero mean and variance dt (the Wiener process, W, is the well-known stochastic process describing Brownian motion). The variance of the fluctuations of I from P, σ_I^2, a measure of the noise intensity, is defined by the magnitude of the coefficient multiplying dW, and thus is fully specified by the receptor parameters k_f, k_r, and N_T for a given $C(t)$. Since W is a stochastic variable, so is I (denoted by bold type). Using eqn (4) it is possible to numerically simulate I over time by specifying k_f, k_r, N_T, and $C(t)$ and sampling one of the infinitely possible realizations of W. To simplify matters, $C(t) = K_d$ is chosen for illustration in Fig. 3, for which $P(t) = P = 1/2$ (i.e. binding equilibrium, where under the ergodic hypothesis, an infinite time-average of I for the N_T receptors would show that half of the receptors are bound), using the above estimate of k_f, k_r, and N_T for the neutrophil–FNLLP case study. The deviations of I from P are obvious. σ_I for these realizations appears to be about 2%, which is the analytical result from eqn (6) below.

It remains to be demonstrated whether receptor noise is important for understanding how well neutrophils orientate in chemoattractant gradients. That, in fact, is quite easy to do given the observation stated in Section 1 that neutrophils exhibit significant orientation bias in gradients of FNLLP with steepness $\varepsilon \approx 1\%$ at $C = K_d$. This true concentration difference ($\approx 1\%$) that the cells sense quite accurately is smaller than the mean fluctuation in receptor-measured concentration ($\approx 2\%$) that each of two receptor populations on either side of the cell would perceive ($N_T = 12\,500$ for each). Under the assumption that the neutrophil sensing mechanism involves a spatial component as implied here and suggested in Section 1, the paradox arises: How can neutrophils sense these spatial chemoattractant gradients so accurately?

This same paradox, first identified for *E. coli* and *D. discoideum* by Berg & Purcell (1977), is due to the assumption thus far that the cell bases its concentration measurement on the *instantaneous* fraction of bound receptors. The paradox is resolvable with the 'time-averaging hypothesis': the cell has some means by which it can time-average the fraction of bound receptors so as to reduce the noise in the associated receptor-measured concentration. The advantage of time-averaging for measurement accuracy follows from the following expression for σ_{CT} (DeLisi, Marchetti & Del Grosso, 1982), where the subscript 'T' denotes time-averaging, again for the case of a homogeneous population of N_T monovalent receptors reversibly binding chemoattractant molecules present at true uniform concentration C_∞ with binding at equilibrium:

$$\left(\frac{\sigma_{CT}}{C_\infty}\right)^2 = \frac{2}{N_T k_r T P}\left[1 - \frac{1-P}{k_r T}\left(1 - \exp\left[-\frac{1-P}{k_r T}\right]\right)\right] \qquad (5)$$

T is the time-averaging period. k_r is a function of P in the general case where receptor binding is limited by diffusion of the chemoattractant molecule to form an encounter complex with the receptor rather than the binding reaction once in the encounter complex (Wiegel, 1983). However, FNLLP binding to neutrophils appears to be reaction-rate limited, so that k_r is a constant independent of P. Dependancies on the parameters are more clearly seen for σ_I^2 and σ_{IT}^2, the related variances of the instantaneous and time-averaged fraction of bound receptors relative to P, respectively (taking the limiting case of $T \gg \tau$ for σ_{IT}^2):

$$\sigma_I^2 = \frac{P(1-P)}{N_T} \qquad (6)$$

$$\sigma_{IT}^2 = \frac{P(1-P)}{N_T}\frac{\tau}{T} \quad \text{where } \tau^{-1} = \tau_f^{-1} + \tau_r^{-1} = k_r(C_\infty/K_d + 1) \qquad (7)$$

The time constant τ is related to the mean times for a free receptor to become bound ($\tau_f = (k_f C)^{-1}$) and a bound receptor to become free ($\tau_r = k_r^{-1}$). In both cases, the mean square deviations decrease with increasing N_T, i.e. the relative significance of receptor binding fluctuations decreases (given that most chemical reaction systems involve numbers of molecules much greater than 12 500, the typical neglect of this type of noise in chemical kinetics is thus justified). For the time-averaged case, deviations also decrease with increasing T and with decreasing τ (e.g. greater k_r for fixed relative concentration C_∞/K_d). These dependancies can be thought of as improvement in measurement accuracy resulting from either more sensors (N_T) or from a greater sampling frequency per sensor ($k_r T$) at a fixed relative concentration.

Although the possibility of a stochastic transformation of a receptor binding event is entirely possible, assessing its quantitative significance is difficult given incomplete knowledge of the molecular basis of the transformation (i.e. the quantitative molecular details of the biochemical transduction pathways and motile force generation events). As suggested in the remarks on magnitude amplification in Section 1, it seems unlikely that any species involved in transduction is present in fewer numbers than the receptors, so that statistical fluctuations associated with finite numbers should not be a significant source of intracellular noise.

3. FIRST GENERATION PROBABILISTIC RECEPTOR MODELS

A series of investigations employed the 'time-averaging hypothesis' to attempt to distinguish whether a particular cell type senses a gradient with either a spatial or a temporal mechanism, and to determine the required value of T. As discussed in Section 1, these mechanisms are probably inappropriate for understanding the chemosensory movement of neutrophils. However, a brief summary and critique of the approaches used in these investigations is worthwhile to highlight their limitations and to justify the more complicated models subsequently discussed at length.

Model descriptions

The seminal work on this problem is due to Berg & Purcell (1977). Besides proposing the time-averaging hypothesis, they also first derived an expression for σ_{CT} (restricted to diffusion-limited binding (i.e. k_r constant) and $T \gg \tau$) and postulated feasibility criteria consistent with spatial and temporal sensing mechanisms. They assumed that a signal to noise (S/N) ratio must exceed unity for a mechanism to be feasible. The signal is defined here by the true concentration difference that the cell is attempting to measure, as consistent with the spatial or temporal mechanism. The noise is defined by the root-mean-square deviation associated with the two receptor-measured concentrations, given by $\sqrt{2}\sigma_{CT}$ upon assumption that the measurements are statistically independent. Thus the feasibility criteria can be stated as:

$$\text{spatial mechanism: } D\frac{dC_\infty}{dx} > \sqrt{2}\sigma_{CT}$$

$$\text{temporal mechanism: } vT\frac{dC_\infty}{dx} > \sqrt{2}\sigma_{CT}$$

where D is the distance separating two hypothetical independently-sensing populations of receptors and v is the component of cell velocity in the direction of the true chemoattractant gradient, dC_∞/dx.

Although Berg & Purcell did not consider neutrophils specifically, DeLisi et al. (1982) did using their more general expression for σ_{CT}, eqn (5). Using experimental estimates for all other (i.e. receptor) parameters in order to calculate T, it was concluded that neutrophils would require days for a spatial mechanism to be feasible but only minutes for a temporal mechanism. In subsequent work, DeLisi & Marchetti (1983) further relaxed the assumption of equilibrium binding for the particular case where at some initial time a cell not previously exposed to chemoattractant is subject to a chemoattractant gradient. This required the introduction of another parameter for the temporal mechanism, the time delay t_1 between two successive time-averaging periods, with an optimum t_1 predicted for minimizing T. The original conclusion that a temporal mechanism must be operative based on a S/N ratio criterion (generalized for non-equilibrium binding) did not change.

Lauffenburger (1982) postulated a different feasibility criterion consistent with a spatial mechanism: a signal threshold (S/Δ) ratio must exceed unity for a neutrophil to exhibit orientation bias in a gradient. The signal is defined here by the (fluctuating) difference in the number of bound receptors across the cell dimension. The threshold, Δ, is some minimum difference that elicits orientation bias. It is a second adjustable parameter in addition to T. Probabilities were computed for all possible signals for a specified gradient with experimental estimates for all other (i.e. receptor) parameters besides assumed values of Δ and T. Comparing each signal to Δ, the total probability for orientation bias up-gradient versus down-gradient was computed (the model cell has only two possible orientation states), random orientation being assumed for sub-threshold signals. For sufficiently small Δ, for example, ten bound receptors, it was found that T of the order of minutes yielded predictions of O_C consistent with the data of Fig. 1(a). Perhaps more significantly, the validity of the S/N criterion was challenged, as the model predicted perfect orientation in some cases where $S/N<1$ and did not predict perfect orientation in all cases where $S/N>1$ (perfect orientation $\equiv Pr$[orientation up-gradient] ≈ 1). This probabilistic model was extended to account for more complicated receptor characteristics, including down-regulation, anterior–posterior asymmetric distribution, and multiple affinity subpopulations (Tranquillo & Lauffenburger, 1986). Also, the nature of the signal used in the S/Δ criterion was changed from a difference in the absolute number of bound receptors to a difference in the *fraction* of bound receptors, which is consistent with a scheme for sensory adaptation (see Section 4 Critique). Interestingly, the dependance of this 'adapting' signal on chemoattractant concentration *without* account for receptor noise is not biphasic like the original 'absolute' signal, seemingly inconsistent with the orientation data (see Fig. 2). However, *with* account for receptor noise, the 'adapting' signal can accommodate significant anterior–posterior receptor asymmetries and yield predictions consistent with the data. The 'absolute' signal cannot.

Critique

Several significant limitations in these models can be identified: (1) They all implicitly assume unrealistic cell movement in order to perform the calculations: the spatial mechanism criterion only rigorously holds if during the time-averaging period the cell is stationary; for the temporal mechanism criterion, the cell must move with constant v, i.e. in the same direction with constant speed. Consequently, there is no account for changes in sensing associated with turning away from (or towards) the direction of the gradient, which seems intrinsic to the chemotactic response (these unrealistic assumptions allows for the simpler probability formalism rather than the stochastic process formalism used in the models beginning with Section 4 which allows for realistic movement), (2) Increasing understanding of the cell biology of neutrophil chemosensory movement supports the more complicated pseudospatial mechanism as discussed in Section 1, (3) None of the models addresses random motility, which seems to be inextricably related to chemotaxis and for which considerable data exists, and (4) Although estimates for T are interesting for comparison to cell movement time scales, the time-averaging period *per se* offers no mechanistic insight.

Despite this last criticism, that the concept of a time-averaging period is somewhat artificial, the concept of time-averaging definitely is not: a receptor signal has a finite lifetime due to its propagation along the transduction and response pathways, and this propagation lifetime is effectively a time-averaging period. Thus, a new model for understanding the consequences of receptor-signal noise was proposed, one which confers on the cell intracellular mechanisms through which receptor signals are generated and from which the change in direction of the cell is determined (Tranquillo & Lauffenburger, 1987). There is no explicit time-averaging period upon which the earlier models are based, the parameters associated with a particular set of mechanisms determining implicitly the extent of time-averaging.

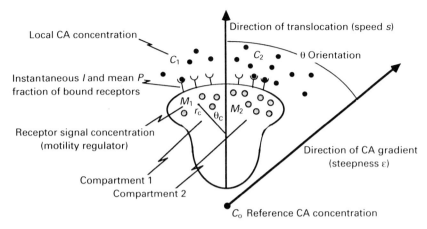

Fig. 4. Second generation model chemosensory cell. See text for description. CA = chemoattractant.

Indeed, the functional form of the mechanisms and their interrelationships determine the characteristics of the implicit time-averaging (e.g. how O_C depends on the parameter values). In addition, the new model cell is capable of autonomous movement in two-dimensional space given a specified chemoattractant concentration field (uniform or gradient), the turning behaviour being determined by a 'pseudo-pseudospatial' mechanism. As seen from these features, the limitations of the earlier models are rectified in this second generation model while maintaining the focus on the certain occurrence of statistical fluctuations in receptor binding.

4. SECOND GENERATION STOCHASTIC RECEPTOR-CELL TURNING MODELS

Model description

The general features are illustrated in Fig. 4. The specific mechanistic assumptions underlying the model equations analysed to date are presented below (Tranquillo & Lauffenburger, 1987; Tranquillo, Lauffenburger & Zigmond, 1988). The cell remains morphologically and behaviourally polarized so that directional change occurs via a continuous 'smooth turn' governed by a stable leading edge. Since the turning behaviour is of primary interest, it is assumed that translocation is uncoupled from turning and is described with a constant speed. A pseudospatial-type mechanism which governs turning is accomplished by modelling the leading edge as two interacting compartments. Assuming a receptor signal is the critical regulator of the motility system, the turning rate of the cell can be related to an

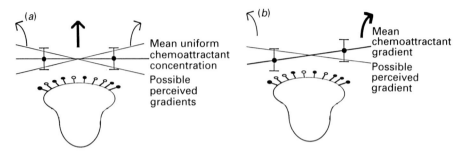

Fig. 5. Consequences of receptor noise on (a) random motility and (b) chemotaxis. The thick lines represent the true chemoattractant gradients and the thin lines represent possible transiently perceived chemoattractant gradients. The receptors on the leading edge of the cell are divided into two populations. Both populations perceive fluctuating chemoattractant concentrations, represented by the error bars, around the true local concentration, indicated by the closed circles, in their receptor measurement of concentration. Thick arrows represent the expected directional responses in the absence of receptor noise and thin arrows the possible responses in the presence of noise.

imbalance of the receptor signal between the two compartments. The generation of the receptor signal in each compartment is related to the stochastic receptor binding process on the cell surface associated with each compartment. In the case of movement in a chemoattractant gradient, a coupling then arises between the movement response (translocation with turning) and the receptor signal since receptor binding depends on chemoattractant concentration, which depends on the cell position and orientation in the chemoattractant gradient.

The model is not entirely consistent with the pseudospatial mechanism described in Section 1 as there is no account for pseudopod extension relative to the cell body, but there is an account for a temporal signal associated with each compartment as it moves through space as the cell translates and turns. Moreover, the issue of global regulation and co-ordination is addressed by the dual compartment structure, albeit in a very superficial way (see Critique). The model also provides a single unifying mechanism to qualitatively explain both random motility and chemotaxis as seen in Fig. 5, consistent with their inextricable relationship. In the case of random motility, the receptor populations of both compartments are constantly subject to the same mean chemoattractant concentration and perceive statistically equivalent fluctuations in chemoattractant concentrations. Thus, the cell perceives fluctuating chemoattractant gradients without a mean reference direction leading to a persistent random walk. In the case of chemotaxis, the cell is subject to a mean chemoattractant gradient with both receptor populations perceiving, in general, statistically different fluctuations from the true local chemoattractant concentrations. At any instant, the cell perceives some deviation from the true gradient (and may even transiently perceive a gradient in the reverse direction) leading to a biased random walk. Note that, for greater (lesser) receptor noise as represented by larger (smaller) error bars, directional persistence in random motility and orientation bias in chemotaxis should both decrease (increase).

The specific mechanisms of receptor binding, signal transduction, and turning response analysed thus far for the stochastic pseudospatial model depicted in Fig. 4 are now considered. Although many details of these processes have recently been elucidated as indicated in Section 1, mechanisms representing a high level of abstraction were initially chosen to facilitate the characterization of a stochastic dynamical system with signal–response structure consistent with the model. These mechanisms represent a set which incorporates the minimal complexity and number of components.

Receptor binding and dynamics

The simplest receptor properties possible have been assumed thus far, those used in generating the realizations of receptor binding fluctuations of Fig. 3: a homogeneous population of monovalent, reversibly binding receptors of constant affinity which do not form ternary complexes, internalize, or diffuse in the membrane (see Critique).

Receptor signal transduction

A single intracellular messenger (i.e. the receptor signal), considered to be the critical regulator of the motility system, is defined with concentration M. In each compartment, the messenger is generated at a rate proportional to the magnitude of some as yet undefined receptor signal R with first-order transduction rate constant k_t, decays according to first-order kinetics with decay rate constant k_d, and is transported between compartments at a rate proportional to the difference in M with diffusive rate constant D.

Cell turning

The turning rate is assumed proportional to the difference in receptor signals between the compartments, $\Delta M = M_1 - M_2$, with the proportionality coefficient κ describing the sensitivity of the cell to the turning signal ΔM. This represents a simplistic force balance normal to the direction of cell movement: resistive force against turning of the cell due to the extracellular environment balanced by the net motile force transmitted by the cell to the substratum via adhesive bonds with the leading edge. It is implicitly assumed that the rate limiting step in activation of the motility system is the generation of M and that the motility components and adhesion sites are uniformly distributed between the compartments. θ_T defines the angle formed by the direction of the cell polarity axis relative to its initial direction and is useful for the characterization of random motility. It has domain $(-\infty, \infty)$ since turning can yield revolutions with respect to the initial direction. The appropriate angle to characterize orientation in chemotaxis is the angle θ defined by the cell polarity axis relative to the direction of the chemoattractant gradient and has domain $(-\pi,\pi)$.

Chemoattractant concentration

The chemoattractant concentration local to the sampling point in each compartment, C, is determined by the movement (translation with rotation) of the cell, given the chemoattractant concentration field (defined by the reference concentration C_0 and gradient steepness ε) and the locations of the sampling points (defined by r and $\pm\theta_c$) (Fig. 4).

In order to proceed, the receptor event R in receptor–signal transduction needs to be specified. The case where the event is the existence of a bound receptor is considered in detail here (see Critique). This is the simplest choice and has been the most common assumption in the interpretation of receptor signal–response data. Making the substitution $R = I \cdot N_T$ completely specifies the binding dynamics of each receptor population. Since all of the variables are coupled together in the associated modelling equations and I is a stochastic variable, they are all stochastic variables. The model can be formalized as a stochastic system with a deterministic transformation of (a) and response to (b) a stochastic signal (c) which is feedback-coupled to the response (d), where (a) = transduction of receptor binding to a receptor signal, (b) = cell translocation with turning, (c) = the receptor binding process, (d) = movement to the chemoattractant gradient. For random motility conditions (uniform chemoattractant concentration), the stochastic signal is not feedback-coupled to the response, which affords considerable simplification as will become evident.

It is instructive to illustrate the model behaviour at this point with some sample paths, i.e. simulated paths of cell movement in the x–y plane for both random motility and chemotaxis obtained by numerical integration of the governing SDE system. Each simulated path of random motility in Fig. 6(a) represents the migration response of the model cell to one realization of binding fluctuations occurring in the receptor population of each compartment, such as presented in Fig. 3. The characteristics of the persistent random walk observed for neutrophil behaviour are evident. The influence of a moderate gradient on each of these paths indicates a biased turning response in the direction of the gradient. These chemotactic responses correspond to the identical driving noise and same cell parameter values used for the random motility simulation. The gradient is of magnitude typically established in Zigmond's visual assay. Also included is the influence of a ten times steeper gradient on each of the random motility paths. A faster turning response with initial oscillatory behaviour occurs. Greater bias up the gradient later in the simulation is also evident. The degree of oscillatory behaviour for this case may be related to the two-compartmental structure of the model; however, a gradient of this magnitude has yet to be established experimentally, so the nature of the response

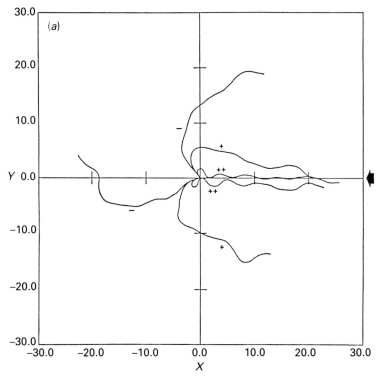

Fig. 6. Simulations of random motility and chemotaxis. (a) Cell paths: two sets of results are indicated. In each set, the cell has the same initial orientation at the origin. For each set, one path is obtained for the cases of random motility ($-$) and chemotaxis in moderate ($+$, $\varepsilon = 1.6\%$) and steep ($++$, $\varepsilon = 16\%$) gradients.

of neutrophils to such steep gradients is unknown. The scatter diagrams of Fig. 6(b) give a different perspective of the model behaviour. In the random motility case, persistence is obvious at the early time and the dispersion characteristic of an unbiased random walk is developing at the later time. For the moderate gradient simulation, the response is consistent with the biased random walk underlying chemotaxis. In the steep gradient simulation, the strongly biased random walk yields a chemotactic wave of migrating cells. These results suggest that the model behaviour is at least qualitatively consistent with the true cell behaviour for neutrophil and motivates the following quantitative analyses.

Analysis and results

Discussion of random motility implies the cell has been subject to a uniform chemoattractant concentration, C, effectively for an infinite time period, so that conditions of binding equilibrium have been established. Then the

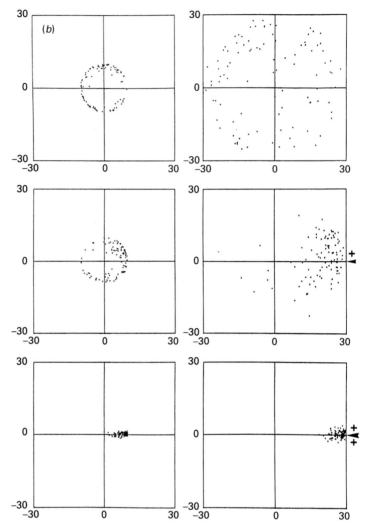

Fig. 6(*b*) Scatter diagrams: the coordinates of the paths of 100 cells at a relatively early time (left column, $t \ll P_T$) and later time (right column, $t \gg P_T$) in the simulations are plotted. As in (*a*), the individual cells have the same initial orientation at the origin and experience the same driving noise for the cases of random motility and chemotaxis in the moderate and steep gradients.

mean fraction of bound receptors is constant and given by $P = C/(K_d + C)$. The model equations are linear in this case and it is possible, in principle, to solve analytically for any desired moment. Since characterizing the directional randomness of movement is of primary interest, the second moment of the orientation process $\langle \theta_T^2 \rangle$ was solved for explicitly. The main result of this analysis is given by:

$$P_T = \lim_{t \to \infty} \frac{2t}{\langle \theta_T^2 \rangle} = \frac{f_s \, \tau_R^4 \, (\rho+1)^3}{N_T \, \tau_D^2 \, \rho} \tag{8}$$

P_T is interpreted as a directional persistence time, the characteristic time that a cell translocates without significant directional change. It serves as an objective measure of the random directional change in random motility. More precisely, P_T is a rotational diffusion coefficient, with the directional persistence time more commonly being defined in terms of the autocorrelation function for the directional process, $(\cos\theta, \sin\theta)$ (Alt, 1990b). f_s is considered to be the sampling frequency for a receptor and given simply by $f_s = k_r$, and N_T is the total number of receptors associated with each compartment as before. ρ is the dimensionless uniform concentration, $\rho = C/K_d$. However, the model only applies for $\rho \leq 1$ which is the chemoattractant concentration range eliciting maximum cell polarity (see Critique). τ_R and τ_D are interpreted as time constants for turning response and turning signal decay, respectively, and are defined in terms of the four parameters associated with the intracellular mechanisms:

$$\tau_R = \frac{1}{(\kappa k_t)^{1/2}} \qquad \tau_D = \frac{1}{2D + k_d} \tag{9}$$

Recalling that ΔM is a turning signal, the random turning behaviour is a consequence of ΔM fluctuating about zero due to I_1 and I_2 fluctuating about P. The result for P_T indicates how, for example, greater directional persistence can be conferred on the model cell by appropriately altering f_s, N_T, τ_R, τ_D: the larger τ_R, the slower the cell responds to a turning signal created by receptor binding fluctuation; the smaller τ_D, the faster the cell eliminates such a turning signal. The dependence of P_T on f_s might be anticipated from σ_{IT}^2 of eqn (7). As f_s increases, the magnitude of the receptor binding fluctuations decreases. The dependence of P_T on N_T is seemingly counterintuitive given that σ_{IT}^2 also states that as N_T increases, the magnitude of the receptor binding fluctuations decreases. However, the rate of receptor signal transduction has been assumed proportional to the total number of bound receptors, $N_T \cdot I$, not to just the fractional number of bound receptors, I (see Critique). Even though the receptor binding fluctuations decrease with increasing N_T, the transduction process is amplifying them in proportion to N_T. The net effect of this trade-off from increasing N_T is a decrease in P_T. Since the results are based on eqn (4) which is valid in the large N_T limit, the conclusion that P_T increases with decreasing N_T may be invalid when N_T becomes sufficiently small. Finally, note that P_T is independent of cell speed, consistent with the model assumption that turning behaviour is uncoupled from translocation.

Having the characterization of the random turning behaviour in a uniform chemoattractant concentration in terms of P_T, the biased turning behaviour in a chemoattractant gradient underlying chemotaxis was examined. The goal was not only to demonstrate that the model cell exhibits chemotactic behaviour consistent with observations of neutrophils, but to elucidate the relationship between the turning behaviour in random motility and chemotaxis as suggested by Fig. 5. In this case, the equations are non-linear and stochastically coupled, reflecting feedback on receptor-sensing associated with translocation and turning in the gradient, and must be solved numerically (i.e. simulated). Orientation bias quantified in the simple form of O_C was chosen to characterize chemotaxis for the purpose of convenient comparison to available data for neutrophils.

The first finding of significance is that simulated cell paths depend only on the parameter groups used to characterize the random motility response (f_s, N_T, τ_R, τ_D) rather than on the individual cell parameters (k_r, N_T, k_t, k_d, D, κ) for a specified chemoattractant gradient (i.e. fixed ρ_0 and ε). If identical initial directions and realizations of the Wiener process are used, identical paths are obtained for any combination of the cell parameters that yield the same values for f_s, N_T, τ_R, and τ_D. Apparently, τ_R and τ_D completely characterize the dynamical characteristics of the intracellular mechanisms. Since there are experimental estimates for f_s and N_T, the adjustable parameter space is reduced to τ_R and τ_D. The parameter space can be further reduced using measured values for the directional persistence time $P = 1$–5 min (Zigmond et al., 1985) for $\rho = 1$. Assuming $P_T \approx P$ in eqn (8) yields the constraint $\tau_R^4/\tau_D^2 = 0.312$–$1.56 \times 10^4$ min^2 (the 25 000 saturable surface receptors for FNLLP at $C = K_d$ appear to be anterior–posterior asymmetrically distributed (Sullivan, Daukas & Zigmond, 1984) – estimating that 40% of the total available surface receptors would be associated with each compartment, then $N_T = 10\,000$; $f_s = k_r = 0.4$ min^{-1} as stated in Section 2).

The next finding of significance is that the percentage of correctly oriented cells (O_C) for the population approaches a quasi-steady value despite population drift to higher chemoattractant concentrations. This allows the dependance of O_C on the steepness of the gradient (ε) to be compared directly to that observed for neutrophils, notwithstanding the complication that the model results do not apply exactly for $\rho = 1$, as do the data, due to the population drift to $\rho > 1$ during the simulation. It is striking from Fig. 7 that the agreement is quantitative as well as qualitative given the variability of cellular dimensions upon which the experimental value of ε is calculated.

Given that validation, the model can be used to examine an intriguing question: what is the relationship between directional persistence (in the absence of a chemoattractant gradient) and orientation bias (in the presence of a chemoattractant gradient) for a cell? The answer was qualitatively suggested in Fig. 5 and is quantitatively provided here in part by the results summarized in Fig. 8. These plots show the dependence of O_C on τ_R^4

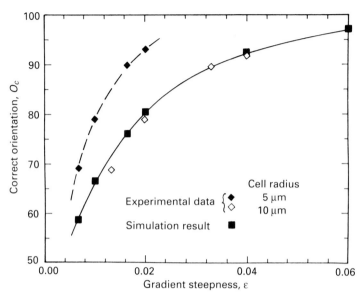

Fig. 7. Dependence of orientation bias on gradient steepness. Experimental orientation data for neutrophils from Fig. 1(a) are replotted (the value of ε against which O_C is plotted depends on the assumed effective cell radius). Simulation results incorporate the previously stated estimates for N_T and f_s and values for k_1, k_d, D, and k which yield $\tau_R = 3.54$ min and $\tau_D = 0.114$ min, determining $P_T = 3.88$ min and $\tau_R^4/\tau_D^2 = 1.21 \times 10^4$ min^2. A population size of 300 with the same initial orientations and realizations of the Wiener process apply to the simulations for the various ε. Other parameter values besides ε need to define the simulation conditions are $r_c = 5$ μm, $\theta_c = \pi/4$, $s = 20$ μm/min.

and τ_D^2. Lines of constant P_T are indicated. In crossing lines from small to large P_T, O_C passes through a relatively shallow maximum, corresponding to an optimal $P_T \approx 3$ min. Following along the line of optimal P_T to smaller values of τ_R^4 and τ_D^2, O_C is seen to increase slightly. So, at least for the values of f_s and N_T applicable to the rabbit neutrophil–FNLLP case study, an optimal P_T is suggested; furthermore, one which reflects small time constants, i.e. a cell that rapidly responds to a turning signal and rapidly eliminates it.

Critique

That receptor binding fluctuations are of consequence for chemoattractant gradient sensing by neutrophils is well supported from the theoretical analyses presented in Section 2. That they can explain simultaneously the component of directional randomness in both random motility and chemotaxis is well supported by this model. It is admittedly surprising that such a simple version of the pseudospatial mechanism yields quantitative as well

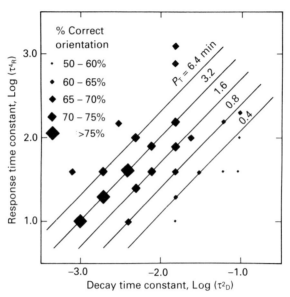

Fig. 8. Relationship between orientation bias and directional persistence. Simulation parameters are the same as Fig. 7 except the population size is 100. $\varepsilon = 1.6\%$ applies to all simulations here.

as qualitative agreement with experimental observations and data given the complexity of the actual biochemical pathways and biophysical events.

One unsatisfying outcome of the model's analysis is with respect to evaluating candidates for the receptor signal, since P_T and O_C depend on the time constants τ_R and τ_D and not on the individual intracellular parameters. Evidently, this is a natural consequence of modelling the cell as an integrated dynamical system. It is expected that any transduction scheme of arbitrary complexity, if it is linear in the model variables (see below), would be reduced to a representation in terms of τ_R and τ_D. However, there is at least one direct test of the particular mechanisms assumed here: the prediction that P_T is inversely proportional to N_T. N_T would have to be modulated in a manner which does not perturb the original state or function of the cell. A competitive inhibitor which binds to the same receptor but neither elicits a signal nor perturbs the cell otherwise would be ideal. Of course, failure of this test would not necessarily invalidate the model structure combining receptor binding fluctuations with a pseudospatial mechanism. It would indicate that the particular mechanisms used to illustrate the general model structure are inadequate or incorrect. Note that if, in fact, the rate of receptor signal transduction is proportional to only the *fraction* of bound receptors (i.e. $R = I$), which is consistent with the simple adaptation mechanism wherein a bound receptor both rapidly produces a non-diffusible generator of the receptor signal and slowly produces a rapidly diffusible degrader of it (Tranquillo & Lauffenburger, 1986), then the random motility analysis yields the same result except that P_T is now directly rather than inversely proportional to N_T. In addition, chemotaxis simulations reveal that a maximum in O_C corresponding to some optimal N_T does not exist for this 'adapting' signal case, as is true for the classical 'absolute' signal case of $R = I \cdot N_T$ presented above; rather, O_C increases monotonically with N_T over the relevant range for N_T. Thus, these alternative transduction schemes can be clearly distinguished by the model's predicted dependence of P_T and O_C on N_T.

The specific equations analysed thus far for receptor signal transduction and turning response are linear in the model variables (i.e. all rates are simply proportional to concentrations, the rate constants for M determining the magnitude amplification of the receptor signal). In reality, there are complex nonlinearities associated with both mechanisms. The justification for the contrived linearity is purely for mathematical tractability: when analysed for random motility, the full stochastic differential equation system is linear (the receptor binding equation, eqn (4), is linear for uniform chemoattractant concentration), yielding the illuminating analytical result for P_T, eqn (8). A major consequence of omitting non-linearities is the inability to amplify small differences in receptor binding between the two compartments into a large difference in the motility response, which is a desirable feature to retain given the occasional dramatic extension or retraction of some region of the leading edge. Generic mechanisms which yield such 'sensitivity amplification' were proposed and analysed by Koshland, Goldbeter & Stock (1982). The simplest example is the response of a co-operative enzyme to its substrate. The Hill coefficient describing the dependence of enzyme velocity on substrate concentration determines the enhancement in velocity given an increase in substrate concentration. A greater Hill coefficient means a greater enhancement because of a greater nonlinearity relating response (velocity) to signal (substrate concentration). This is very similar to the argument behind neutrophils having the greatest gradient sensitivity near $C = K_d$, assuming they are using a spatial mechanism. It is near $C = K_d$ that the equilibrium binding curve is most non-linear, so that small changes in chemoattractant concentration yield the largest changes in receptor occupancy. Even though P_T as well as O_C may have to be determined from simulated cell paths upon incorporating more complicated non-linear transduction functions, the result of this extension of the second generation model should be illuminating.

In light of the receptor dynamics which are now known or speculated (reviewed in Section 1), the assumptions made in this initial characterization of the second generation model, that receptors are immobile in the plasma membrane and only undergo a simple reversible binding reaction to a single transducing state, is clearly oversimplified. In order to investigate the consequences of receptor dynamics on gradient sensing and the orientation

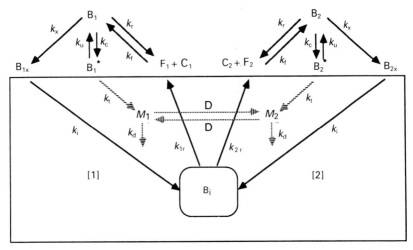

Fig. 9. Extension of the second generation model to receptor dynamics. [1] and [2] represent the two compartments of the cell's leading edge depicted in Fig. 4. M is generated by a reaction involving a rapidly dissociating CA-receptor-G-protein ternary complex B^* formed reversibly from the bound receptor B and G-protein. B^* is the signalling receptor state in the model (though the actual signalling species involves the activated G-protein subunit which dissociates from B^*). The bound receptor can also become irreversibly cytoskeletally associated, represented as B_X, the inactive signalling state targeted for internalization into a putative intracellular pool (B_i), and after sorting from the CA molecule is reinserted back to the cell surface with pseudo-first-order kinetics. To account for a spatial bias in receptor recycling, the recycling rate constants, k_{1r} and k_{2r}, will be trial functions of M. The receptor network is indicated by solid arrows, while that for the signalling second messenger M is shown by dashed arrows. The other aspects of the model are as described for Fig. 4.

response, we are extending the second generation model as illustrated in Fig. 9. There are two main features of receptor dynamics which are accommodated: a transient signalling state of the receptor and receptor recycling. As before, the fluctuating numbers of receptors in the various states are accounted for, as determined by the interconverting rate constants and the total receptor number. One consequence of the transient signalling state is the introduction of a new fast time-scale, although the consequences for f_s, τ_R, and τ_D are not obvious. The implications of receptor recycling include a partial account for adaptation via down-regulation and the intriguing possibility that the location for receptor reinsertion in the plasma membrane is regulated by the sensing and response history of the cell. We are assessing the dependance of P_T and O_C on different functions for (lateral) asymmetric receptor reinsertion into the two compartments. For example, since the receptor signal regulates motility, or equivalently, the state of the cytoskeleton and cortex, it is reasonable to propose that the fraction of receptors recycling to a compartment from the intracellular endocytic pool is given by the instantaneous fraction of M in that compartment, i.e. $k_{1r} = k_r\, M_1/(M_1 + M_2)$ and $k_{2r} = k_r\, M_2/(M_1 + M_2)$. This is effectively another type of positive feedback in the model, in addition to the turning towards the compartment having greater M already employed.

The prospect of using the extended model to elucidate which receptor reinsertion function optimizes the chemotactic response raises the issue of what optimization criterion should be applied. It is known that neutrophils express separate receptors for several chemoattractants (listed in Section 2) all of which may be present in the *in vivo* environment simultaneously. In fact, each of these receptors appears to signal via a G-protein and share the same transduction pathways. Thus, the chemotactic response for which the cell is optimized may include the response to multiple chemoattractant gradients. Unfortunately, there is no orientation data yet for this case, and one must resort to optimization based on maximizing O_C to a single gradient, despite the likelihood that robustness of the response to multiple gradients exists. Of particular interest is the possible positive feedback when one product of DAG metabolism, LTB_4, a potent chemoattractant itself, is released upon stimulation with other chemoattractants.

An inherent limitation of the second generation model is that it applies only for chemoattractant concentrations where the cell retains its morphological and behavioural polarity (i.e. $C \leqslant K_d$). Although this model based on the limit of high polarity yields significant insight into the relationships between random motility and chemotaxis and provides a convenient framework for investigating the consequences of more complicated receptor dynamics and transduction mechanisms, the notion that neutrophil directional changes can be described as turning associated with rotation is quite unsatisfying given the actual process of lamellipodial extension and retraction occurring locally around the leading edge, the resultant of which determines the direction of translocation. However, modelling the mechanics of lamellipodial activity, even for the high polarity limit where they are largely restricted to the leading edge, is extremely complicated, especially since the relevant information, namely, the nature of motility forces and the identity and regulatory mechanisms of the associated molecular species, is again complex and incomplete. The situation is even more complicated when the polarity is not stable (i.e. significant cytoskeletal rearrangements occur), characteristic for $C \gg K_d$.

A number of investigators have recently proposed models for the extension of a single idealized pseudopod. The models differ with respect to the assumed forces and regulatory mechanisms as well as the level of abstraction (Oster & Perelson, 1985; Zhu, Skalak & Schmid-Schonbein, 1989; Alt, 1990a). None of them has treated the case of simultaneous, competing lamellipodia, which is central to the pseudospatial sensing mechanism. This forms the basis of the third generation of models.

5. THIRD GENERATION STOCHASTIC RECEPTOR-LAMELLIPODIAL ACTIVITY MODEL

Model description

The key distinction of the third generation models it that it accounts for dynamic lamellipodial activity occurring locally around the cell with associated cortical flow, whose resultant determines the direction and speed of translocation (Tranquillo & Alt, unpublished studies). The equations modelling the cell mechanics (response component) are based on the theory developed by Dembo, Harlow & Alt (Dembo, Harlow & Alt, 1984; Alt, 1988; Dembo, 1989) which models the cortical cytoplasm as a reactive, contractive, two-phase fluid, *reactive* meaning a dynamic assembly and disassembly of actin filaments, *contractive* meaning contraction forces associated with actomyosin complexes in the reactive cortical actin network, and *two-phase fluid* meaning cytosol flowing through a filamentous actin network. The equations modelling receptor binding (sensory component) are based on an extension of the stochastic methods underlying eqn (4) to account for a generalized spatial description of free and bound receptor distributions (as opposed to the coarse spatial description based on two compartments), including receptor mobility in the membrane and resultant receptor diffusion. The link between receptor binding and lamellipodial activity (signal–response coupling) is made by making a parameter (or parameters) describing the properties of the reactive, contractive fluid a function of a receptor-mediated second messenger, the motility regulator in the second generation model for example.

Before briefly discussing the elements of the cell mechanics theory, some general features of the preliminary model for lamellipodial activity based on such a fluid are noted (Fig. 10). We assume that the cell is initially nonstimulated, i.e. spherical and nonpolar with a uniform distribution of cortical fluid (i.e. cytosol and F-actin) in a subplasmalemmal cortex with a uniform distribution of receptors on the plasma membrane. The remainder of the cell (i.e. the cell body and all organelles) is assumed to be of secondary importance, justified by the observation that cytoplasts exhibit chemotaxis. The idealized cell is assumed to be suspended or completely nonadherent to the substratum so that upon exposure to chemoattractant, shape changes are not influenced by adhesion. For simplicity we assume only lamellipodial extensions in a single plane, consistent with a cell being constrained between two surfaces as in Zigmond's visual assay chamber (i.e. the two-dimensional analogue of a neutrophil stimulated in suspension). The goals of this version of the third generation model are to predict the spatial and temporal correlations of the lamellipodial activity which ensues when the non-stimulated idealized circular cell with outer annulus of reactive, contractive fluid is subject to a chemoattractant field (lamellipodial activity being the dynamic changes in radius around the perimeter), to interpret the correlations given the assumed receptor signal generation and motility response mechanisms, and to quantitatively relate these correlations to the parameter values in the mathematical terms modelling these mechanisms. For the model to be acceptable, it should admit the development of a global anterior–posterior cell polarity which is largely stable over time, the occasional occurrence of simultaneous lamellipodia, and the rare occurrence of an inversion of polarity due to dominance of a lamellipodium developing away from the leading edge. Of course, the surface receptor distribution should conform to available data.

In the two-phase fluid model of cortical cytoplasm developed by Dembo, Harlow & Alt, molecular details, such as the microstructure of the actin network, are ignored and the mean local volume fraction of only a network phase (F-actin and accessory proteins) and aqueous phase (the remaining solution of the cortical cytoplasm, assumed organelle-free) are considered, as well as their respective velocities. A mass balance equation for the network phase, including assembly–disassembly kinetics, defines the volume fractions and a force balance equation for each phase defines the two velocities and the effective hydrostatic pressure of the aqueous phase. Alt has applied this theory to the case of extension and retraction of a single lamellipodium (Alt, 1990a). The force balance equations include terms representing viscous shear, frictional drag, swelling, pressure, and contraction forces. Except for the last term which models the forces associated with a distribution of actomyosin complexes throughout the network, the other terms are characteristic of a viscous solution flowing through a dynamic polymeric network. Assuming linear assembly–disassembly kinetics around an equilibrium value for the network volume fraction, that the network can 'slip' with respect to the membrane at the tip of the lamellipodium (since accessory proteins which connect the network to the membrane are likely to be mobile in the membrane), and that the hydrostatic pressure inside the tip is related to the membrane curvature there with surface tension proportional to the volume fraction of network, Alt shows that the modelling equations yield numerical solutions consistent with a lamellipodium *autonomously* exhibiting cycles of extension and retraction for certain combinations of parameter values. The '*internal* driving force' for extension is the hydrostatic pressure in the cell body when it exceeds that at the tip where the network has detached ('slipped'); that for retraction is actomyosin contraction force when the network is continuous from the cell body to the tip. Autonomously means that an '*external* driving force', such as receptor-mediated chemoattractant stimulation, is not necessary to evoke the observed dynamics. Clearly, however, this coupling exists in neutrophils, and a way to account for it is discussed below.

In order to address the problem of a cell exhibiting multiple lamellipodia and the related problem of their global co-ordination, we assume that the cortical network in the idealized annulus flows in response to the lamellipo-

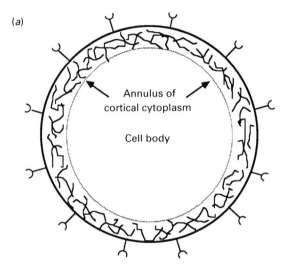
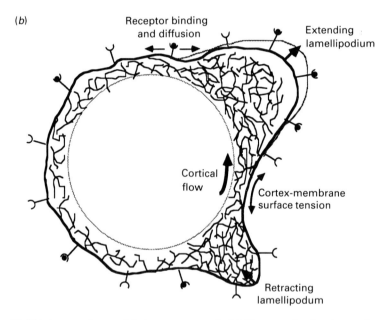

Fig. 10. Third generation model chemosensory cell. (*a*) The non-stimulated nonpolar state: the idealized circular cell has an outer annulus of reactive, contractive cortical fluid, representing the cell cortex, which is assumed to be spatially homogeneous. Chemoattractant receptors are located uniformly on the plasma membrane. (*b*) The chemoattractant-stimulated state: lamellipodia locally extend and retract in response to local fluctuating receptor signals, individually obeying a radial force balance. The cortical network concentration at the base of each lamellipodium which contributes to the radial force balance varies due to cortical flow, which obeys a tangential force balance. A cortex tension related to the local curvature from a lamellipodium is proposed to contribute to the tangential force balance, counteracting the extension of a lamellipodium. See text for more discussion of operative forces.

dial activity, satisfying a force balance in the tangential direction (Fig. 10b). The terms appearing in this equation are the same as for the network phase force balance in the radial direction for a lamellipodium, as is necessary since the networks in the cortex and lamellipodia are assumed to have the same properties, except that an additional force is added to model a 'cortex–membrane surface tension', i.e. the resistance of the cortex to curvature associated with evaginations like lamellipodia (and invaginations) because of network and perhaps other cytoskeletal attachments to the plasma membrane. Further co-ordination between lamellipodia is associated with the hydrostatic pressure term in the radial force balance for each lamellipodium, since it is modulated by cell volume changes, or equivalently, the changing mean radius for the cell. However, in order to illustrate the model behaviour here, we assume in what follows that changes in radius do not occur, i.e. that lamellipodium formation is incipient, being inferred from the distribution of cortical network in the annulus. Further, it is assumed that viscous shear force is negligible compared to drag, swelling, and contraction forces in the network, and that there is no net assembly of the network. In this special case the tangential force balance equation can be combined into the mass balance equation to eliminate the cortical network velocity, yielding a single partial differential equation which describes the temporal and angular dependence of the distribution of the network around the annulus:

$$\underset{\text{network accumulation}}{\frac{\partial a}{\partial t}} + \underset{\substack{\text{friction} \\ }}{\frac{1}{\phi}\frac{\partial^2}{\partial x^2}} \left\{ \underset{\text{swelling}}{\sigma \ln\left(1 - \frac{a}{a_{\max}}\right)} + \underset{\text{contraction}}{\frac{\Psi}{2} a^2} + \underset{\text{cortical tension}}{\nu \frac{\partial}{\partial x}\left(a \frac{\partial a}{\partial x}\right)} \right\} = 0 \qquad (10)$$

(network convection)

$a(t,x)$ is the network concentration, t is time, and x is the angular coordinate. ϕ is the drag coefficient describing the magnitude of the frictional force on the network due to the flowing cytosol, σ describes the magnitude of the swelling force as the network becomes concentrated (a approaches $a_{\max}$), Ψ describes the magnitude of the contraction force based on a pairwise interaction law for actomyosin complexes, and v describes the magnitude of the cortical tension force depending on the local cortex curvature (here inferred from the local spatial gradient of a since the radius is assumed constant), with tension proportional to the local value of a (note that the cortical tension term in eqn (10) is the linearized version of a more general form).

Discussion is now briefly turned back to receptor signalling and signal–response coupling. To illustrate the model behaviour, very simple assumptions are made once again: the chemoattractant receptors are a homogeneous population of monovalent, reversibly binding receptors with single affinity which do not internalize nor form ternary complexes, as for the illustration of the second generation model, but *can* diffuse in the membrane (with free and bound receptors having different diffusion coefficients). The full spatial description of the motility mechanism embodied in eqn (10) makes an account for receptor diffusion possible, which is not the case for the spatially coarse two compartment structure of the second generation models. Using an extension of the stochastic methods used to derive eqn (4), it is possible to numerically simulate the fluctuating density of free and bound receptors at any location around the cell membrane as the receptors bind and dissociate chemoattractant and diffuse in the membrane. The simulation involves dividing the annulus into a large number, n, of pair-wise connected compartments, where the larger the value of n, the better the approximation to spatially-continuous receptor densities. As before, the magnitudes of the fluctuations are fully defined by receptor parameters. Signal–response coupling is accounted for in the simplest possible fashion in this illustration, by making a parameter in eqn (10) linearly proportional to the local bound receptor density (implying the dynamics of the transduction pathway connecting a bound receptor to the action of a second messenger are rapid relative to the effect or response elicited by the second messenger). In particular, we choose here the contraction parameter Ψ. A biological interpretation is the local, relatively rapid activation of myosin light chain kinase associated with chemoattractant binding.

The results of Fig. 11(*a*) show representative results from numerically solving eqn (10) for one realization of stochastic receptor binding and diffusion around the cell annulus for the case of a uniform chemoattractant concentration. In presentation of the simulation results, the annulus has been cut and unfolded onto a straight line, so the cell coordinates $x=0$ and 1 represent the same positions on the annulus. The network concentration is scaled with respect to the initial uniform concentration (with the assumption here of no net assembly, the total amount of network remains constant), the bound receptor density deviation is scaled to the mean density, and time is scaled to the mean time a receptor remains bound as in Fig. 3. Initially, the network and receptors are taken to be uniformly distributed with $I=P$ in all compartments, thus $a=1$, $b=0$ at all positions for $\tau=0$.

The key to interpreting the results is to recognize that the network will tend to accumulate ($a>1$) where the contraction coefficient is highest, i.e. where there are sustained positive deviations in bound receptor density ($b>0$). As can be seen, the initially nonstimulated non-polar cell develops a bipolar morphology at early times, but then returns to a nonpolar state by $\tau=1$, reminiscent of the periodic nature of cell locomotion often reported for neutrophils. If the region of network accumulation is interpreted as the leading edge ($x\approx0.8$), the reason for the loss of polarity is clear from

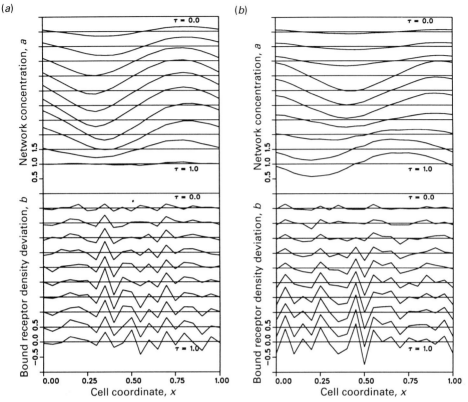

Fig. 11. Simulation of polarity development and lamellipodial activity. The spatial profile of network concentration, a, is plotted in the upper panel and the spatial profile of deviations in the density of bound receptors from the mean value, b, is plotted in the lower panel over $\tau = 0$–1 in ten increments of $\tau = 0.1$. The scale on the ordinate applies directly to the $\tau = 1$ profiles, and to the other profiles upon translation down to $a = 1$ or $b = 0$. (a) and (b) are separate simulations of eqn (10) based on different realizations of the Wiener process. Values of receptor parameters for the neutrophil–FNLLP case study are used with $n = 20$. See text for more discussion.

inspecting the bound receptor deviations: a significant positive deviation in bound receptor density is sustained in the uropod ($x \approx 0.3$) over the course of the simulation which creates a focus of contraction force there, leading to a successful competition with the established focus at the leading edge. Fig. 11(b) is another simulation for a different realization of stochastic receptor binding and diffusion (i.e. W). Again a bipolar state evolves from the uniform initial state, but here the receding leading edge is reinforced and repositioned due to the influence of a sustained positive deviation in bound receptor density again located away from the leading edge ($x \approx 0.45$), suggestive of speed variation with directional change in a neutrophil exhibiting random motility.

It is not apparent from comparing the network concentration and bound receptor density deviation profiles why the leading edge develops at the observed positions on the annulus. Fig. 12 shows the profiles corresponding to Fig. 11(b) at values of τ two (first column) and one (second column) orders of magnitude smaller than those displayed in Fig. 11(b) (third column). The correspondence between regions of network accumulation and sustained positive deviations in bound receptor density is clear at the earliest times, $\tau = 0.001-0.004$. The orientation of polarity (i.e. the locations of the emerging leading edge and uropod) is rapidly established due to a relatively small localized fluctuation in receptor binding (cf the magnitude of the fluctuations at later times), suggesting that the mechanical properties of the network confer significant sensitivity amplification. The correspondence becomes far less obvious for $\tau \geqslant 0.005$, but, as can be seen during the $\tau = 0.1-1.0$ sequence, the consequences of receptor binding fluctuations are still critical in determining the network distribution. In passing, we note that, by adjusting the network property parameters, it is possible to obtain multiple regions of network accumulation, suggestive of the occasional occurrence of multiple competing lamellipodia.

Critique

Interpreting the region of network accumulation as the leading edge is arguable since it arises as a focus of actomyosin contraction. An interpretation of that region as the uropod is certainly tenable and underscores a difficulty in applying the current cell mechanics theory: myosin is not considered as a separate species from what has been referred to above as 'network'. The form of the contraction term in eqn (10) implies that the concentration of actomyosin complexes is proportional to the network concentration, e.g. network represents F-actin and myosin is everywhere present in excess. The observation that myosin II is localized in the uropod of neutrophils makes such an interpretation doubtful. This ambiguity will need to be resolved. Also, the simulations presented here assume no net assembly of network. Clearly, future simulations of polarity development and lamellipodial activity must make proper account for the rapid and extensive assembly of network (conversion of G-actin into F-actin) which correlates with lamellipod formation.

Despite these deficiencies in this preliminary version of the third generation model, there is finally an integration of receptor signal generation and realistic motility. The properties ascribed to our caricature of the cortical network embodied in eqn (10) make it a morphogenetic medium, i.e. capable of generating spatial patterns. The wealth of spatial patterns that eqn (10) can predict is augmented with complex dynamical properties by tying the value of a cortical network parameter to the fluctuating value of a receptor signal, a desirable and seemingly necessary feature of any model for lamellipodial activity. The model also provides a framework

THEORIES AND MODELS OF GRADIENT PERCEPTION

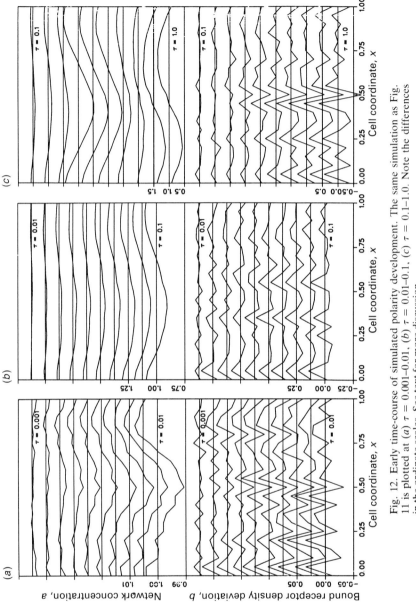

Fig. 12. Early time-course of simulated polarity development. The same simulation as Fig. 11 is plotted at (a) $\tau = 0.001-0.01$, (b) $\tau = 0.01-0.1$, (c) $\tau = 0.1-1.0$. Note the differences in the ordinate scales. See text for more discussion.

for accommodating receptor dynamics and transduction cascades of arbitrary complexity. Receptor dynamics schemes which account for affinity conversion, ternary complex formation, internalization, and recycling such as depicted in Fig. 9, can occur locally within each of the n compartments. The main deficiency of the third generation model is not being able to predict cell translocation, since there is currently no account for adhesion of lamellipodia to a substratum or anchorage around fibrils. This may define the fourth generation model, the model that is "made as simple as possible" for understanding the certainty of (chemo)taxis.

ACKNOWLEDGEMENTS

The author is indebted to Doug Lauffenburger and Sally Zigmond for their many contributions to these ideas and their development and to Wolfgang Alt for his ongoing collaboration. Support for the ongoing project is from an NSF Research Initiation Award NSF/EET–8809439 and an NSF Presidential Young Investigator Award NSF/BCS–8957736.

REFERENCES

Alt, W. (1988). Models of cytoplasmic motion. In *From Chemical to Biological Organization*, eds. Markus, M., Muller, S. C., and Nicolis, G., vol. 39, Series in Synergetics, pp. 235–47. Berlin: Springer.

Alt, W. (1990a). Mathematical models and analysing methods for the lamellipodial activity of leukocytes. In *Biomechanics of Active Movement and Deformation of Cells*. Springer (NATO ASI Ser. H) (in press).

Alt, W. (1990b). Correlation analysis of two-dimensional moving paths. In *Proceedings of the Workshop on Biological Motion*. Springer (Lecture Notes in Biomathematics) (in press).

Berg, H. C. & Purcell, E. M. (1977). Physics of chemoreception. *Biophysical Journal*, **20**, 193–219.

Cassimeris, L. & Zigmond, S. H. (1990). Chemoattractant stimulation of polymorphonuclear leukocyte locomotion. In *Seminars in Cell Biology*, vol. 1, pp. 125–34. Saunders Press.

DeLisi, C., Marchetti, F. & Del Grosso, G. D. (1982). A theory of measurement error and its implications for spatial and temporal gradient sensing during chemotaxis. *Cell Biophysics*, **4**, 211–29.

DeLisi, C. & Marchetti, F. (1983). A theory of measurement error and its implications for spatial and temporal gradient sensing during chemotaxis – II. The effects of non-equilibrated ligand binding. *Cell Biophysics*, **5**, 237–53.

Dembo, M. (1989). Field theories of the cytoplasm. *Comments in Theoretical Biology*, **1**, 159–77.

Dembo, M., Harlow, F. J. & Alt, W. (1984). The biophysics of cell surface motility. In *Cell Surface Dynamics: Concepts and Models*. Perelson, A., DeLisi, C. & Wiegel, F., eds. pp. 495–543. New York: Marcel Dekker.

Devreotes, P. N. & Zigmond, S. H. (1988). Chemotaxis in eukaryotic cells: A focus on leukocytes and *Dictyostelium*. *Annual Reviews in Cell Biology*, **4**, 649–86.

Gardiner, C. W. (1985). *Handbook of Stochastic Methods for Physics, Chemistry and the Natural Sciences*. New York: Springer-Verlag.

Koshland, D. E., Goldbeter, A. & Stock, J. B. (1982). Amplification and adaptation in regulatory and sensory systems. *Science*, **217**, 220–5.
Lackie, J. M. (1986). *Cell Movement and Cell Behaviour*. London: Allen & Unwin.
Lauffenburger, D. A. (1982). Influence of external concentration fluctuations on leukocyte chemotactic orientation. *Cell Biophysics*, **4**, 177–209.
Lauffenburger, D. A., Farrell. B. E., Tranquillo, R. T., Kistler, A. & Zigmond, S. H. (1987). Commentary: Gradient perception by neutrophil leukocytes, continued. *Journal of Cell Science*, **88**, 415–16.
Omann, G. M., Allen, R. A., Bokoch, G. M., Painter, R. G., Traynor, A. E. & Sklar, L. A. (1987). Signal transduction and cytoskeletal activation in the neutrophil. *Physiological Reviews*, **67**, 285–321.
Oster, G. F. & Perelson, A. S. (1985). Cell spreading and motility. *Journal of Mathematical Biology*, **21**, 383–8.
Singer, S. J. & Kupfer, A. (1986). The directed migration of eukaryotic cells. *Annual Reviews in Cell Biology*, **2**, 337–65.
Sklar, L. A. & Omann, G. M. (1990). Kinetics and amplification in neutrophil activation and adaptation. In *Seminars in Cell Biology*, vol. **1**, pp. 115–24. Saunders Press.
Sullivan, S. J., Daukas, G. & Zigmond, S. H. (1984). Asymmetric distribution of the chemotactic receptor on polymorphonuclear leukocytes. *Journal of Cell Biology*, **99**, 1461–7.
Sullivan, S. J. & Zigmond, S. H. (1980). Chemotactic peptide receptor modulation in polymorphonuclear leukocytes. *Journal of Cell Biology*, **85**, 703–11.
Tranquillo, R. T. (1990). Models of chemical gradient sensing by cells. In *Proceedings of the Workshop on Biological Motion*. Springer (Lecture Notes in Biomathematics) (in press).
Tranquillo, R. T. & Lauffenburger, D. A. (1986). Consequences of chemosensory phenomena for leukocyte chemotactic orientation. *Cell Biophysics*, **8**, 1–46.
Tranquillo, R. T. & Lauffenburger, D. A. (1987). Stochastic model of chemosensory cell movement. *Journal of Mathematical Biology*, **25**, 229–62.
Tranquillo, R. T., Lauffenburger, D. A. & Zigmond, S. H. (1988). Stochastic model for leukocyte random motility and chemotaxis based on receptor binding fluctuations. *Journal of Cell Biology*, **106**, 303–9.
Wiegel, F. W. (1983). Diffusion and the physics of chemoreception. *Physics Report*, **95**, 283–319.
Zhu, C., Skalak, R. & Schmid-Schonbein, G. W. (1989). One-dimensional steady continuum model of retraction of pseudopod in leukocytes. *Journal of Biomechanical Engineering*, **111**, 69–77.
Zigmond, S. H. (1977). Ability of polymorphonuclear leukocytes to orient in gradients of chemotactic factors. *Journal of Cell Biology*, **75**, 606–16.
Zigmond, S. H. (1981). Consequences of chemotactic peptide receptor modulation for leukocyte orientation. *Journal of Cell Biology*, **88**, 644–7.
Zigmond, S. H. (1982). Polymorphonuclear leucocyte response to chemotactic gradients. In *Cell Behavior*. Bellairs, R., Curtis, A. and Dunn, G., eds. pp. 183–202. Cambridge University Press.
Zigmond, S. H. & Sullivan, S. J. (1979). Sensory adaptation of leukocytes to chemotactic peptides. *Journal of Cell Biology*, **82**, 517–27.
Zigmond, S. H., Klausner, R., Tranquillo, R. T. & Lauffenburger, D. A. (1985). Analysis of the requirements for time-averaging of the receptor occupancy for gradient detection by polymorphonuclear leukocytes. In *Membrane Receptors and Cellular Regulation*, pp. 347–56. New York: Alan R. Liss.

GENETICS, STRUCTURE, AND ASSEMBLY OF THE BACTERIAL FLAGELLUM

ROBERT M. MACNAB

Department of Molecular Biophysics and Biochemistry, Yale University, New Haven, Connecticut 06511, USA

INTRODUCTION

One of the most complex structures in the bacterial cell is its organelle of motility, the flagellum. Some idea of its complexity can be gained from the number of genes that are involved in flagellar assembly and function, about 40. To put this number in perspective, we may note that even complex biosynthetic pathways involve far fewer genes (for example, 14 in the case of histidine biosynthesis, Sanderson & Roth (1988). Indeed, of all the gene systems, only the ribosomal system (with about 90 genes, Bachmann (1983) is more complex than that of the flagellum. Why does the flagellum need such a large genetic underpinning? Part of the reason is certainly the intrinsic complexity of the structure and the functions that it has to perform. Another factor is that the structure has to be integrated in a specific fashion into the various components of the cell surface, namely, the cell membrane, the periplasm, the peptidoglycan layer, and the outer membrane (DePamphilis & Adler, 1971*b*). And, finally, there is the fact that most of the mass of the flagellum (notably the filament) is external to the cell, so that assembly is a complicated logistical process rather like the construction of a space station.

A system as complex as this has many interesting features: genetic regulation, biosynthesis, export, assembly, and function. One thing that makes the flagellar system especially attractive as a subject for study is the fact that flagellation and motility, though undoubtedly of great value to the cell under natural conditions, is not essential for viability. Therefore absolute (even null) mutants with defects in flagellar genes are readily obtained and do not need any special care for their maintenance. This contrasts with the situation for many other complex gene systems, which are involved in vital cellular functions such as DNA replication or protein synthesis, where one is forced to work with conditional mutants. For reviews of the bacterial flagellum, see Iino (1985), Macnab (1987*a,b*), Macnab & DeRosier (1988), Eisenbach (1990), and Jones & Aizawa (1990).

Before going into some more detailed considerations of flagellar structure and assembly, it is necessary to give a brief overview of the main features of the organelle. The descriptions in this chapter will be confined to the two closely related bacteria, *Escherichia coli* and *Salmonella typhimurium*.

Mechanism of propulsion

How does the flagellum propel the cell? The answer is an elegantly simple one: Construct a helical filament and then rotate it by a motor at its base (Silverman & Simon, 1974a). As has been pointed out, it is not a trivial matter to achieve such a helical structure from subunits of a single protein, however (Calladine, 1982; Kamiya, Hotani & Asakura, 1982). The intersubunit interactions require a rather specific type of symmetry breaking, without which the filament would be straight.

It is interesting that the bacterial flagellum seems to be the only mechanoenzyme to have evolved with a rotary mechanism and that nature has not employed such mechanisms more widely.

Energy source

The next obvious question is the source of energy. In contrast to virtually all other mechanoenzymes that have been studied, the bacterial flagellum does not use the free energy of ATP hydrolysis but rather takes advantage of its location at the cell surface and employs ionic transmembrane potentials. In many of the bacterial species that have been examined, the ion used is the proton (Larsen *et al.*, 1974a; Manson *et al.*, 1977), which has been actively pumped out of the cell by the electron transport chain (or the proton ATPase under anaerobic conditions). In several species of alkalophilic and marine bacteria, the ion used is sodium (Imae & Atsumi, 1989), pumped out of the cell by sodium/proton antiporters or an electron transport chain; typically, such bacteria have relatively low proton potentials and therefore, if their motors were proton driven, they would (for the same proton stoichiometry) be able to perform much less work than those of neutrophilic species like *E. coli*. The fact that there are sodium-driven motors as well as proton-driven ones narrows the mechanism of rotation down to one involving electrostatic interactions and ion association/dissociation events, rather than one involving hydrogen bonding.

Switching

Another important feature of the flagellar motor is that it switches its direction of rotation back and forth between counterclockwise and clockwise (Silverman & Simon, 1974a). This switching goes on all the time, and is responsible for the motility pattern of alternate swimming and tumbling that takes the cell on a random zig-zag trajectory through the medium. However, the probability of the motor being in counterclockwise vs. clock-

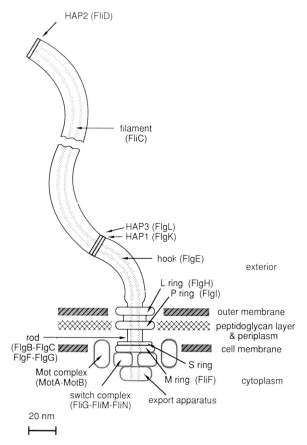

Fig. 1. Schematic illustration of the bacterial flagellum of *S. typhimurium*. Morphological features such as rings, rod, etc. are indicated, together with the gene products from which they are constructed. The filament is much longer than shown here. The origin of the S ring is unknown. The locations of the MotA and MotB proteins, the three switch proteins, and the proteins that comprise the flagellum-specific export apparatus are based on a variety of lines of indirect evidence. The channel through which the protein subunits travel to reach their site of assembly is shown in dotted outline. HAP, hook-associated protein.

wise rotation can be modulated by sensory information in a manner that results in a bias in the trajectory and hence the behavioural response known as chemotaxis (Larsen *et al.*, 1974*b*).

FLAGELLAR STRUCTURE

The overall structure of the flagellum is illustrated in Figure 1. The known structure consists of (i) the helical filament or propeller, (ii) a hook that is thought to be a universal joint, (iii) various junction and capping proteins for the hook and filament (collectively known as the hook-associated proteins, or HAPs (Homma et al., 1984b), (iv) a rod that is thought to be the transmission shaft, (v) outer rings that are thought to be a bushing, and (vi) inner rings that may be directly or indirectly involved in motor rotation. The rod–ring complex is commonly known as the basal body (DePamphilis & Adler, 1971a).

In addition, there is strong circumstantial evidence that there must also be structures associated with energy transduction (conversion of proton motive force into torque) (Block & Berg, 1984; Blair & Berg, 1988; Khan, Dapice & Reese, 1988), switching (Parkinson et al., 1983; Yamaguchi et al., 1986a,b), export of external components (Kuwajima et al., 1989), and anchoring to the cell surface. Before going on to consider flagellar structure and assembly, I shall look at the underlying genetics, since it is largely through this that an understanding of the flagellar system has developed.

THE FLAGELLAR REGULON

The flagellar regions

As has already been mentioned, there are some 40 flagellar genes (Bachmann, 1990; Sanderson & Roth, 1988). There are also six chemotaxis genes involved in signal transduction, and a substantial number of receptor genes involved in the initial generation of sensory information. The flagellar and chemotaxis genes are highly clustered on the chromosome (Fig. 2). Until recently it was thought that there were three such clusters in *E. coli* (Kutsukake et al., 1988) plus, in *S. typhimurium*, a fourth small one containing a second allele of the gene encoding the filament protein, flagellin. However, a more careful scrutiny of flagellar region III has revealed that it is in fact two regions with a substantial stretch of DNA unrelated to flagellar function in between (I. Kawagishi, V. Müller, A. Williams & R. Macnab, unpublished data). There are therefore four major regions, termed I, II, IIIa and IIIb, which we will now consider in more detail (Table 1). (In the literature prior to 1989, the nomenclature for flagellar genes was different for *E. coli* and *S. typhimurium*, and had arisen more or less randomly. We follow here the new unified nomenclature as drawn up by Iino et al. (1988)).

Region I genes

Region I contains three operons. The first is *flgA*, consisting of a single gene of unknown function. Then follows the largest flagellar operon, *flgB*, which contains many of the known structural genes, including those for four rod proteins, the hook protein, and the proteins for the outer two rings of the basal body (Komeda, Silverman & Simon, 1978; Homma et al., 1987c; Jones, Homma & Macnab, 1987; Jones, Homma & Macnab, 1989; Okino et al., 1989; Homma, DeRosier & Macnab, 1990a; Homma

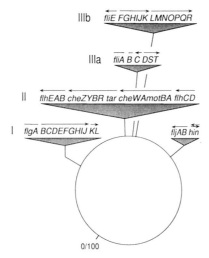

Fig. 2. Flagellar and related genes of *S. typhimirium*. The genes are highly clustered into four main regions, two of which (IIIa and IIIb) have only recently been recognized as being distinct (I. Kawagishi et al., unpublished data). There is also a small region containing a second allele of the flagellin gene (*fljB*) and control genes for alternate expression of *fljB* and the other flagellin allele, *fliC*. The *flhE* gene (M. Kim & P. Matsumura, personal communication) and the *fliS* and *fliT* genes (I. Kawagishi et al., unpublished data) have only recently been discovered. Operons are shown by overlining, with an arrowhead where the polarity is known. The map has been rotated to place the flagellar genes at the top; the origin of the map is at 0/100.

et al., 1990*b*); it also contains a gene, *flgD*, that participates in a poorly understood process of 'rod modification' during the flagellar assembly process (Suzuki *et al.*, 1978; Suzuki & Komeda, 1981; Iino, 1985), and a gene of unknown function, *flgJ*.

The *flgB* operon is interesting in that it is punctuated between *flgG* and *flgH* by an unusually large stretch of noncoding DNA (54 bp) containing a transcription-terminator-like sequence, but no obvious internal promoter (Jones *et al.*, 1989). The genes before this point are associated with what have been termed axial components of flagellar structure (such as rod and hook; Homma *et al.*, 1990*a,b*), while those immediately after it encode the outer rings (Jones *et al.*, 1989). These structures are exported and assembled in a very different fashion, and so the punctuation of the operon may reflect a need for different patterns of gene expression. However, the role of this non-coding sequence has not yet been examined experimentally.

The last operon in region I is *flgK*, and contains the genes encoding the two HAPs that exist at the hook-filament junction (Homma, Kutsukake & Iino, 1985; Homma & Iino, 1985*a*; Ikeda *et al.*, 1987; Homma *et al.*, 1990*a*).

Table 1. *Flagellar and motility operons in* S. typhimurium *and* E. coli

Region I (middle flagellar structure)
  flgA operon (class 2; unknown function)
    FlgA (unknown function)

  flgB operon (class 2; hook and basal-body structure)
    FlgB (rod protein)
    FlgC (rod protein)
    FlgD (rod modification)
    FlgE (hook protein)
    FlgF (rod protein)
    FlgG (rod protein)
    FlgH (basal-body L-ring protein)
    FlgI (basal-body P-ring protein)
    FlgJ (unknown function)

  flgK operon (class 3a; hook-filament junction structure)
    FlgK (HAP1; first junction protein)
    FlgL (HAP3; second junction protein)

Region II (flagellar regulation; motility; chemotaxis)
  flhB operon (class 2; flagellar export and assembly?)
    FlhB (unknown function)
    FlhA (flagellum-specific export?)
    FlhE (unknown function)

  tar operon (class 3b; sensory reception and transduction)
    Tar (aspartate/maltose receptor)
    Tap (dipeptide receptor; established in *E. coli* only)
    CheR (receptor methylating enzyme)
    CheB (receptor demethylating enzyme)
    CheY (modulation of flagellar switch bias)
    CheZ (deactivation of CheY)

  motA operon (class 3b; energy transduction and sensory transduction)
    MotA (motor rotation)
    MotB (motor rotation)
    CheA (phosphorylation of CheY and CheB)
    CheW (modulation of CheA activity)

  flhD operon (class 1; master regulatory operon)
    FlhD (sigma factor for class 2 and 3a operons?)
    FlhC (sigma factor for class 2 and 3a operons?)

Table 1 *continues opposite*

Region I is therefore seen to contain predominantly genes involved in hook and basal-body structure.

Region II genes

Region II cannot be described in terms of genes of any single overall category. It contains the *flhD* operon, which encodes two proteins that are needed for the expression of all of the other flagellar operons (see below). It also contains another flagellar operon, *flhB*, which contains three genes of poorly understood function (M. Kim & P. Matsumura, personal commu-

Table 1 *continued*

Region IIIa (late flagellar structure and assembly)
    fliA operon (class 2; transcriptional initiation)
        FliA (sigma factor for class 3a and 3b operons)
    fliB operon (class 3b? flagellin modification)
        FliB (flagellin methylating enzyme; established in *S. typhimurium* only)
    fliC operon (class 3b; filament structure)
        FliC (flagellin)
    fliD operon (class 3a; filament assembly)
        FliD (filament capping protein)
        FliS (negative regulation of expression of class 3a and 3b operons?)
        FliT (negative regulation of expression of class 3a and 3b operons?)

Region IIIb (early flagellar structure)
    fliE operon (class 2; basal body structure)
        FliE (basal-body protein)
    fliF operon (class 2; basal-body structure; energy transduction and switching; export and assembly?)
        FliF (basal-body M-ring protein)
        FliG (flagellar switch)
        FliH (flagellum-specific export?)
        FliI (flagellum-specific export?)
        FliJ (unknown function)
    fliL operon (class 2; energy transduction and switching)
        FliL (unknown function)
        FliM (flagellar switch)
        FliN (flagellar switch)
        FliO (unknown function)
        FliP (unknown function)
        FliQ (unknown function)
        FliR (unknown function)

---

nication); for one of these genes, *flhA*, a role in the process of flagellum-specific export is suggested by experiments that will be described below.

The remaining two region II operons contain a mixture of different types of genes. The *motA* (also known as *mocha*) operon (Silverman & Simon, 1976) includes the two motility genes, *motA* and *motB*, whose products are not needed for flagellar assembly, but are needed for its rotation (Block & Berg, 1984; Blair & Berg, 1988). Yet it also contains two of the chemotaxis genes, *cheA* and *cheW*, neither of whose products are thought to have any direct interaction with the flagellum. Thus these two pairs of gene products have a rather indirect functional relationship to each other; however, they do share the common property that they are not needed until the flagellum has been assembled and their existence in the same operon is consistent with the need for similar regulation of expression. Finally, there is the *tar* (or *meche*) operon (Krikos *et al.*, 1983; Mutoh & Simon, 1986), which contains the remaining four central sensory genes, as well as two of the receptor genes. The rest of the receptor genes, such as *tsr*, exist in isolated regions of the chromosome, unlinked to other flagellar or chemotaxis genes.

Thus we can describe region II as a potpourri of flagellar genes (none of which have so far been demonstrated to be structural), all of the known motility and chemotaxis genes, and some receptor genes.

Region IIIa genes

Region IIIa may be viewed as one dedicated to genes that are related to the expression and structure of the later components of the overall system, and most notably to the filament protein, flagellin. It contains: (i) *fliA*, which encodes a flagellum-specific sigma factor that operates on class 3a and class 3b operons (Kutsukake, Ohya & Iino, 1990; Ohnishi et al., 1990); (ii) *fliB*, which encodes an enzyme that carries out post-translational modification of flagellin by methylating certain of its lysine residues (Stocker, McDonough & Ambler, 1961; Konno et al., 1976); (iii) *fliC*, which encodes flagellin itself (Yamaguchi et al., 1984); and (iv) the *fliD* operon. Until very recently, it was thought that the latter operon contained only *fliD*, which encodes the filament capping protein, HAP2 (Homma et al., 1985; Kutsukake et al., 1990); however, cloning and DNA sequence analysis has shown that *fliD* is followed by two small genes, *fliS* and *fliT*, whose function has not yet been established (I. Kawagishi, V. Müller & R. Macnab, unpublished data).

Region IIIb genes

In contrast to region IIIa, which is associated with the last events in the flagellar assembly process, region IIIb seems to be associated with earlier events. It contains three operons. The first, *fliE*, contains a single gene (Kutsukake et al., 1990), whose product is a basal-body protein (V. Müller & C. Jones, unpublished data). The location and function of FliE within the basal body is not yet known. Since related genes tend to be organized into multi-gene operons, the fact that *fliE* is in an operon by itself may indicate that its product plays a unique structural role. The *fliF* operon includes the M-ring structural gene (Homma et al., 1987a; Jones et al., 1989), one of the three switch genes (which are also needed for energy transduction) (Yamaguchi et al., 1986b; Kihara et al., 1989), a gene (*fliK*) whose product is involved in controlling the length of the flagellar hook (Silverman & Simon, 1972; Patterson-Delafield et al., 1973), and three genes of poorly understood function (*fliH*, *fliI*, and *fliJ*). Two of the latter may be involved in the process of flagellum-specific export (A. Vogler & R. Macnab, unpublished data), which will be discussed below; the *fliK*-mediated process of hook length control may also be related to flagellum-specific export. The last operon in region IIIb, *fliL*, contains seven genes,

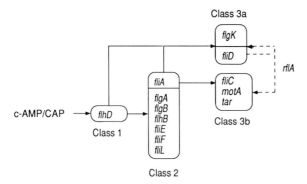

Fig. 3. Hierarchy of operons within the flagellar regulon of *S. typhimurium*. The sole operon is class 1, *flhD*, activates operons in classes 2 and 3a. The *fliA* operon in class 2 encodes a flagellum-specific sigma factor that recognizes promoters in classes 3a and 3b, so that class 3a is under dual control from classes 1 and 2. The other operons in class 2 (*flgA*, etc) encode many of the known flagellar structural genes, but these genes are also necessary for expression of class 3b and to a lesser extent class 3a genes (see text). In class 3a, the *fliD* operon exerts a repressor-like activity (*rflA*) on class 3a and 3b operons (see text). Based on data described by Iino (1985), Kutsukake *et al.* (1990), and Ohnishi *et al.*, 1990.

two of which (*fliM* and *fliN*) encode switch components (Yamaguchi *et al.*, 1986b; Kihara *et al.*, 1989), while the remaining five are of unknown function.

Regulation of expression of flagellar genes

As we have seen, within each region there are several operons, ranging in size from those that contain only a single gene to the largest (the *flgB* operon), which contains nine genes. In total there are at least 14 operons, which together constitute a unit of gene expression, or regulon (Kutsukake *et al.*, 1990). At the head of the hierarchy is the *flhD* operon, whose two products, FlhD and FlhC, are both absolutely required for expression of the remaining 13 operons. Activation of the *flhD* operon itself is under the control of c-AMP via the regulatory protein CAP (Komeda *et al.*, 1975; Silverman & Simon, 1974b), and so flagellar synthesis is sensitive to the catabolic state of the cell.

On the basis of operon fusions to *lacZ*, a rather complex hierarchy of six operon classes was described for *E. coli* (Komeda, 1982). Some features of this scheme seemed to contradict data concerning the assembly process, and in a more recent operon fusion study in *S. typhimurium* a simpler scheme has been presented, involving just three operon classes (Kutsukake *et al.*, 1990). There is biochemical evidence, however, that suggests matters are slightly more complicated, and we will use here a scheme with four classes of operons (Fig. 3), similar to that used by Iino (1985). Class 1 consists of only one operon, *flhD*, whose expression is independent of the status of the other operons. The products of the *flhD* operon, FlhD and FlhC, act directly on class 2 operons, which include the *fliA* operon (encoding a flagellum-specific sigma factor), and other operons that contain genes encoding many of the known structural components of the hook and basal body, as well as a variety of genes of less well understood function. *All* of the genes in class 2 operons are needed for expression of class 3b operons, which encode flagellin, the motility proteins MotA and MotB, and the chemotaxis proteins. In the *lac*-fusion study, class 2 genes were also found to be necessary for detectable expression of class 3a operons, which encode the HAPs. However, to some degree, expression of class 3a operons must also occur under direct control of the *flhD* operon, since *fliA* mutants do synthesize and assemble the HAPs (but not flagellin) (Homma *et al.*, 1984*b*). Negative regulation of class 3b operons, and also the *flgK* operon, by the product(s) of the *fliD* operon was indicated by the abnormally high levels of β-galactosidase activity that were detected in *fliD* mutants (Kutsukake *et al.*, 1990); this regulatory activity of the *fliD* operon has been termed *rflA*.

How should this hierarchy of operons be interpreted? The structural components encoded by class 3a or 3b operons either are incorporated into the structure at a later stage of the assembly pathway than components encoded by class 2 operons (as is the case with the HAPs and flagellin), or serve no useful purpose for the cell unless the flagellar apparatus has been assembled (as is the case with the chemotaxis proteins). Within the class 3a and 3b operons, class 3a may be regarded as the earlier, since (see below) HAPs are added before flagellin. Thus the broad features of the hierarchy make logistical sense, although it is not clear whether the hierarchy generates a sequential order of gene expression in the cell during the normal course of flagellar assembly, or whether it is there primarily to avoid waste under environmental conditions (such as catabolite repression) or genetic conditions (mutation) such that flagellar assembly cannot proceed to completion.

What is it about flagellar operons that distinguishes them from all others on the chromosome and allows them to be placed under separate control? As DNA sequence information about the flagellar gene system accumulated, it became evident that there was very little evidence for the use of the promoter consensus that is recognized by the primary sigma factor of RNA polymerase: TTGACA-N_{17}-TATAAT. Instead a distinctly different consensus was noted by Helmann & Chamberlin (1987): TAAA-N_{15}-GCCGATAA. This led them to propose that the flagellar system employs a special sigma factor. On the basis of sequence similarity (albeit not very strong) between both FlhC and FlhD and a known sigma factor (sigma-28 from *Bacillus subtilis*), plus the known role of FlhC and FlhD as positive regulators of the flagellar regulon, they suggested that one or both of these proteins might be the factor. Subsequently, the flagellum-specific sigma factor activity (Arnosti & Chamberlin, 1989) has been found to reside, not in either FlhC or FlhD, but in another protein, FliA (Ohnishi et al., 1990). It has also become clear that, although all flagellar operons (except for the master operon, *flhD*) display some version of the GCCGA-TAA portion of the flagellum-specific consensus, many do not display the TAAA portion (Bartlett, Frantz & Matsumura, 1988; Kutsukake et al., 1990). The relation between those that do (classes 3a and 3b) and do not (class 2) can be understood in terms of the hierarchy of operons within the regulon, which was discussed above. It is reasonable to assume that FliA requires the full flagellum-specific consensus, because the promoters for the class 3a and 3b operons are under FliA control. FlhD or FlhC, or both, might also be flagellum-specific sigma factors, but ones that recognize primarily the -10 portion of the consensus, i.e. the promoters for class 2 operons. For the *flhD* operon itself, incidentally, the nature of the promoter (and hence the means of operon expression) is problematical since there is not clear evidence for either a primary or a flagellum-specific consensus.

A truly remarkable aspect of the overall regulatory scheme is the fact that a large number of gene products, many of them known to be components of flagellar structure, are (at least in the formal sense) also positive regulators of gene expression, since mutants with defects in any of these genes do not express the class 3b operons and express the class 3a operons weakly. It is very hard to imagine that a multitude of flagellar structural proteins exercise control of gene expression in a direct fashion, as DNA binding proteins. A more plausible explanation would be that in some way the wild-type flagellar structure either sequesters or otherwise inactivates a negative regulator, or releases or otherwise activates a positive regulator; mutant flagellar structures would fail in this regard, and as a consequence class 3a and 3b operons would be repressed. Among the class 2 operons, therefore, the *fliA* operon should probably be viewed quite differently from all the others in that it is the only one that is *directly* involved in gene expression.

Another puzzle is the negative regulatory role (*rflA* activity; Kutsukake et al., 1990) of the *fliD* operon, in view of the fact that the FliD protein is a structural component that caps the filament (Ikeda, Asakura & Kamiya, 1985; Ikeda et al., 1987) and thus is located at the most physically remote point of the entire cell. It is difficult to see how it could exert control on gene expression from that location. Cloning and DNA sequence analysis of the *fliD* operon (I. Kawagishi et al., unpublished data) has recently established that it contains not only *fliD* but two additional flagellar genes (*fliS* and *fliT*). It may be that these, rather than *fliD*, are the ones involved in the *rflA* regulatory role, although the deduced amino acid sequences do not reveal any of the structural motifs that are characteristic of DNA binding proteins. Thus the mechanism of *rflA* regulation is completely obscure at present.

FLAGELLAR ASSEMBLY

We now turn to the question of flagellar assembly. In order to understand any assembly process, it is clearly necessary to know the structure that is being assembled, and it must be acknowledged that there are still many aspects of flagellar structure that are not well understood. For example, the components responsible for the switching of the flagellar motor have not yet been detected morphologically, even though there is genetic evidence (for interaction with a cytoplasmic protein, CheY) (Parkinson *et al.*, 1983; Yamaguchi *et al.*, 1986*a*) that places them at the location indicated in Figure 1. There is morphological evidence that supports the idea that the Mot proteins surround the basal body (Khan *et al.*, 1988), but this has not been proved directly. It is not known how the flagellum is anchored to the cell surface in order to permit torque to be transmitted to the external filament. Finally, although there is evidence in favour of a flagellum-specific export apparatus (see below), there is no morphological information about it and only tentative evidence concerning its genetic and biochemical identity. Thus our statement of the assembly process will largely be confined to the dozen or so proteins whose location and function in the flagellum are fairly well understood.

The morphogenetic pathway

Analysis of the assembly process has accumulated from electron microscopic analysis of partial structures, immuno-electron microscopy, and most recently from radiolabelling analysis of structures assembled by conditional mutants (Suzuki *et al.*, 1978; Suzuki & Komeda, 1981; Homma & Iino, 1985*a*; Iino, 1985; Ikeda *et al.*, 1985; Ikeda *et al.*, 1987; Jones & Macnab, 1990). The overall conclusions regarding the order of assembly of the various known substructures are given in Figure 4. The order is for the most part from proximal to distal: first the M ring and other associated structures in the cell membrane, then successive rod segments, the P and L rings, the hook, the hook–filament junction proteins and the filament capping protein, and finally the filament. There are some exceptions to this generalization of proximal-to-distal assembly: (i) the P and L rings are not distal to the entire rod; (ii) the filament is not distal to the filament cap; (iii) the Mot proteins can be synthesized and inserted into the membrane and function perfectly well after all the rest of the flagellar apparatus is complete (Block & Berg, 1984; Blair & Berg, 1988), although this may not be the normal temporal order of events in a wild-type cell; and (iv)

THE STRUCTURE OF THE BACTERIAL FLAGELLUM

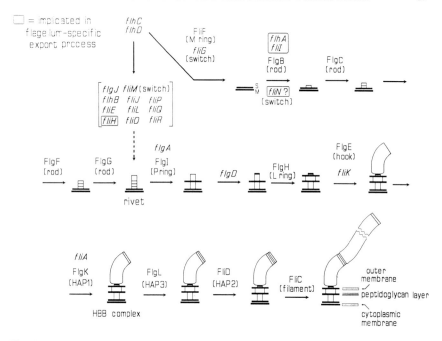

Fig. 4. Morphogenetic pathway for the flagellum of *S. typhimurium*. Where a gene product has been identified biochemically in the structure and its encoding gene has been established, the product is indicated (as, e.g. FlgE); otherwise, the gene is indicated (as, e.g. *flgD*). The genes in brackets are needed prior to the rivet structure, but it is not known where prior to that structure they participate. The order of addition of the proximal rod proteins (FlgB, FlgC, and FlgF) is tentative. HBB, hook-basal body. Based on the results of Suzuki *et al.* (1978), Suzuki & Komeda (1981), Homma & Iino (1985a), Ikeda *et al.* (1985), Ikeda *et al.* (1987), and Jones & Macnab (1990). Figure modified from Jones & Macnab (1990).

there are likely to be a number of peripheral structures such as the switch complex that presumably cannot assemble at the cytoplasmic face of the cell membrane until the M ring is complete.

The general progression from proximal to distal substructures is probably reflected even at the level of the individual subunit; in other words, a given substructure develops by addition of subunits at its distal end. This has been proved for the major external substructure, the filament (Iino, 1969; Emerson, Tokuyasu & Simon, 1970). It also applies (*in vitro* at least) to the hook (Iino, 1974), and it seems likely that it will apply to the rod and the HAPs.

A point worth noting is that in the flagellar assembly process there do not seem to be any examples of modular substructures being joined together. Although the various partial structures seen with mutants seem well-defined (Suzuki *et al.*, 1978; Suzuki & Komeda, 1981), this is presumably because they are blocked at a well-defined point, and it is probable that the route from one partial structure to the next in the normal assembly process involves addition of individual protein subunits to a single structure of ever increasing complexity. The reason that this type of assembly process, rather than a modular one, has evolved is probably because much of the assembly is taking place beyond the cell membrane. The logistics of exporting a modular structure, or of joining together modules in an extracellular location, seem much more daunting than simple subunit export and addition.

The description of the assembly process given thus far is empirical, and does not give any indication of the mechanism or mechanisms by which both export and assembly occur. It will be helpful first to categorize the various substructures according to their cellular location. At least five distinct locations are involved: (i) peripheral to the cell membrane (switch complex? export apparatus?), (ii) integral to the cell membrane (M ring, Mot proteins), (iii) periplasmic (rod and P ring), (iv) integral to the outer membrane (L ring), and (v) extracellular (hook, hook-filament junction proteins, filament, and filament cap).

Peripheral structures are probably capable of self-assembly whenever the necessary integral membrane structure (M ring and possibly other structures) is available for assembly onto. Integral membrane flagellar proteins presumably insert in the conventional (although not by any means fully understood) fashion using the Sec system (Saier, Werner & Müller, 1989), and then nucleate onto each other to give substructures such as the M ring.

What about the components that lie beyond the cell membrane? Do they employ the primary cellular export pathway, involving signal peptide cleavage (Oliver, 1985)? For two of the components, the proteins of the P and L rings, the answer is yes (Homma et al., 1987b; Jones et al., 1989; Jones et al., 1990). In principle, there could be flagellum-specific components that interface with the Sec system, but the fact that nonflagellate minicells were still competent to export the L- and P-ring proteins makes this unlikely. Use of the primary export system would seem to make sense for these proteins, in view of (i) the finite size of the destination compartments (the periplasm and the outer membrane, respectively), such that dilution losses are tolerable, and (ii) the fact that although eventually they are incorporated into the macromolecular structure, they undoubtedly have to be exported through the Sec system as monomers and continue to exist as such, independent of any other flagellar structure, until they can nucleate onto a recently completed rod. The conventional export pathway seems entirely adequate for this situation. The same argument might also seem to apply to the subunits of another periplasmic structure, namely, the rod. However, as we now discuss, there exists a flagellum-specific export pathway, which is used for all of the external structures such as the hook and the filament, but is also well suited for use by the rod.

Flagellum-specific export

The need for a specialized pathway becomes evident when it is recognized not only how much biomass of the flagellum lies beyond the outer membrane (roughly 20 000 subunits of flagellin per flagellum, plus smaller amounts of the hook and other proteins), but how far beyond the membrane it extends. The tip of an average-sized flagellum is about 10 μm away from the cell. Recalling that filament growth proceeds at the tip (Fig. 4), one realizes what an exceedingly small target this would provide if the protein subunits were to be exported into the bulk medium.

It is interesting to ask why a mechanism involving distal addition should have evolved rather than the apparently simpler one of progressive insertion of subunits at the base. Part of the reason may be the segmented character of the structure, such that it consists of a series of substructures of which the filament is the last to be assembled; thus hook protein addition has to be distal to the rod, and flagellin addition has to be distal to the hook. Given this, distal addition to the nascent substructure itself (e.g. hook subunits to nascent hook) is the simplest strategy. There may also be a mechanical reason why distal addition is advantageous. During motor rotation, which goes on continuously during filament elongation, the proximal end bears the heaviest torsional load, while the tip bears no load at all; ongoing insertion of subunits at the former, with the temporary dislocation that this event must inevitably entail, would probably be structurally catastrophic.

Therefore, given distal addition, how is the cell to avoid intolerable dilution losses of the exported subunits? The obvious solution is to reduce the export process from a three-dimensional to a one-dimensional one, such that all exported proteins encounter their intended target at the growing tip. In principle, one might imagine such a process involving diffusion along the external surface of the filament, but there are many objections to this 'SpiderMan' model, and there is no experimental evidence in its favour. Specifically, neither light nor electron microscopy indicate any external decoration of filament.

The most likely physical pathway therefore is through the nascent structure itself (Fig. 1). In this way the affinity problem disappears. Also, the process becomes much more efficient, since now the proteins are constrained to move unidirectionally and in single file, with any energy supplied in the initial transport event now being effectively transmitted to the entire column of subunits (like a pea-shooter filled with peas).

What is the evidence for a flagellum-specific export process and pathway? The evidence against use of the primary export pathway (aside from the horrendous losses it would entail) is simply that none of the rod proteins, hook protein, hook-associated proteins, or filament protein have a cleaved signal peptide (Joys & Rankis, 1972; Joys, 1985; Homma *et al.*, 1990*a,b*;

Jones *et al.*, 1990). Another piece of evidence, very clear in the case of the filament as a result of a recent X-ray fiber diffraction study (Namba, Yamashita & Vonderviszt, 1989), is that there is a central channel of suitable size (*c.* 5 nm diameter) to accommodate subunits of the proteins that are to be exported. This channel can also be seen in transverse sections of reconstituted filaments (Jones & Aizawa, 1990).

In the case of the hook, although image reconstruction indicates that a channel exists (Wagenknecht *et al.*, 1982), its diameter is not as clearly defined as in the case of filament for technical reasons. However, there is every reason to suppose that the channel will be of adequate size, since the hook and filament are organized along very similar structural lines. Reconstructed images of basal bodies (Stallmeyer *et al.*, 1989) are too noisy near the particle axis to say whether there is a channel through the rod. Comparisons of the amino acid sequences of the rod proteins with those of the hook, hook-associated, and filament proteins indicate that all these proteins belong to a structurally related family (Homma *et al.*, 1990a,b). In light of this, it seems very likely that the rod too will be hollow.

We consider next whether the exported proteins self-assemble once they reach their destination or whether they require participation of other structures. Nothing is known about this for the rod proteins. Purified flagellin can be reconstituted *in vitro* to form indefinite filamentous structures indistinguishable from the native ones (Asakura, Eguchi and Iino, 1966), and the structures once formed are stable indefinitely, even when placed in fresh buffer lacking monomer. A similar situation exists for hook protein (Kato, Aizawa & Asakura, 1982), except that in this case the reconstituted hook structure has an indefinite length, whereas the native structure in a wild-type cell has a fixed length. Also, HAP3 and HAP2 can add to the proximal and distal ends of filament, respectively, and HAP1, HAP3 and flagellin can successively assemble onto hooks (Ikeda, Asakura & Kamiya, 1989). Thus self-assembly is certainly possible by exogenous addition of proteins, provided these are present at fairly high concentrations.

In vivo, where the monomers are being supplied from within the structure, the situation may be different, since now each monomer when it reaches its potential assembly site must quickly form a stable association with the nascent structure, or be lost for ever into an essentially infinite sink. The situation for the rod proteins, where the sink is presumably the periplasmic space, may be less severe. At least for the case of flagellin, the unaided balance between assembly and loss is heavily in favour of loss (Homma *et al.*, 1984a), and this situation has to be redressed by the presence of a cap, made out of HAP2, at the tip of the filament (Ikeda, Asakura & Kamiya, 1985; Ikeda *et al.*, 1987). Whether this cap acts as a scaffolding enzyme to enhance the intrinsic rate of assembly of flagellin into filament, or whether it simply acts to contain the protein and thus block the loss process, is not known. Surprisingly, the cap does not block the passage of HAPs, which continue to be excreted into the medium even while filament elongation is underway (Homma & Iino, 1985*b*). Continuous export of HAP2 is probably essential, in order that a newly broken filament can be recapped and resume elongation.

Hook elongation, in contrast to filament elongation, seems to be able to proceed without the aid of a capping protein; although in the complete flagellum HAP1 is distal to the hook (as HAP2 is to the filament), it is not needed for hook assembly (Homma *et al.*, 1984*b*). It is not obvious why assembly of these two filamentous structures should differ in this regard. Perhaps the intrinsic rate of hook protein addition is higher than that of flagellin, although *in vitro* hook growth is actually slower that filament growth (Kato *et al.*, 1982): However, the data also suggest that the binding constant for hook protein is higher, and so the slow elongation rate may indicate a rate-limiting step between the addition of one monomer and the next rather than a slow association rate. It is also possible that hook protein addition involves accessory proteins that have not yet been detected or are discarded after use.

Hook protein and flagellin addition differ in another major regard: the hook is subject to length control but the filament is not. The process of hook protein addition and the process of hook length control may in some way be linked.

In our discussion of the flagellum-specific export pathway, we have so far only considered the physical conduit for flagellar proteins once they have crossed the plane of the cell membrane. We have not considered the apparatus that transfers the protein subunits from the cytoplasm into that conduit, i.e. the export apparatus itself. This apparatus may be expected to have a number of properties: (i) it has to recognize the relevant flagellar proteins; (ii) it has to transfer them across a pore or gate in the cell membrane, presumably within the centre of the flagellar base itself, either by facilitated diffusion or, more likely, by active transport; (iii) it may have to transport the various proteins in a particular order (rod proteins, then hook protein, etc); (iv) it may have to export subunits with a defined stoichiometry; (v) it must fail to recognize all other proteins, and also physically prevent their passage; (vi) it must do the same for all of the small molecules and ions in the cell. There may also be a need to maintain the proteins in a export-competent state, i.e. to provide a chaperonin-like function.

Until very recently, there was no information concerning the components of this export apparatus. However, we have carried out experiments that at least point to possible candidates (A. Vogler & R. Macnab, unpublished data). We have found that, when conditional mutants are grown at the permissive temperature and then have their flagella sheared off, with few exceptions the flagella grow, even at the restrictive temperature. The exceptions are mutants defective in three genes of previously unknown function (*flhA*, *fliH*, and *fliI*), and in one of the flagellar switch genes (*fliN*). We suspect that the products of these genes are either directly or indirectly involved in the process of flagellar export: This hypothesis is further supported by the stage at which they are needed in the assembly pathway, after the M ring but before the rod (Fig. 4).

A likely location for such an export apparatus is peripheral to the M ring, which raises the interesting problem of how the M ring avoids being an open, nonspecific pore during the earliest stages of assembly before the export apparatus is in place. Perhaps other proteins are associated with the M ring in a way that prevents this nonspecific leakage from happening. In our assembly analysis (Jones & Macnab, 1990) we found at least one protein (currently known only by its apparent molecular mass, 23 kD, and not by its genetic origin) that is involved in the very earliest stages of flagellar assembly, before rod assembly has commenced, and might be involved in sealing the M-ring structure against non-specific leakage. On the other hand, it may simply be that the inner surface of the M-ring annulus is hydrophobic, and that the central channel can be effectively sealed by lipid molecules until such time as the components of the export apparatus become available.

We now turn to the questions of how the export apparatus identifies its substrates, how the export process proceeds in a sequential fashion, and how the number of subunits within a substructure is determined.

Substrate recognition for export

What is it about exported flagellar proteins that enables them to be distinguished from other proteins? In an attempt to address this, we have asked which part of the exported protein (specifically, flagellin) is needed for it to be recognized, and have found that the N-terminal one-third of flagellin suffices (Kuwajima *et al.*, 1989). We are currently attempting to refine this information and establish what information within the N-terminus is responsible for recognition. In another approach, we have compared the deduced amino acid sequences of all nine of the known exported proteins (four rod proteins, hook protein, three HAPs, and flagellin) (Joys, 1985; Homma *et al.*, 1990*a,b*), and have found that they do share similarities, especially near both their N and C termini (Fig. 5(*a*)). However, interpretation of the data is not as straightforward as it might seem, since there is strong reason to presume that proteins such as the rod, hook, hook-associated, and filament proteins are structurally related as well as sharing a common export pathway. As an illustration of this, we have noted that the proteins have hydrophobic residues at intervals of 7 at both termini, which is suggestive of amphipathic α-helical structure (Fig. 5(*b*)). The exis-

(a)

```
Rod  (FlgB)                              MLDRLDAALRFQQ EAL NLRAQRQ EILAANIAN ADTPGYQARD...
Rod  (FlgC)                              MALLNIFDIAG SAL AAQSKRL NVAASNLAN ADSVTGPDGQ...
Rod  (FlgF)                              MDHAIYTAM GAA SQTLNQQ AVTASNLAN ASTPGFRAQL...
Rod  (FlgG)                              MISSLWIAK TGL DAQQTNM DVIANNLAN VSTNGFKRQR...
Hook (FlgE)                              MSFSQAV SGL NAAATNL DVIGNNIAN SATYGFKSGT...
Flagellin (FliC)  MAQVINTNSLSLLTQNNLNKSQSGTAIERLS SGL RINSAKD DAAGQAIAN RFTANIKGLT...
HAP1 (FlgK)                              MSSLINHAM SGL NAAQAAL NTVSNNINN YNVAGYTRQT...
HAP3 (FlgL)                 MRISTQMMYEQNM SGI TNSQAEW MKLGEQMST GKRVTNPSDD...

HAP2 (FliD)   MASISSLGVGSNLPLDQLLTDLTKNEKGRLTPIT KQQ SANSAKL TAYGTLKSA LEKFQTANTA...
```

(b)

N-terminus

```
                                              SG  L              ANN  L  AN
Rod (FlgB)                              MLDRLDAA L RFQQEA L NLRAQR Q EILAAN I ANADTPGYQARDID
Rod (FlgC)                              MALLNI F DIAGSA L AAQSKR L NVAASN L ANADSVTGPDGQPY
Rod (FlgF)                              MDHA I YTAMGA A SQTLNQ Q AVTASN L ANASTPGFRAQLNA
Rod (FlgG)                              MISS L WIAKTG L DAQQTN M DVIANN L ANVSTNGFKRQR..
Hook (FlgE)                             MS F SQAVSG L NAAATN L DVIGNN I ANSATYGFKSG...
Flagellin (FliC)  MAQVINTNSLSLLTQNNLNKSQSALGTA I ERLSSG L RINSAK D DAAGQA I ANRFTANIKGL...
HAP1 (FlgK)                             MSSL I NHAMSG L NAAQAA L NTVSNN I NNYNVAGYTRQ...
HAP3 (FlgL)                MRISTQMM Y EQNMSG I TNSQAE W MKLGEQ M STGKRVTNPSD...
```

C-terminus

```
                                ELVN  M  I
Rod (FlgB)          ...YRVPDQPS L DGNTVD M DRERTQ F ADNSLK Y QMGLTV L GSQLKG M MNVLQGGN*
Rod (FlgC)          ...PLADANGY V KMPNVD V VGEMVN T MSASRT Y QANIEV L NTVKSM M LKTLTLGQ*
Rod (FlgF)          ...SIRIMSGV L EGSNVK P VEAMTD M IANARR F EMQMKV I TSVDEN E GRANQLLSMS*
Rod (FlgG)          ...AGLLYQGY V ETSNVN V AEELVN M IQVQRA Y EINSKA V STTDQM L QKLTQL*
Hook (FlgE)         ...FGKLTNGA L EASNVD L SKELVN M IVAQRN Y QSNAQT I KTQDQI L NTLVNLR*
Flagellin (FliC)    ...VNNLSSAR S RIEDSD Y ATEVSN M SRAQIL Q QAGTSV L AQANQV P QNVLSLLR*
HAP1 (FlgK)         ...VKQLYKQQ Q SVSGVN L DEEYGN L QRYQQY Y LANAQV L QTANAL F DALLNIR*
HAP3 (FlgL)         STLDSLGSDRA L GQKLQM S NLVDVD W NSVISS Y VMQQAA L QASYKT F TDMQGMSLFQLNR*
HAP2 (FliD)         QDNVNATLKSLTKQ Y LSVSNS I DETVAR Y KAQFTQ L DTMMSK L NNTSSY L TQQFTAMNKS*
```

Fig. 5. Comparison of the deduced amino acid sequences of the axial flagellar proteins, which are exported by a flagellum-specific export pathway. (a) Localized consensus sequences that could be involved in recognition by this pathway, or could reflect structural similarities that are important in the quaternary interactions among subunits in these related structures. (b) Existence of hydrophobic residues at intervals of 7 at the N and C termini of the axial proteins, which probably reflect amphipathic α helices (see text) that are important in the final structure rather than in the export process. Based on results in Homma et al. (1990a,b); figure modified from Homma et al. (1990a).

tence of such structure is experimentally supported by the X-ray fiber diffraction referred to above (Namba et al., 1989), as well as by spectral studies (Fedorov, Kostyukova & Pyatibratov, 1988; Vonderviszt et al., 1990). Thus we are not inclined to view that particular motif as having a role in recognition for flagellum-specific export, since presumably there are many other proteins in the cell that have an amphipathic helical structure.

Control of order of export

Is it necessarily the case that *export* (as opposed to assembly) of the various components proceeds sequentially? Flagellar proteins might get sent out in random order, but only the ones appropriate to the current stage of morphogenesis might add on. In the case of the rod proteins as a group, there is no evidence for or against sequential export. In the case of the HAPs and flagellin, the evidence suggests that once hook assembly is complete all four can be exported independently of each other, since (i) mutants defective in a given HAP still export the other HAPs and flagellin (Homma *et al.*, 1984a; Homma & Iino, 1985b), (ii) a wild-type cell excretes HAPs even while in the process of filament elongation (Homma & Iino, 1985b), and (iii) a broken filament resumes elongation (Iino, 1969), presumably as a result of re-addition of the cap. An implication of this random order of export, incidentally, is that there cannot be strict control of the number of subunits of a given protein such as HAP1 that are *exported*, even though the stoichiometry of subunits in the assembled structure may be well-determined.

In the case of hook protein, the evidence for the most part argues against excretion of hook protein once the hook has reached its normal length (Homma & Iino, 1985b). The sole exception is that hook protein is detected in the culture medium of mutants defective in the filament cap gene, *fliD*. The explanation of this result is not clear. As we shall see, the hook provides a critical test of theories regarding the process of export and assembly of flagellar proteins.

Stoichiometry of flagellar components

Are the stoichiometries of the various proteins well-defined, and if so what is the mechanism by which this happens? The context of subunit stoichiometry varies from case to case. For the annular substructures, the stoichiometries (c. 26 subunits in each of the M-, P-, and L-rings; Jones *et al.*, 1990) are probably self-determined by the closed set of bonding interactions and by other constraints such as the diameter of the rod around which they are assembled.

The other established flagellar substructures are either known (hook and filament) (O'Brien & Bennett, 1972; Wagenknecht *et al.*, 1982) or suspected (rod and HAPs) (Homma *et al.*, 1990a,b) to be organized with helical symmetry. [*Helical* here refers to the disposition of the individual subunits and not to the shape of the assembled substructure, which may be either helical (hook and filament) or straight (rod).] Substructures with such symmetry can in principle be of indefinite stoichiometry, just as a spiral staircase can go on for ever. This is effectively the situation with the filament, whose length seems to be determined only by the rate of growth versus the rate of breakage, as the filament though quite rigid is also quite brittle.

The rod, hook, and HAP structures, however, seem to be of fairly closely defined stoichiometry (Homma *et al.*, 1984*b*; Ikeda *et al.*, 1987; Jones *et al.*, 1990) even though, as we noted above, the stoichiometry of their export may be ill-defined. For some of these components, specifically the HAPs and the proximal rod proteins, where the values are low enough to represent either one or two turns of helix, the inter-subunit interactions themselves may fix the stoichiometry (Jones *et al.*, 1990). Indeed, *in vitro* prolonged incubation of hook with monomeric HAPs does not increase the assembled HAP zone beyond its normal extent (Ikeda *et al.*, 1989).

It is hard to use the self-limiting type of argument for a structure such as the hook, which has a rather closely defined length, but one sufficient to consist of roughly 24 turns of helix (Wagenknecht *et al.*, 1982; Jones *et al.*, 1990). The distal portion of the rod presents a similar problem.

The mechanism for control of hook length therefore represents one of the most intriguing aspects of the assembly process. How does the export system know when to stop export of rod proteins and start export of hook protein, how long to continue such export, and finally how to switch specificity of export from hook protein to the HAPs and flagellin?

Hook length control

The key to an understanding of the process of hook assembly and hook length control has to be a gene known as *fliK*, which if lacking causes two phenotypic defects: (i) a failure to stop hook elongation, so that hooks of indefinite length ('polyhooks') result (Patterson-Delafield *et al.*, 1973; Silverman & Simon, 1974*b*; Suzuki & Iino, 1981), and (ii) a failure to start HAPs assembly (Homma *et al.*, 1984*b*; Aizawa *et al.*, 1985) and, as a consequence, filament assembly. We will consider what is known about FliK and what constraints the information places upon the mechanism of hook length control. It must be admitted at the start that there are more questions than answers at this stage, and that no model seems completely satisfactory. The following discussion is primarily intended to provoke thinking on the subject.

One well-studied mechanism of length control is the *molecular ruler* mechanism as exemplified by the tail of phage lambda (Katsura, 1987) (Fig. 6(a)). However, the mechanism for FliK as a ruler cannot be directly analogous to this, or else the absence of FliK would result in a zero-length hook, rather than a polyhook as is observed. An inverse mechanism could in principle apply, whereby some feature of FliK would be responsible for actively inhibiting further elongation (Fig. 6(b)), rather than failing to enable it. There is currently no experimental evidence to support the hypothesis that FliK is a ruler, and there are a number of arguments against it: (i) FliK is not found in isolated hook-basal body complexes or hook-filament complexes (Aizawa *et al.*, 1985), although it could be present

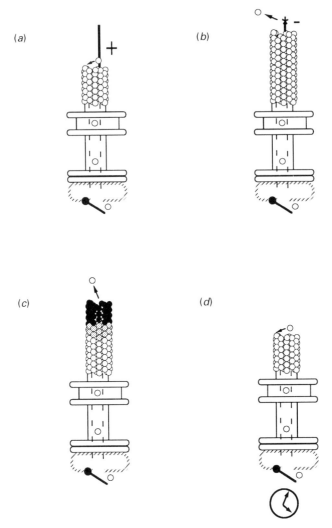

Fig 6. Possible mechanisms for assembling a flagellar hook of defined length (see the text for discussion of the various mechanisms). (a) *Positive ruler*: A molecule, whose length matches that of a complete hook, catalyzes hook subunit addition along its length. (b) *Negative ruler*: Hook subunits are self-assembling, but the tip of the ruler contains an inhibitory site. (c) *Accumulated strain*: Hook subunits are self-assembling, but subunit addition has a slight negative cooperativity, so that as the hook elongates the binding constant gets progressively weaker (indicated schematically by white → grey → black subunits) until association and dissociation rates are in balance and so the hook has reached its final length. (d) *Control of export or synthesis*: The cell (globally) or the flagellum (locally) undergoes a switch from the export (or synthesis) of rod proteins to hook protein and later from hook protein to HAPs and flagellin. The extent of the phase involving hook protein has to be controlled in order to produce a hook of fixed length. The control could be intracellular, involving a timing mechanism (indicated here schematically as a clock) or a counting mechanism; or it could be extracellular, involving a structural feedback to the cell from the hook structure itself or from a ruler molecule.

during assembly and then discarded, (ii) it is located predominantly in the cytoplasm in minicell experiments (Homma, Iino & Macnab, 1988), and (iii) it shows no sequence similarity to the other exported flagellar proteins (I. Kawagishi & R. Macnab, unpublished data). While its molecular size ($c.$ 400 residues) would make it a suitable ruler as a single α-helix with an overall length similar to that of the hook (55 nm), its deduced amino acid sequence, with a large number of proline residues in the central portion, makes this most unlikely. Interestingly, in an extended state it would be about 120 nm long, which would correspond roughly to the distance from the M ring to the hook tip.

In spite of the absence of evidence in favour of a ruler mechanism, it remains conceptually the easiest way to determine with fairly high precision the length of what would otherwise be a structure of indefinite length. It may be that one of the flagellar genes of currently unknown function is involved in such a role. FlgD, which is responsible for a process called rod modification that must occur before hook assembly can initiate (Suzuki et al., 1978), might also be involved as a ruler in hook length control. Incidentally, if there is a ruler and it is not FliK, then FliK must have another function which, if defective, causes the ruler mechanism to be defeated and hook assembly to proceed indefinitely.

An accumulated strain model, such that the affinity of subunits for the hook structure decreases with hook length (Fig. 6(c)), seems very unlikely since (i) it should provide a rather broad spectrum of lengths, (ii) it implies that the distal end of the hook would not be strong, and (most compelling of all) (iii) wild-type hook protein can under some circumstances assemble into indefinitely long 'polyhook' structures.

Another class of mechanism for hook length control would be one where FliK operates from within the cell, to control either synthesis or export of hook protein versus later proteins like the HAPs and flagellin (Fig. 6(d)). Control of synthesis will be considered first. The gene for the hook protein is in a different class (class 2) from those for HAPs and flagellin (classes 3a and b, respectively), and so the opportunity for differential control of transcription does exist, although this would not be possible for the rod genes versus the hook gene, which are not only in the same transcriptional class, but in the same operon. Studies of transcriptional control do not implicate *fliK* in the process (Komeda, 1982; Kutsukake et al., 1990). Control at the translational level is also a possibility.

Control of synthesis would have to be either global (cell-wide) or local (at the level of the individual flagellum). It would be possible for some steps of the synthesis or assembly process to be global, perhaps tied to the cell cycle or some other triggering event, but it cannot be true for the final stage, namely, HAPs and flagellin synthesis and assembly, for the following reasons: (i) cells in a non-synchronized culture, whose flagellar filaments have been sheared off mechanically by passage of the cells through a hypodermic needle, all regrow filaments within minutes, whereas *de novo* synthesis takes much longer (A. Vogler, unpublished data); (ii) elongation of existing filaments (and therefore HAPs and flagellin synthesis and export) proceeds in a continuous fashion over many cell cycles, while in the meantime *de novo* assembly of other flagella has been initiated (Jones & Aizawa, 1990). However, this still leaves open the possibility that *hook* protein synthesis is temporally controlled at a global level. Temporal or local control of synthesis at the level of the individual flagellum would certainly permit a sequential order of assembly, but implies a dedicated transcriptional or translational apparatus at each flagellum, for which there is no experimental evidence. Regardless of whether it is global or local, however, temporal control of synthesis does not seem well-suited to precise control of length.

Finally we consider mechanisms whereby FliK operates from within the cell to control the specificity of *export* (Fig. 6(*d*)), rather than of synthesis. Since mutants defective in the HAP genes, specifically that for HAP1, which is adjacent to the hook, still control hook length normally (Homma *et al.*, 1984*b*), switching on of HAP export cannot be responsible for switching off of hook export. FliK might, however, control both the switching off of hook export *and* the switching on of HAPs/flagellin export. The deduced amino acid sequence of FliK suggests there may be at least two distinct domains (I. Kawagishi & R. Macnab, unpublished data), which would be consistent with a hypothesis of dual function. Another fact supporting the hypothesis is the existence of a special type of allele known as Phf$^+$ (polyhook–filament), where the hook length control is abnormal but addition of HAPs and flagellin is more or less normal (Suzuki & Iino, 1981). Also, under special circumstances (hook protein overproduction), polyhooks are formed but are competent to grow filaments at their tips (Ohnishi *et al.*, 1987). No *fliK* alleles have been found that give the inverse phenotype, namely, normal hook length but failure of HAPs and filament export and assembly.

It is all very well to talk about control of specificity of export, but this begs the question as to what signals the point in the process at which to perform the switch in specificity from export of hook protein to export of HAPs and flagellin (a similar question arises for switching specificity of synthesis). There are several possibilities, based on time, number, size, or structure. In all cases, there has to be an initiating event that is communicated to the export apparatus, say, addition of the L ring to the newly completed rod.

For a mechanism based on time, the initiating event would set in motion some process, perhaps, a covalent modification of the regulator of specificity, that takes a time interval matching that needed for export and assembly of the appropriate number of hook monomers. But it seems unlikely that the kinetics of the two processes could be made to match under different growth conditions, where hook length remains invariant. It is also questionable whether the narrow length distribution of a population of hooks could be achieved by this mechanism.

For a mechanism based on number, the initiating event would start an actual counting of the hook subunits as they were being exported. This implies a quantized parameter, matching the number of hook monomers in the mature hook, that is to be decremented to zero (or incremented to the final number). How this would be achieved in molecular terms is obscure; I am not aware of any precedent for such a mechanism in biology.

For a mechanism based on size (if by size we mean the size of the hook itself), we find ourselves back at a version of the ruler model, the only difference being that now the role of the ruler is to send back a signal to stop hook protein export rather than acting to directly control the assembly process.

A mechanism based on hook structure (in the absence of a ruler) implies that the structure in some way changes with length, and that this information can be communicated back to the cell. Setting aside problems associated with the latter aspect, the fact that wild-type hook protein can (in *fliK* mutants, and in other circumstances as well) grow to indefinite length makes this mechanism seem very unlikely.

It should be obvious from the above discussion that the mechanism of hook length control remains very much a mystery but one which, if it can be solved, may give insight into the general problem of length measurement in cellular structures.

CONCLUDING COMMENTS

Not too many years ago, although many flagellar genes had been identified, there were relatively few of them for which a structure or function had been assigned. This situation has changed considerably, with the components of most of the known structure now identified both biochemically and genetically. Components of a number of presumed structures have also been identified. Also, almost all of the flagellar genes have been cloned and sequenced, and their products identified. It is getting to the stage where structure and function needs to be analysed in detail. For example, to say a protein is a switch protein is merely to say that it is responsible for determining the direction of rotation of the flagellar motor, and does not give any clue as to how it does so. Answering questions like this will be the real challenge in the future study of this fascinating system, which has the potential for becoming a paradigm for the genetics, assembly and function of a cellular organelle.

ACKNOWLEDGEMENTS

I am grateful to members of my laboratory, and to David DeRosier of Brandeis University, for stimulating discussions on many aspects of flagellar assembly; to P. Matsumura, C. Jones and S.-I. Aizawa for making available results prior to publication; and to the United States Public Health Service for support of the research in my laboratory under grants AI12202 and GM40335.

REFERENCES

Aizawa, S.-I., Dean, G. E., Jones, C. J., Macnab, R. M. & Yamaguchi, S. (1985). Purification and characterization of the flagellar hook-basal body complex of *Salmonella typhimurium*. *Journal of Bacteriology*, **161**, 836–49.
Arnosti, D. N. & Chamberlin, M. J. (1989). Secondary σ factor controls transcription of flagellar and chemotaxis genes in *Escherichia coli*. *Proceedings of the National Academy of Sciences, USA*, **86**, 830–4.
Asakura, S., Eguchi, G. & Iino, T. (1966). *Salmonella* flagella: *In vitro* reconstruction and over-all shapes of flagellar filaments. *Journal of Molecular Biology*, **16**, 302–16.
Bachmann, B. J. (1990). Linkage map of *Escherichia coli* K-12, Edition 8. *Microbiological Reviews*, **54**, 130–97.
Bartlett, D. H., Frantz, B. B. & Matsumura, P. (1988). Flagellar transcriptional activators FlbB and FlaI: Gene sequences and 5' consensus sequences of operons under FlbB and FlaI control. *Journal of Bacteriology*, **170**, 1575–81.
Blair, D. F. & Berg, H. C. (1988). Restoration of torque in defective flagellar motors. *Science*, **242**, 1678–81.
Block, S. M. & Berg, H. C. (1984). Successive incorporation of force-generating units in the bacterial rotary motor. *Nature, London*, **309**, 470–2.
Calladine, C. R. (1982). Construction of bacterial flagellar filaments, and aspects of their conversion to different helical forms. In *Prokaryotic and Eukaryotic Flagella*, ed. W. B. Amos & J. G. Duckett, pp. 33–51. Cambridge: Cambridge University Press.
DePamphilis, M. L. & Adler, J. (1971a). Fine structure and isolation of the hook-basal body complex of flagella from *Escherichia coli* and *Bacillus subtilis*. *Journal of Bacteriology*, **105**, 384–95.
DePamphilis, M. L. & Adler, J. (1971b). Attachment of flagellar basal bodies to the cell envelope: Specific attachment to the outer, lipopolysaccharide membrane and the cytoplasmic membrane. *Journal of Bacteriology*, **105**, 396–407.
Eisenbach, M. (1990). Functions of the flagellar modes of rotation in bacterial motility and chemotaxis. *Molecular Microbiology*, **4**, 161–7.
Emerson, S. U., Tokuyasu, K. & Simon, M. I. (1970). Bacterial flagella: Polarity of elongation. *Science*, **169**, 190–2.
Fedorov, O. V., Kostyukova, A. S. & Pyatibratov, M. G. (1988). Architectonics of a bacterial flagellin filament subunit. *FEBS Letters*, **241**, 145–8.
Helmann, J. D. & Chamberlin, M. J. (1987). DNA sequence analysis suggests that expression of flagellar and chemotaxis genes in *Escherichia coli* and *Salmonella typhimurium* is controlled by an alternative σ factor. *Proceedings of the National Academy of Sciences, USA*, **84**, 6422–4.
Homma, M., Aizawa, S.-I., Dean, G. E. & Macnab, R. M. (1987a). Identification of the M-ring protein of the flagellar motor of *Salmonella typhimurium*. *Proceedings of the National Academy of Sciences, USA*, **84**, 7483–7.
Homma, M., DeRosier, D. J. & Macnab, R. M. (1990a). Flagellar hook and hook-associated proteins of *Salmonella typhimurium* and their relationship to other axial components of the flagellum. *Journal of Molecular Biology*, **213**, 819–32.
Homma, M., Fujita, H., Yamaguchi, S. & Iino, T. (1984a). Excretion of un-

assembled flagellin by *Salmonella typhimurium* mutants deficient in the hook-associated proteins. *Journal of Bacteriology*, **159**, 1056–9.
Homma, M. & Iino, T. (1985a). Locations of hook-associated proteins in flagellar structures of *Salmonella typhimurium*. *Journal of Bacteriology*, **162**, 183–9.
Homma, M. & Iino, T. (1985b). Excretion of unassembled hook-associated proteins by *Salmonella typhimurium*. *Journal of Bacteriology*, **164**, 1370–2.
Homma, M., Iino, T. & Macnab, R. M. (1988). Identification and characterization of the products of six region III flagellar genes (*flaAII.3* through *flaQII*) of *Salmonella typhimurium*. *Journal of Bacteriology*, **170**, 2221–8.
Homma, M., Komeda, Y., Iino, T. & Macnab, R. M. (1987b). The *flaFIX* gene product of *Salmonella typhimurium* is a flagellar basal body component with a signal peptide for export. *Journal of Bacteriology*, **169**, 1493–8.
Homma, M., Kutsukake, K., Hasebe, M., Iino, T. & Macnab, R. M. (1990b). FlgB, FlgC, FlgF and FlgG. A family of structurally related proteins in the flagellar basal body of *Salmonella typhimurium*. *Journal of Molecular Biology*, **211**, 465–77.
Homma, M., Kutsukake, K., Iino, T. & Yamaguchi, S. (1984b). Hook-associated proteins essential for flagellar filament formation in *Salmonella typhimurium*. *Journal of Bacteriology*, **157**, 100–8.
Homma, M., Kutsukake, K. & Iino, T. (1985). Structural genes for flagellar hook-associated proteins in *Salmonella typhimurium*. *Journal of Bacteriology*, **163**, 464–71.
Homma, M., Ohnishi, K., Iino, T. & Macnab, R. M. (1987c). Identification of flagellar hook and basal body gene products (FlaFV, FlaFVI, FlaFVII and FlaFVIII) in *Salmonella typhimurium*. *Journal of Bacteriology*, **169**, 3617–24.
Iino, T. (1969). Polarity of flagellar growth in *Salmonella*. *Journal of General Microbiology*, **56**, 227–39.
Iino, T. (1974). Assembly of *Salmonella* flagellin *in vitro* and *in vivo*. *Journal of Supramolecular Structure*, **2**, 372–84.
Iino, T. (1985). Genetic control of flagellar morphogenesis in *Salmonella*. In *Sensing and Response in Microorganisms*, ed. M. Eisenbach & M. Balaban, pp. 83–92. Amsterdam: Elsevier Science Publishers B.V.
Iino, T., Komeda, Y., Kutsukake, K., Macnab, R. M., Matsumura, P., Parkinson, J. S., Simon, M. I. & Yamaguchi, S. (1988). New unified nomenclature for the flagellar genes of *Escherichia coli* and *Salmonella typhimurium*. *Microbiological Reviews*, **52**, 533–5.
Ikeda, T., Asakura, S. & Kamiya, R. (1985). 'Cap' on the tip of *Salmonella* flagella. *Journal of Molecular Biology*, **184**, 735–7.
Ikeda, T., Asakura, S. & Kamiya, R. (1989). Total reconstitution of *Salmonella* flagellar filaments from hook and purified flagellin and hook-associated proteins *in vitro*. *Journal of Molecular Biology*, **209**, 109–14.
Ikeda, T., Homma, M., Iino, T., Asakura, S. & Kamiya, R. (1987). Localization and stoichiometry of hook-associated proteins within *Salmonella typhimurium* flagella. *Journal of Bacteriology*, **169**, 1168–73.
Imae, Y. & Atsumi, T. (1989). Na^+-driven bacterial flagellar motors. *Journal of Bioenergetics and Biomembranes*, **21**, 705–16.
Jones, C. J. & Aizawa, S.-I. (1990). The bacterial flagellum and flagellar motor: Structure, assembly, and function. *Advances in Microbial Physiology* (in press).
Jones, C. J., Homma, M. & Macnab, R. M. (1987). Identification of proteins of the outer (L and P) rings of the flagellar basal body of *Escherichia coli*. *Journal of Bacteriology*, **169**, 1489–92.
Jones, C. J., Homma, M. & Macnab, R. M. (1989). L-, P-, and M-ring proteins

of the flagellar basal body of *Salmonella typhimurium*: Gene sequences and deduced protein sequences. *Journal of Bacteriology*, **171**, 3890–900.

Jones, C. J. & Macnab, R. M. (1990). Flagellar assembly in *Salmonella typhimurium*: Analysis with temperature-sensitive mutants. *Journal of Bacteriology*, **172**, 1327–39.

Jones, C. J., Macnab, R. M., Okino, H. & Aizawa, S.-I. (1990). Stoichiometric analysis of the flagellar hook-(basal body) complex of *Salmonella typhimurium*. *Journal of Molecular Biology*, **212**, 377–87.

Joys, T. M. (1985). The covalent structure of the phase-1 flagellar filament protein of *Salmonella typhimurium* and its comparison with other flagellins. *Journal of Biological Chemistry*, **260**, 15758–61.

Joys, T. M. & Rankis, V. (1972). The primary structure of the phase-1 flagellar protein of *Salmonella typhimurium*. I. The tryptic peptides. *Journal of Biological Chemistry*, **247**, 5180–93.

Kamiya, R., Hotani, H. & Asakura, S. (1982). Polymorphic transition in bacterial flagella. In *Prokaryotic and Eukaryotic Flagella*, ed. W. B. Amos & J. G. Duckett, pp. 53–76. Cambridge: Cambridge University Press.

Kato, S., Aizawa, S. & Asakura, S. (1982). Reconstruction *in vitro* of the flagellar polyhook from *Salmonella*. *Journal of Molecular Biology*, **161**, 551–60.

Katsura, I. (1987). Determination of bacteriophage λ tail length by a protein ruler. *Nature, London*, **327**, 73–5.

Khan, S., Dapice, M. & Reese, T. S. (1988). Effects of *mot* gene expression on the structure of the flagellar motor. *Journal of Molecular Biology*, **202**, 575–84.

Kihara, M., Homma, M., Kutsukake, K. & Macnab, R. M. (1989). Flagellar switch of *Salmonella typhimurium*: Gene sequences and deduced protein sequences. *Journal of Bacteriology*, **171**, 3247–57.

Komeda, Y. (1982). Fusions of flagellar operons to lactose genes on a Mu *lac* bacteriophage. *Journal of Bacteriology*, **150**, 16–26.

Komeda, Y., Silverman, M. & Simon, M. (1978). Identification of the structural gene for the hook subunit protein of *Escherichia coli* flagella. *Journal of Bacteriology*, **133**, 364–71.

Komeda, Y., Suzuki, H., Ishidsu, J.-I. & Iino, T. (1975). The role of cAMP in flagellation of *Salmonella typhimurium*. *Molecular & General Genetics*, **142**, 289–98.

Konno, R., Fujita, H., Horiguchi, T. & Yamaguchi, S. (1976). Precise position of the *nml* locus on the genetic map of *Salmonella*. *Journal of General Microbiology*, **93**, 182–4.

Krikos, A., Mutoh, N., Boyd, A. & Simon, M. I. (1983). Sensory transducers of *E. coli* are composed of discrete structural and functional domains. *Cell*, **33**, 615–22.

Kutsukake, K., Ohya, Y. & Iino, T. (1990). Transcriptional analysis of the flagellar regulon of *Salmonella typhimurium*. *Journal of Bacteriology*, **172**, 741–7.

Kutsukake, K., Ohya, Y., Yamaguchi, S. & Iino, T. (1988). Operon structure of flagellar genes in *Salmonella typhimurium*. *Molecular & General Genetics*, **214**, 11–15.

Kuwajima, G., Kawagishi, I., Homma, M., Asaka, J.-I., Kondo, E. & Macnab, R. M. (1989). Export of an N-terminal fragment of *Escherichia coli* flagellin by a flagellum-specific pathway. *Proceedings of the National Academy of Sciences, USA*, **86**, 4953–7.

Larsen, S. H., Adler, J., Gargus, J. J. & Hogg, R. W. (1974a). Chemomechanical coupling without ATP: The source of energy for motility and chemotaxis in bacteria. *Proceedings of the National Academy of Sciences, USA*, **71**, 1239–43.

Larsen, S. H., Reader, R. W., Kort, E. N., Tso, W. -W. & Adler, J. (1974b). Change in direction of flagellar rotation is the basis of the chemotactic response in *Escherichia coli*. *Nature, London*, **249**, 74–7.
Macnab, R. M. (1987a). Flagella. In *Escherichia coli and Salmonella typhimurium: Cellular and Molecular Biology*, vol. 1, ed. F. C. Neidhardt, J. Ingraham, K. B. Low, B. Magasanik, M. Schaechter & H. E. Umbarger, pp. 70–83. Washington, DC: American Society for Microbiology Publications.
Macnab, R. M. (1987b). Motility and chemotaxis. In *Escherichia coli and Salmonella typhimurium: Cellular and Molecular Biology*, vol. 1, eds. F. C. Neidhardt, J. Ingraham, K. B. Low, B. Magasanik, M. Schaechter & H. E. Umbarger, pp. 732–59. Washington, DC: American Society for Microbiology Publications.
Macnab, R. M. & DeRosier, D. J. (1988). Bacterial flagellar structure and function. *Canadian Journal of Microbiology*, **34**, 442–51.
Manson, M. D., Tedesco, P., Berg, H. C., Harold, F. M. & van der Drift, C. (1977). A protonmotive force drives bacterial flagella. *Proceedings of the National Academy of Sciences, USA*, **74**, 3060–4.
Mutoh, N. & Simon, M. I. (1986). Nucleotide sequence corresponding to five chemotaxis genes in *Escherichia coli*. *Journal of Bacteriology*, **165**, 161–6.
Namba, K., Yamashita, I. & Vonderviszt, F. (1989). Structure of the core and central channel of bacterial flagella. *Nature, London*, **342**, 648–54.
O'Brien, E. J. & Bennett, P. M. (1972). Structure of straight flagella from a mutant *Salmonella*. *Journal of Molecular Biology*, **70**, 133–52.
Ohnishi, K., Homma, M., Kutsukake, K. & Iino, T. (1987). Formation of flagella lacking outer rings by *flaM*, *flaU* and *flaY* mutants of *Escherichia coli*. *Journal of Bacteriology*, **169**, 1485–8.
Ohnishi, K., Kutsukake, K., Suzuki, H., & Iino, T. (1990). Gene *fliA* encodes an alternative sigma factor specific for flagellar operons in *Salmonella typhimurium*. *Molecular and General Genetics*. **221**, 139–47.
Okino, H., Isomura, M., Yamaguchi, S., Magariyama, Y., Kudo, S. & Aizawa, S.-I. (1989). Release of flagellar filament-hook-rod complex by a *Salmonella typhimurium* mutant defective in the M ring of the basal body. *Journal of Bacteriology*, **171**, 2075–82.
Oliver, D. (1985). Protein secretion in *Escherichia coli*. *Annual Reviews in Microbiology*, **39**, 615–48.
Parkinson, J. S., Parker, S. R., Talbert, P. B. & Houts, S. E. (1983). Interactions between chemotaxis genes and flagellar genes in *Escherichia coli*. *Journal of Bacteriology*, **155**, 265–74.
Patterson-Delafield, J., Martinez, R. J., Stocker, B. A. D. & Yamaguchi, S. (1973). A new *fla* gene in *Salmonella typhimurium–flaR* and its mutant phenotype–superhooks. *Archiv für Mikrobiologie*, **90**, 107–20.
Saier, M. H. Jr., Werner, P. K. & Müller, M. (1989). Insertion of proteins into bacterial membranes: Mechanism, characteristics, and comparisons with the eucaryotic process. *Microbiological Reviews*, **53**, 333–66.
Sanderson, K. E. & Roth, J. R. (1988). Linkage map of *Salmonella typhimurium*, Edition VII. *Microbiological Reviews*, **52**, 485–532.
Silverman, M. R. & Simon, M. I. (1972). Flagellar assembly mutants in *Escherichia coli*. *Journal of Bacteriology*, **112**, 986–93.
Silverman, M. & Simon, M. (1974a). Flagellar rotation and the mechanism of bacterial motility. *Nature, London*, **249**, 73–4.
Silverman, M. & Simon, M. (1974b). Characterization of *Escherichia coli* flagellar mutants that are insensitive to catabolite repression. *Journal of Bacteriology*, **120**, 1196–203.

Silverman, M. & Simon, M. (1976). Operon controlling motility and chemotaxis in *E. coli*. *Nature, London*, **264**, 577–80.

Stallmeyer, M. J. B., Aizawa, S.-I., Macnab, R. M. & DeRosier, D. J. (1989). Image reconstruction of the flagellar basal body of *Salmonella typhimurium*. *Journal of Molecular Biology*, **205**, 519–28.

Stocker, B. A. D., McDonough, M. W. & Ambler, R. P. (1961). A gene determining presence or absence of ε-N-methyl-lysine in *Salmonella* flagellar protein. *Nature, London*, **189**, 556–8.

Suzuki, T. & Iino, T. (1981). Role of the *flaR* gene in flagellar hook formation in *Salmonella* spp. *Journal of Bacteriology*, **148**, 973–9.

Suzuki, T., Iino, T., Horiguchi, T. & Yamaguchi, S. (1978). Incomplete flagellar structures in nonflagellate mutants of *Salmonella typhimurium*. *Journal of Bacteriology*, **133**, 904–15.

Suzuki, T. & Komeda, Y. (1981). Incomplete flagellar structures in *Escherichia coli* mutants. *Journal of Bacteriology*, **145**, 1036–41.

Vonderviszt, F., Kanto, S., Aizawa, S.-I. & Namba, K. (1989). Terminal regions of flagellin are disordered in solution. *Journal of Molecular Biology*, **209**, 127–33.

Wagenknecht, T., DeRosier, D. J., Aizawa, S.-I. & Macnab, R. M. (1982). Flagellar hook structures of *Caulobacter* and *Salmonella* and their relationship to filament structure. *Journal of Molecular Biology*, **162**, 69–87.

Yamaguchi, S., Aizawa, S.-I., Kihara, M., Isomura, M., Jones, C. J. & Macnab, R. M. (1986a). Genetic evidence for a switching and energy-transducing complex in the flagellar motor of *Salmonella typhimurium*. *Journal of Bacteriology*, **168**, 1172–9.

Yamaguchi, S., Fujita, H., Ishihara, A., Aizawa, S.-I. & Macnab, R. M. (1986b). Subdivision of flagellar genes of *Salmonella typhimurium* into regions responsible for assembly, rotation, and switching. *Journal of Bacteriology*, **166**, 187–93.

Yamaguchi, S., Fujita, H., Sugata, K., Taira, T. & Iino, T. (1984). Genetic analysis of H2, the structural gene for phase-2 flagellin in *Salmonella*. *Journal of General Microbiology*, **130**, 255–265.

TRANSDUCERS: TRANSMEMBRANE RECEPTOR PROTEINS INVOLVED IN BACTERIAL CHEMOTAXIS

G. L. HAZELBAUER, R. YAGHMAI, G. G. BURROWS, J. W. BAUMGARTNER, D. P. DUTTON AND D. G. MORGAN

Biochemistry/Biophysics Program, Washington State University, Pullman, Washington 99164–4660, USA

INTRODUCTION

The essence of receptor function is transduction of a stimulus into an informational signal and transmission of that signal from the site of recognition to the appropriate effector within the cell. We do not understand how any transmembrane receptor transduces ligand recognition into a signal or how that signal is transmitted across the membrane. The receptor proteins that mediate chemotaxis in *Escherichia coli* are the focus of intensive studies aimed at providing such an understanding. In this chapter we will summarize information available about these chemoreceptors, often called transducers, describe some recent observations from our own laboratory and suggest plausible models for some aspects of receptor structure and function.

BACTERIAL BEHAVIOUR

A short summary of chemotactic behaviour by *E. coli* is provided here as background to our consideration of transducers. Primary references for much of the basic information presented in this article can be found in a number of early reviews (Adler, 1975; Berg, 1975; Springer, Goy & Adler, 1979; Koshland, 1981; Boyd & Simon, 1982). Discussions of current understanding of aspects of bacterial chemotaxis besides receptor function can be found in other chapters in this volume and in several recent reviews (Stewart & Dahlquist, 1987; Macnab & De Rosier, 1988; Hazelbauer, 1988; Stock, Ninfa & Stock, 1989; Stock, Stock & Mottonen, 1990; Eisenbach, 1990).

E. coli moves in a three-dimensional random walk. Cells swim in gentle curves, called runs, punctuated every second or two by brief episodes of unco-ordination, called tumbles, that result in new, randomly chosen directions of swimming. In a spatial gradient of attractant or repellent, the sensory system alters the pattern of swimming by extending runs in favourable directions, thus creating a biased random walk that produces net progress.

Transducers, specific transmembrane receptor proteins, detect the amino acids and sugars that are the most effective attractants for *E. coli*. The bacterium senses a spatial gradient by detecting the temporal gradient created as it swims. This implies that the cell has a way to measure current concentration, a means of storing a record of past concentration and a mechanism for comparing the two values. For wild-type cells, co-ordinated swimming occurs when the left-handed helices of bacterial flagella rotate counterclockwise and tumbling occurs when they rotate clockwise. The sensory system directs cellular migration by controlling the switch that determines the direction of flagellar rotation.

The bacterial response to a temporal gradient is transient. Exposure of cells to attractant results in suppression of tumbles and exposure to repellent increases tumbling. However, after a time ranging from seconds to minutes, depending on the compound and the magnitude of the gradient, cells resume their initial behavioural patterns of runs and tumbles, even though the attractant or repellent is still present. Therefore, like most sensory cells, bacteria adapt. The biochemical mechanism of adaptation involves covalent modification of the transducer protein through which the stimulus has passed. Specifically, several glutamyl residues on the cytoplasmic domain of the transducer can be enzymatically methylated to create carboxyl methyl esters. An appropriate change in extent of methylation is mechanistically linked to adaptation.

COMPONENTS OF THE CHEMOTACTIC SYSTEM

The chemotactic system is composed of four transducer proteins, Tsr, Tar, Trg and Tap; six cytoplasmic proteins, CheA, CheW, CheY, CheZ, CheR and CheB, and three components of the flagellar switch. CheA is a protein kinase that phosphorylates CheY and CheB (Hess *et al.*, 1987; 1988*b*; Oosawa, Hess & Simon, 1988*a*, Matsumura *et al.*, this volume). It is presumably phospho-CheY that interacts with the flagellar motor to induce clockwise rotation (Eisenbach, 1990) and phospho-CheB is the activated form of the methylesterase (Stewart *et al.*, 1988; Lupas & Stock, 1989). Both phospho-CheY and phospho-CheB are unstable. CheZ appears to be a phosphatase that increases the rate of hydrolysis of phospho-CheY. Transducers control flagellar rotation and activity of the esterase by influencing the level of phospho-CheA, a process in which CheW participates (Borkovich *et al.*, 1989; Liu & Parkinson, 1989). The CheA kinase transfers phosphate from ATP to one of its own histidine residues (Hess, Bourret & Simon, 1988*a*) and subsequently to an aspartyl residue on CheY and presumably on CheB (Sanders *et al.*, 1989; Bourret, Hess & Simon, 1990). CheR and CheB are the enzymes catalyzing methylation and demethylation, respectively, and are thus central to adaptation. S-adenosylmethionine is

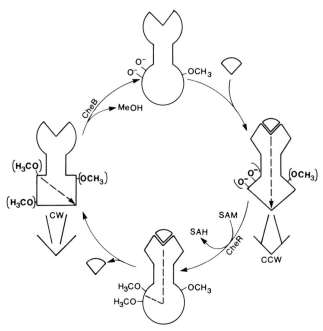

Fig. 1. Transducer states in the cycle of excitation and adaptation. See text for explanation. SAM and SAH are S-adenosylmethionine and S-adenosylhomocysteine, respectively. CCW and CW represent counterclockwise and clockwise signals, respectively. Adapted from Fig. 2, Hazelbauer (1988).

required as a methyl donor in adaptation and ATP is the initial source of phosphate in the transduction cascade. ATP may have an additional role (Smith et al., 1988), perhaps related to the involvement of acetyladenylate in controlling the direction of flagellar rotation (Wolfe, Conley & Berg, 1988).

THE ROLE OF TRANSDUCERS

Events thought to occur at the transducer can be summarized by a model for a cycle of excitation and adaptation (Fig. 1). 1. In the unstimulated state (top drawing), the ligand-binding site is unoccupied and the methyl-accepting sites (four or five depending on the transducer) are continually methylated and demethylated, producing an average steady-state level of approximately one methyl group per protein. 2. Addition of ligand creates a positive stimulus by interaction at a stereospecific site in the periplasmic domain of the transducer (right-hand drawing). Binding results in a shift in the bias of the flagellar switch to counterclockwise. Transduction of information from binding site to flagellum can be considered in two steps: intramolecular, through the transducer from periplasmic domain to cytop-

lasmic domain, and intermolecular, through the cytoplasm from transducer to the flagellar switch by means of phosphotransfer and protein–protein interactions. Until this point in the cycle, methylation does not play a role. However, intramolecular transduction activates the methyl-accepting sites for a net increase in methylation (symbolized in the figure by brackets around the methyl-accepting groups), thus setting the stage for adaptation.
3. The system adapts to the continued presence of the attractant (lower drawing). A new steady-state level of methylation is achieved, balancing the effect of ligand binding such that the cytoplasmic domain regains a null signalling state, functionally indistinguishable from the state prior to stimulation. Saturation of a ligand-binding site is balanced by a two- to three-fold increase in the average number of methyl groups per transducer (Stock, Clarke & Koshland, 1984; Terwilliger, Wang & Koshland, 1986).
4. Loss of ligand creates a negative stimulus (left-hand drawing), causing the periplasmic domain to assume its unoccupied conformation and thus eliminating the excitatory input from periplasmic to cytoplasmic domain. The change induced by methylation now has nothing to balance against and thus causes the cytoplasmic domain to assume an excitatory conformation that generates an intracellular signal for clockwise rotation. In this conformation, methyl esters are activated for net demethylation. 5. The adapted state is established in the unoccupied transducer by demethylation, restoring the cytoplasmic domain to the null signalling state.

The properties necessary for detecting a temporal gradient are a measurement of current concentration, a record of recent concentration and a means for comparison. As diagrammed in Figure 1, it appears that the transducer protein is directly involved in all three functions. Occupancy of the ligand-binding site is a measure of current concentration and covalent modification appears to be involved in information storage. The modification reactions are slow on the time-scale of changes in occupancy, therefore, in an environment of changing concentration, the current extent of modification is the level that compensated for the level of ligand occupancy a few moments previously. Thus the extent of modification is a record of prior concentration and the postulated balancing of ligand occupancy and modification, is a comparison of current and past concentrations. In fact, cells appear to compute running averages of present and past concentrations. The result obtained for the past second is compared to the concentrations in the previous three seconds (Segall, Block & Berg, 1986). As long as the two values are matched, behaviour persists in the null state; when one value exceeds the other, an excitatory signal affects the flagellar switch and activities are set in motion to reestablish the balance.

The notion that transducer methylation is crucial for detection of temporal gradients and thus for the ability of cells to make net progress in spatial gradients had been challenged by J. Stock and colleagues (see Stock & Stock, 1987 for a summary of the arguments). However, examination

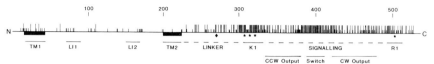

Fig. 2. Primary structure of chemotactic transducers from *E. coli*. The horizontal line represents the aligned sequences of Tar, Tsr, Tap and Trg as shown in Hazelbauer (1988). The sequences are 553, 536, 535 and 537 residues, respectively. In the alignment, the first ten positions contain only Trg residues and the final 15 only Tar residues. No internal gaps were placed in the Tsr sequence; two gaps of one residue each occur in Tar, one gap of four residues in Tap and one gap of two residues in Trg, all in the periplasmic domain. The tall and short vertical lines indicate the identical amino acid at that position for all four or three of the four transducer sequences, respectively. Notable features common to all four sequences are indicated below the line. See text for explanations.

of behaviour of single cells stimulated with attractant pulses of physiologically relevant magnitude (Segall et al., 1986), of cells lacking modification enzymes migrating in defined, preformed gradients (Weiss & Koshland, 1988) and of cells with transducers lacking methyl-accepting glutamyls migrating in metabolically created gradients (Hazelbauer, Park & Nowlin, 1989) provides overwhelming evidence for the crucial involvement of transducer modification in the fundamental mechanism of gradient detection. The apparently contradictory evidence may, in large part, reflect potentially misleading artefacts possible in migration of bacteria in semisolid agar (Wolfe & Berg, 1989) and in movement in certain kinds of gradients (Weiss & Koshland, 1988).

OVERVIEW OF TRANSDUCER STRUCTURE

The deduced amino acid sequences (Fig. 2) of the transducers of *E. coli* (Boyd, Kendall & Simon, 1983; Krikos et al., 1983; Russo & Koshland, 1983; Bollinger et al., 1984) revealed a family of four closely related proteins all with a molecular mass of approximately 60,000. Figure 2 provides a diagram of amino acid identities among the four sequences. In the diagram, two stretches of hydrophobic amino acids, 27 and 24 residues, respectively, are marked with black boxes and labelled transmembrane 1 (TM1) and transmembrane 2 (TM2), methyl-accepting sites are marked with stars, and particular deduced common features are indicated below the line representing the primary sequences. The pattern suggests a simple model for disposition of the homologous polypeptides across the cytoplasmic membrane (Fig. 3) in which the two segments of hydrophobic amino acids span the membrane, creating a periplasmic domain of approximately 155 amino acids and a cytoplasmic domain of approximately 320. A large body of genetic and biochemical evidence has supported and refined this model. Binding sites for attractants have been located in specific regions (LI-1 and LI-2) of the periplasmic domain as the result of evidence discussed in detail in a later section. The postulated orientation of the two membrane-

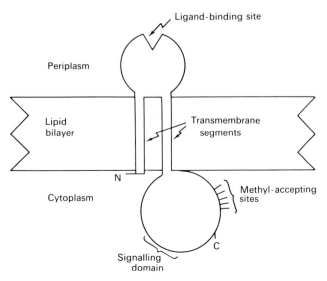

Fig. 3. A model for disposition of transducer proteins across the cytoplasmic membrane. Adapted from Fig. 1, Hazelbauer (1988).

spanning segments has been demonstrated directly by labelling experiments utilizing specific cysteine substitutions in Tar probed with membrane-impermeant reagents (Falke et al., 1988). The specific methyl-accepting positions in the cytoplasmic domain of Tsr (Kehry & Dahlquist, 1982), Tar (Terwilliger et al., 1986) and Trg (Nowlin, Bollinger & Hazelbauer, 1987) have been identified as either glutamates or as glutamines that are deamidated by the CheB esterase to create methyl-accepting glutamates. Four of these sites are at identical positions in the aligned sequences (Fig. 2). The sites occur in two regions: K1 which contains three, or for Trg and Tsr, four sites, and R1 which contains a single identified site. Because of these characteristic covalent modifications, the transducers are also called methyl-accepting chemotaxis proteins. A signalling domain within the cytoplasmic domain has been defined by a high frequency of mutational substitutions that lock the transducer in a clockwise or counterclockwise signalling mode (Ames & Parkinson, 1988; Park & Hazelbauer 1986b; Mutoh, Oosawa & Simon, 1986). This region, located between the K1 and R1 segments, contains primarily conserved amino acids, some of which presumably constitute common sites of interaction with CheA and CheW, the proteins responsible for transmitting the signal to the flagella. The distribution of subclasses of *tsr* mutations that create transducers with locked signal outputs has lead to the suggestion that the signalling domain contains a region of counterclockwise output near the K1 segment, a region of clockwise output near the R1 region and a 'switch' region separating the two (Ames & Parkinson, 1988). In addition, occurrence of some locked mutations and many mutations that create constitutive signalling without being dominant identify the 'linker' region between TM2 and the K1 segment as a potential participant in intramolecular transduction from the periplasmic domain to the signalling domain (Ames & Parkinson, 1988).

Three of the four transducers recognize polypeptide ligands, in each case specific periplasmic binding proteins in their ligand-occupied conformations. The Trg transducer mediates taxis to galactose and to ribose by binding the two respective sugar-binding proteins; Tar mediates maltose taxis by recognizing the maltose-binding protein; and Tap mediates response to dipeptides through a dipeptide-binding protein. In addition, Tar binds aspartate directly and the fourth transducer, Tsr, binds serine but apparently no polypeptide. Each of the transducers also mediates the response to one or more repellents (Yamamoto, Macnab & Imae, 1990 and references cited therein). In a number of cases, the effects of repellents on transducers appear to be via relatively nonspecific interactions that do not have the character of binding to stereospecific recognition sites.

Transducers exist as dimers (Milligan & Koshland, 1988) but the functional significance of dimeric organization is not yet understood. The three-dimensional organization of the dimeric structure has been investigated by elegant use of a family of altered Tar molecules, each with a single cysteine placed at a specific point within the primary sequence (Falke & Koshland, 1987; Falke *et al.*, 1988). Measurements of accessibility to sulphydryl reagents and of rate of oxidative cross-linking between cysteines on associated monomers indicated that the two transmembrane segments of a single polypeptide chain are close together at the periplasmic side of the membrane and that the first membrane-spanning regions of the two monomers in a dimer are near each other at the central axis of the dimeric structure. The pattern of cross-linking between cysteines at various positions and the differing functional activities of the resulting cross-linked proteins indicate a striking dynamic flexibility of the transducer. The effects of ligand on patterns of cross-linking indicate a global change in structure when a ligand binds.

Circular dichroic spectra of purified Tsr (Foster *et al.*, 1985) and Trg (Burrows, Newcomer & Hazelbauer, 1989) both indicate a content of approximately 80% α-helix, 20% random coil and no detectable β-sheet. The similarity of the two spectra is consistent with a common content of secondary structure for the two proteins. A single structural organization for the cytoplasmic domains of the four transducers is likely since there is a high degree of residue identity among the primary structures (Fig. 2). Although there is very little conservation of sequence in the four periplasmic domains, a number of observations discussed in a later section suggest a common structure. Thus it is reasonable to assume that the four transducers in *E. coli* are variations on one structure, and that this structure involves interactions of secondary structure units that are primarily, if not exclusively, α-helices.

Like many polypeptides found in the periplasm, the periplasmic domains of transducers are relatively resistant to proteolysis, indicating a compact structure providing little access for proteases to the polypeptide backbone. In contrast, the cytoplasmic domain contains a site in the linker region approximately 50 residues past TM2 (marked by a diamond in Fig. 2) that is extremely sensitive to proteolysis, implying that this region of the linker is relatively exposed, perhaps as the result of a loop or hinge. The 31 kD, carboxy-terminal fragment created by cleavage at this site is itself rather sensitive to further proteolysis, indicating a less compact structure than the amino-terminal fragment. Proteolytic sensitivity of the intact protein is greatly reduced by glycerol at concentrations 10% or higher, presumably because a more compact conformation is induced. This effect may be related to the general sensory effect glycerol has on all transducers, acting as a repellent (Oosawa & Imae, 1983, 1984) at concentrations corresponding to those at which a more compact structure is induced *in vitro*. Observations about proteolytic sensitivity of transducers were first documented by Mowbray, Foster & Koshland (1985) for Tar. In this laboratory we have noted an almost identical behaviour for Trg.

The 31 kD carboxy-terminal fragment exhibits some features of a functionally and structurally independent domain. Hybrid proteins between Tsr and Tar fused using an *Nde*I cut very near the codon corresponding to the proteolytic cleavage site are fully functional (Krikos *et al.*, 1985) as are fusions of Tsr or Tar to the more distantly related Trg (J. W. Baumgartner, unpublished data). The proteolytically released domain has some residual ability to be covalently modified *in vitro* (Mowbray *et al.*, 1985). *In vivo*, carboxy-terminal domains coded by truncated wild-type genes or truncated genes containing locked signalling mutations exhibit some appropriate covalent modification and flagellar signalling activities (Oosawa, Mutoh & Simon, 1988). These observations indicate that at least some features of the active protein are maintained in the absence of connection to the membrane. It is clear, however, that the fragment is not in a native state. This may be due in part to the lack of interaction with membrane-embedded components. Some observations imply that, in the intact transducer, the cytoplasmic domain is not completely separated from lipid bilayer, a feature not represented in simple models like Figure 3. For instance, after proteolysis *in vitro*, detergent is required to release the carboxy-terminal fragment from the membrane (Mowbray *et al.*, 1985) and the defective phenotype created in Tar by substitution of a lysine in TM1 is suppressed by several different compensating changes in a short segment of residues just past the site of proteolysis (Oosawa & Simon, 1986). These observations imply a functional and perhaps physical interaction between linker and transmembrane regions. The linker may loop into the lipid bilayer. Interestingly, the cytoplasmic fragment created by proteolysis (Mowbray et al., 1985) or truncation of the gene (Kaplan & Simon, 1988) is prone to aggregation in the absence of detergent but behaves as a homogenous species in the presence of detergent, implying that some segment of the fragment is stabilized by a hydrophobic environment. It is interesting that in the absence, but not in the presence, of detergent a cytoplasmic fragment of Tar containing a locked signalling mutation exhibited a different pattern of aggregation to the wild-type protein (Kaplan & Simon, 1988).

RELATED RECEPTORS IN OTHER SYSTEMS

Methyl-accepting taxis proteins are present in many bacterial species besides *E. coli* and *Salmonella typhimurium*. Genes for homologs of *tsr* and *tar* in the relatively closely related species *Enterobacter aerogenes* have been cloned and sequenced, and it appears that transducer genes are also present in the related but non-motile species, *Klebsiella pneumoniae* (Dahl, Boos & Manson, 1989). Part of the sequence of a gene important in aggregation by *Myxococcus xanthus* is strikingly similar to the section of transducer genes coding for the cytoplasmic domain (McBride, Weinberg & Zusman, 1989). It appears that the protein product would contain methyl-accepting sites and indeed methylation has been measured in this system (see chapter 9 in this volume for details of this unusual system). Methyl-accepting proteins have been identified as part of the tactic systems of *Bacillus subtilus* (Ullah & Ordal, 1981), *Pseudomonas aeruginosa* (Craven & Montie, 1983), *Spirochaeta aurantia* (Kathariou & Greenberg, 1983), *Caulobacter crescentus* (Shaw et al., 1983), a thermophilic bacterium PS-3 (Hirota, 1984, 1986), *Rhodospirillum rubrum* (Sockett, Armitage & Evans, 1987), *Agrobacterium tumefaciens* (Shaw et al., 1988), *Pseudomonas putida* (Harwood, 1989) and *Halobacterium halobium* (Alam et al., 1989). These proteins appear to share structural as well as functional features with the well-characterized transducers of the enteric bacteria since antiserum raised to Trg immunoprecipitates the proteins from *B. subtilis* and *S. aurantia* (Nowlin et al., 1985) and sera raised to Trg or to other enteric transducers react in an immunoblot with methyl-accepting proteins from *C. crescentus* (Gomes & Shapiro, 1984), *A. tumefaciens* (C. Shaw, personal communication), *P. pudita* (C. Harwood, personal communication), *H. halobium* (D. G. Morgan & G. L. Hazelbauer, unpublished data) and *Rhizobium meliloti* (D. G. Morgan & G. L. Hazelbauer, unpublished data). Antiserum raised to the methyl-accepting proteins of the thermophilic PS-3 recognizes transducers of *E. coli* (Hirota, 1986). A substantial number of other bacterial species representing a wide range of taxonomic classes as determined by analysis of rRNA (Woese, 1987) have been surveyed by immunoblotting with anti-Trg serum and many contain specific cross-reacting proteins of apparent molecular weights between 60 kD and 90 kD (D. G. Morgan, unpublished data). The sum of these observations indicate that homologous transducer proteins are components of tactic systems throughout the range of eubacterial and archebacterial species. This implies that a primordial transducer gene originated prior to the divergence of the ancestors of modern-day archaebacteria and eubacteria, over three billion years ago (Woese, 1987).

In addition to the wide taxonomic distribution of transducer-related proteins that presumably are components of homologous tactic systems, it has recently become evident that the components of the chemotactic system of *E. coli* are themselves representatives of a wider family of sensory systems in bacteria (Ronson *et al.*, 1987; Stock *et al.*, 1989). There are kinase–substrate pairs homologous to the CheA–CheY pair in many systems that control specific gene expression in response to environmental stimuli (Stock *et al.*, 1990). The systems include ones sensitive to phosphate, osmotic pressure, and aerobic/anaerobic balance in *E. coli*; nitrogen in nitrogen-fixing bacteria; components that induce virulence or colonization for bacteria that interact with animal or plant cells; and signals that induce sporulation in *B. subtilus*. For many of these systems, the kinase is a transmembrane protein with a deduced transmembrane disposition identical to chemotactic transducers. It appears that the component corresponding to the cytoplasmic domain of a chemotactic transducer has been eliminated and a CheA-related kinase is fused directly to the ligand-recognition domain via TM2. For systems controlling gene expression the response should persist as long as the stimulus is present and thus there is no need for the sophisticated methylation system that mediates adaptation and creates sensitivity to changes in stimulus intensity. Instead there can be persistent response to absolute levels of stimulant molecules. The protein substrates for the receptor/kinases that control gene expression contain a segment with substantial sequence identity to CheY but in addition they also have a segment that exerts an effect on transcription. In some cases it has been documented that control involves direct binding of the phosphorylated substrate protein to appropriate segments of DNA (Ninfa, Reitzer & Magasanik, 1987; Aiba *et al.*, 1989; Igo, Ninfa & Silhavy, 1989). Existence of a common mechanism for transmembrane signalling in the chemotactic transducers and the transmembrane receptor/kinases has been strikingly demonstrated by activation of appropriate gene expression with a hybrid protein. This protein has the ligand-recognition domain and transmembrane regions of the chemotactic transducer Tar fused, at the same *Nde*I site used to create hybrid transducers, to the kinase domain of EnvZ, a transmembrane protein that mediates a response to the osmotic environment of *E. coli* (Utsumi *et al.*, 1989). Activation of appropriate gene expression by this chimeric protein required the presence of a ligand (aspartate) for Tar as well as the substrate protein for phosphorylation by EnvZ. Thus it seems that occupancy at the ligand-binding site of Tar can be effectively transduced across the membrane to activate the fused kinase domain of EnvZ, implying that the two receptors, Tar and EnvZ, function by the same intramolecular transduction mechanism. In collaboration with the group of M. Inouye, who characterized the Tar-EnvZ fusion, we have investigated an analogous Trg–EnvZ hybrid protein and observed that it also can induce appropriate gene expression.

There are some interesting parallels between models for structural organization of chemotactic transducers and for the transmembrane disposition of eukaryotic receptors of growth factors and related hormones. The deduced amino acid sequences of receptors for insulin, epidermal growth factor and platelet-derived growth factor, for putative receptors for unknown ligands like the product of the *neu* oncogene, and for putative receptors involved in development such as the products of the *Drosophila* genes *sevenless* (Schejter et al., 1986) and *torso* (Sprenger, Stevens & Nüsslein-Volhard, 1989) all suggest an organization in which large, hydrophilic domains on each side of the membrane are connected by a single segment of hydrophobic residues that span the membrane. Details of these systems and references to original observations not cited above can be found in reviews by Carpenter (1987), Yarden & Ullrich (1988), and Williams (1989). Oligomer formation is a common feature of many receptors in this group and appears to be an important factor in signal transduction (reviewed in Yarden & Ullrich, 1988). The active forms of those receptors therefore contain two or more transmembrane segments. For biochemically characterized receptors in this family, the intracellular domains have been shown to possess tyrosine kinase activity that is activated by ligand binding in the extracellular domain. Kinase activity appears to be a crucial step in intracellular transduction. The possibility that the features shared by these eukaryotic receptors and chemotactic transducers might extend to common mechanisms of action has been investigated by making hybrid proteins with the exterior domain from the insulin receptor and the interior domain from the Tar transducer or vice versa. For the former combination, the hybrid protein bound insulin but there was no indication of transmembrane signalling (Ellis et al., 1986). In the latter combination, aspartate binding appeared to alter the activity of the internal tyrosine kinase domain (Moe, Bollag & Koshland, 1989). This tantalizing observation implies that eukaryotic and prokaryotic receptors related by similar transmembrane organization may share common mechanisms of intramolecular transduction.

INTRAMOLECULAR TRANSDUCTION

Intramolecular transduction begins with ligand binding. Characterization of mutations that specifically affect this transducer function for Trg (Park & Hazelbauer, 1986a), Tar (Wolff & Parkinson, 1988; Kossman, Wolff & Manson, 1988; Lee & Imae, 1990) or Tsr (Lee, Mizuno & Imae, 1988)

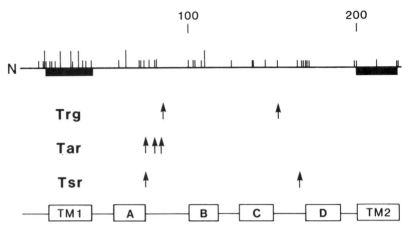

Fig. 4. Positions of mutational substitutions that affect interaction of transducer with ligand. The sequence alignment for the periplasmic domain is reproduced from Fig. 2. The arrows indicate position of mutational substitutions characterized in the respective transducers. The changes in Trg are R85H and G151D (Park & Hazelbauer, 1986a), in Tar R64C (two independent isolations), R69C, R69H (four independent isolations), R73W (two independent isolations) and R73Q (two independent isolations) (Wolff & Parkinson, 1988) and in Tsr R64C, R64H, T156I (two independent isolations) and T156P (Lee et al., 1988). The lower line indicates positions of amphipathic helices that could pack to create a four-helix bundle (Moe & Koshland, 1986).

indicate that two small segments at similar positions in the periplasmic domains of each transducer are important for effective ligand recognition (Fig. 4). We will refer to these two regions as ligand-interaction regions 1 and 2 (LI-1 and LI-2). The similar placement in different transducers of mutations that affect ligand interaction suggests that even though few residues in the periplasmic domain are conserved among the four transducers, the overall structure of the domains may exhibit a common organization. In fact, the suggestion by Moe & Koshland (1986) that the periplasmic domains are all organized as analogous four-helix bundles places the two ligand-interaction regions in close proximity in the postulated three-dimensional structure (Fig. 5). We will base our considerations of the organization and function of the periplasmic domains of transducers on the idea of a common three-dimensional structure and use the information gained from the study of different transducers to create a unified model of structure and function.

It is not known how ligand binding in the periplasmic domain affects the activity of the cytoplasmic domain. It seems likely that ligand interaction induces local perturbations that initiate conformational changes within the periplasmic domain. A ligand-occupied, conformationally altered periplasmic domain must in turn alter the state of the cytoplasmic domain. This might be accomplished either by an induced conformational change

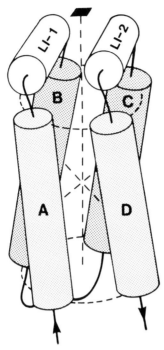

Fig. 5. A model of the periplasmic domain of transducers. After Moe & Koshland (1986) with indication of ligand-interaction regions 1 and 2. The figure is adapted from Fig. 1 of Sheridan, Levy & Salemme (1982). The diameter of the cylinders, representing helices, correspond to the dimension of the polypeptide backbone and the labels on the four helices of the bundle correspond to those in Fig. 4.

in the membrane-spanning segments that is transmitted directly to the cytoplasmic domain or by indirect effects resulting from dimer structure in which altered interactions of ligand-occupied periplasmic domains cause alterations of the cytoplasmic domains as a result of alternative oligomeric states. Recent studies of the Tar protein indicate, however, that the transducer is essentially always found as a dimer and it is dimeric whether or not ligand is bound (Milligan & Koshland, 1988). The persistence of the dimeric state may very well reflect an extensive interface of monomer–monomer interaction. Cleavage of Tar at the protease-sensitive loop separates the cytoplasmic domains from the periplasmic plus membrane-spanning domains, yet both halves of the transducer remain dimeric (Mowbray et al., 1985). Observations discussed in a later section indicate that the transmembrane segments are specifically involved in monomer–monomer interactions. Despite this extensive interface of interaction, it has been shown that monomer exchange does occur and is drastically inhibited in the ligand-occupied state (Milligan & Koshland, 1988; Falke et al., 1988), however, the significance of these observations is not yet understood. It appears that transmembrane transduction is not linked to gross changes in monomer-dimer equilibria. Transducers are found only as dimers, with no apparent pool of monomers. If changes in oligomeric state are mechanistically involved in transduction, they must be relatively subtle alterations. In contrast, several observations implicate the transmembrane regions in intramolecular signalling. Introduction of a lysine in the middle of TM1 of Tar eliminated sensory function without perturbing membrane insertion (Oosawa & Simon, 1986). Some mutational changes that suppress this defect are substitutions in TM2, implying that the two transmembrane regions are in proximity and that they must do more than just span the membrane to allow normal transducer function. One dominant mutation out of many that lock Tsr in a signalling mode is the substitution of a serine in place of Phe-18, in the midst of TM1 (Ames & Parkinson, 1988). In addition, there are mutational substitutions in TM1 and TM2 of Tsr and Trg that send constitutive signals to the cytoplasmic domain without causing a dominant phenotype (Ames *et al.*, 1988; J. W. Baumgartner, unpublished data). These observations suggest that the transmembrane segments may be part of the pathway of intramolecular transduction that would then continue through the linker region to the signalling domain.

LIGAND INTERACTION

The site of ligand interaction within the periplasmic domain of a transducer must perform two roles: i. physical interaction with the ligand to produce stereospecific recognition of certain molecules but not others and ii. initiation of intramolecular transduction by a conformational change created by ligand interaction. Specific residues could be important for binding, for the ability of binding to initiate transduction, or for both. Mutations affecting each of those three classes of residues might have different phenotypes. For example, a *trg* mutuation in LI-1, in which Arg-85 is replaced by a His, reduces the effectiveness in signalling but not the apparent affinity of ligand interaction, while a mutation in LI-2, G151D, reduces both parameters (Park & Hazelbauer, 1986*a*).

Two regions of the periplasmic domain important for ligand interaction (LI-1 and LI-2) have been identified by the striking clustering of twenty mutations, obtained in three different transducer genes (Fig. 4). Most of the mutants were identified as likely to be altered specifically in ligand interaction by a phenotype in which the response to one compound was reduced drastically while responses to other compounds sensed by that transducer (alternative attractants or repellents) were little affected (Park & Hazelbauer, 1986*a*, Wolff & Parkinson, 1988; Lee *et al.*, 1988). The mutational changes fell into two groups, characterized by different location along the primary sequence and by different phenotypic characteristics. In LI-1, all substitutions were at arginine residues and caused a reduced response to all attractants recognized by the particular transducer, albeit to different degrees. In the collection of *tar* mutations obtained by Wolff & Parkinson (1988), identical substitutions occurred independently several times implying that few other substitutions could create the identifying phenotype. In LI-2, substitutions were not at arginine residues, and for Trg G151D and Tar T154I (a substitution created in Tar to mimic the Tsr T156I mutant, Lee & Imae, 1990) response to one attractant (ribose or maltose, respectively) was indistinguishable from that of wild-type while the response to the other attractant (galactose or aspartate, respectively) was drastically defective. Thus LI-2 contains residues crucial for interaction with one ligand and not another, while the mutated residues in LI-1 disrupt functions required for full response to any ligand. Residues in LI-1 could be important for effective intramolecular transduction upon binding of any ligand as documented for Trg R85H (see above). However, it is also attractive to postulate that the three arginines in LI-1, present in both Tsr and Tar, provide a charged pocket for binding the carboxyl group of serine and aspartate, respectively (Wolff & Parkinson, 1988; Lee *et al.*, 1988) and that the conserved threonine in LI-2 would be involved in hydrogen bonding to another feature common to both amino acids (Lee & Imae, 1990). In any case, it is clear that the arginine pocket must play a role besides binding to amino acids since substitutions at the arginines in Tar alter the effectiveness of interaction with maltose-binding protein. In the arginine mutants, response mediated by wild-type maltose-binding protein is reduced while responses mediated by some chemotaxis-defective maltose-binding proteins are improved (Kossmann *et al.*, 1988).

These ideas are being investigated by oligonucleotide-directed, random mutagenesis of the codons corresponding to LI-1 and LI-2 (R. Yaghmai, unpublished data). We have already generated a collection of Trg proteins with amino acid substitutions throughout LI-1. Many of these substitutions have no discernible effect on transducer function, some result in substantial susceptibility to proteases presumably because of disrupted structure, and others alter functional properties of the protein without substantial alterations of stability. The third group of mutational defects can be placed into two distinct classes by a number of phenotypic characteristics. A striking observation is that the two classes are linearly segregated along the twenty-codon target of mutagenesis (residues 69 to 88), with position 79 as the overlapping border. Class A includes substitutions at positions 71 to 78 and a substitution of leucine for Gln 79; Class B includes proline in place of Gln 79 and substitutions at positions 81 to 87. In Class A, the pattern of covalent modification of the mutant transducer in the absence of attractant corresponds to the increased methylation observed when wild-type Trg protein has adapted to the continued presence of saturating amounts of galactose or ribose. Thus it appears that Class A substitutions mimic the presence of ligand in a way that causes a characteristic change in the extent of covalent modification in the cytoplasmic domain, presumably by way of the normal pathway of intramolecular, transmembrane signalling. The implication is that the mutated residues are involved in, or can induce, the conformational changes that constitute the initial step of intramolecular transduction. This constitutive signalling does not appear to saturate either the intramolecular signalling pathway or the methylation capacity of the mutant proteins. Thus, when produced at normal levels in an otherwise wild-type cell, the mutant proteins mediate chemotaxis to galactose and to ribose. Class B mutations have a more drastic effect on Trg function. The presence of attractant results in little or no increased methylation, and chemotactic migration mediated by the mutant proteins is significantly defective. Substitutions in this region appear to reduce or eliminate effective interaction of ligand with Trg. This could be the result of disruption of residues and structures that physically interact with the binding proteins. Alternatively, if Class B alterations were the converse of Class A changes, inducing intramolecular signalling corresponding to loss of ligand occupancy, then the effects of ligand occupancy might be reduced or even eliminated. A hint of such a situation comes from examination of mutations in *tsr* that are related to these *trg* changes. Ames *et al.*, (1988) describe *tsr* mutations that cause the transducer to generate constitutive intracellular signals, either clockwise or counterclockwise, of moderate intensity. Only a few of these mutations cause substitutions in the periplasmic domain of Tsr, but three of them are in LI-1, one at position 73 in the aligned sequences, corresponding to a Class A mutation and two at positions (81 and 83) corresponding to Class B mutations. Consistent with the phenotypes of the two classes in *trg*, the *tsr* mutation in the Class A region is counter-clockwise-biased and the two in the Class B segment are clockwise-biased. The second Class B mutation alters one of the arginine positions in Tsr at which other substitutions greatly reduced the effectiveness of serine binding.

We are developing three-dimensional structural models of LI-1 with the hope of gaining insights into how the residues might be involved in both the physical interactions that bind the ligand to the transducer and in the conformational changes induced by that binding. Residues in the region are certainly involved in both phenomena and some may be central to both. An interesting model is created by investigating the way in which the bridging helix (labelled LI-1 in Fig. 5) could be connected to helix A. A type-1 turn (Richardson, 1981) between helix A and the bridging helix brings three arginines of Trg together in a pocket even though two of the three arginines are at different positions on the primary sequence when compared with the three arginines postulated to form an arginine pocket in Tsr and Tar. A particularly tantalizing observation is that alternative hydrogen bonding in the turn provides two plausible stable structures with different angles between the two helices, suggesting alternative structures that could be involved in conformational changes that are induced by binding and in turn generate transductional signals. The glutamine at the border between Class A and Class B mutations in LI-1 of Trg is within the turn, placing the two classes on adjoining helices. The possibility that the mutational substitutions would favour one or the other helix–helix angle is currently being investigated.

Three of the four transducers recognize specific, ligand-occupied binding proteins. Thus, in these cases ligand recognition is actually a protein–protein interaction. The situation is somewhat puzzling, since the best estimates using physiological data, of the dissociation constants for these protein–protein complexes are in the order of 100 to 1000 μM (Manson et al., 1985), in contrast to transducer-small molecule dissociation constants which are in the 1 to 10 μM range (Clarke & Koshland, 1979). The low estimates of affinity reflect primarily the apparently very high concentration (100 to 1000 μM) of binding protein in the periplasm of appropriately induced cells. It is not clear how the association of occupied binding protein and transducer can be as weak as apparently required yet be quite specific. At least one region of the galactose-binding protein has been found to be crucial for interaction with its transducer, Trg. Substitution of an aspartyl residue in place of Gly74 (Scholle et al., 1987) in the binding protein eliminates chemotactic activity but not galactose binding or transport (Hazelbauer & Adler, 1971). Comparison of the three-dimensional structure of mutant and wild-type binding proteins reveals a difference only in a shallow groove between two α-helices. The groove is partly occupied by the aspartyl group and thus might be blocked sterically to prohibit correct placement of a segment of Trg in the groove (Vyas, Vyas & Quiocho, 1988). If the aspartyl group interfered with correct placement of a segment of polypeptide backbone from Trg that usually interacted with several residues in the groove, then single amino acid substitutions in Trg could not easily compensate for the alteration in the binding protein. This notion would explain the lack of success experienced by several groups in attempts to obtain changes in *trg* that suppressed the G74D substitution in the binding protein.

Three mutational substitutions in the maltose-binding protein reduce chemotactic activity of the protein but have little effect on sugar binding or transport (Manson & Kossmann, 1986). The tactic defects for two of the substitutions are suppressed by *tar* mutations at the crucial arginines in LI-1 (Kossmann et al., 1988). The loci of those mutational changes in the binding protein, separated by only one residue in the primary structure, define a segment of the polypeptide that is likely to participate in transducer interaction. Unfortunately, the three-dimensional structure of maltose-binding protein is not generally available, so it is not yet possible to make an unambiguous comparison of the position of these mutational substitutions with the analogous one in galactose-binding protein.

Both Trg and Tar recognize two different ligands using non-identical binding sites. For Trg, the sites for galactose- and ribose-binding proteins overlap functionally, that is saturation with one ligand greatly reduces response to the other (Strange & Koshland, 1976; Park & Hazelbauer, 1986a), but at least some residues involved in recognition are different (Park & Hazelbauer, 1986a). For Tar, saturation with one ligand reduces but does not entirely block response to the other (C. Wolff, Diploma thesis, University of Konstanz, Federal Republic of Germany, 1983; Mowbray & Koshland, 1987), and some mutational substitutions affect response to one ligand more than the other (see above). In Trg, the two ligands may bind to nonidentical but overlapping sets of residues, while Tar may contain physically separated but functionally linked sites (Wolff & Parkinson, 1988). The data for the galactose-binding protein suggest that transducers recognize binding proteins by intercalation of segments of structure, for example, one or both bridging helices on the top of the four-helix bundle of the transducer could fit into one or more shallow grooves on a face of the binding proteins. Specific residues in the bridging helices involved in the physical interaction could be different for the two different transducer-binding protein pairs and thus the interaction faces on the two sugar-binding proteins could include quite different residues, even though the same region of Trg was involved in recognizing both binding proteins. Could binding of serine and aspartate by Tsr and Tar, respectively, involve structural units related to those involved in recognition of binding proteins? Clearly, sites for aspartate and maltose-binding protein on Tar are not identical, but the clustering in the three transducer sequences of mutational loci that affect recognition of both small molecules and polypeptides suggests there might be an intimate relationship. Could it be that segments of the transducer that change conformation as part of transductional alterations induced by recognition of binding proteins have come to serve as specific binding sites for certain amino acids? For instance the arginine pocket might function solely as a tranductional element in Trg but also serve as part of the binding site for amino acids in Tsr and Tar.

THE MEMBRANE-SPANNING REGIONS

A native transducer dimer contains four transmembrane segments, two in each monomer, each is likely to be an α-helix. The residues in these segments are exclusively hydrophobic or neutral amino acids. Yet particular sequences are, however, different for each of the four transducers and very few positions have identical residues in even three out of four cases. Is each of the helices completely surrounded by hydrocarbon chains of membrane lipid or are there protein–protein interactions among the helices? If such interactions exist, are they within monomers, between monomers, or both? We have begun to explore these issues with a family of Trg proteins produced from mutated *trg* genes, each altered to contain a single cysteine codon at a position corresponding to a residue in a transmembrane segment (G. G. Burrows & D. P. Dutton, unpublished data). The strategy is an extension of that used successfully by Falke & Koshland (1987) and Falke *et al.* (1988) in characterizing cysteines substituted in hydrophilic domains of Tar. Each mutant Trg protein is being examined in its native dimer form for the degree of reactivity with the sulphydryl reagent N-ethylmaleimide (NEM) and for the extent to which the cysteines on the two monomers of a dimer can be oxidatively cross-linked. Our initial work has been done with a family of seven proteins, each with a single cysteine in a limited segment of TM1. We found that sulphyldryls at some positions react readily with ^{3}H-NEM, at other positions are not modified by the reagent and at others exhibit limited but detectable labelling. The extent of reactivity does not correlate with the distance from the hydrophilic boundaries of the transmembrane segment and thus reactivity does not appear to be a simple function of distance from an aqueous environment. In fact, NEM is known to react with sulphydryls in hydrophobic environments of lipid bilayers and detergent micelles (Altenbach *et al.*, 1989). We conclude that low reactivity indicates a residue is shielded from the hydrocarbon environment by a stable interaction with another segment of protein. Since the two transmembrane regions each span the membrane only once, it is unlikely that residues on the same transmembrane segment pack on each other to an extent that would block reaction with NEM. Instead, we suspect that the residues with low or no reactivity are packed on residues from another transmembrane segment. Arrangement of TM1 of Trg as a helix places the two residues of lowest accessibility to NEM at non-adjacent positions around the circumference of the helix, separated by residues exhibiting substantial accessibility. We suspect that two faces of the helix, defined by the positions of the inaccessible residues, are packed on two other helical segments of Trg, either TM2 of the same molecule or one of the helices in the other molecule of the dimer.

A complementary approach is to explore interactions between the pair of molecules contained in a dimer using disulfide cross-linking between those monomers via the single cysteine position in each monomer and to relate the pattern of cross-linking to the packing of helices deduced from accessibility experiments. The results of these experiments are striking. At some positions in the first transmembrane region, cysteines can effectively cross-link one monomer to the other and at other positions crosslinking hardly occurs. The experiments are still in progress, therefore not all the possibilities have been explored for the seven cysteine positions, but already interesting conclusions can be made. For one position, oxidative cross-linking is extremely fast and that rate is not reduced by ten-fold dilution of detergent-solubilized dimers, implying that the first membrane-span-

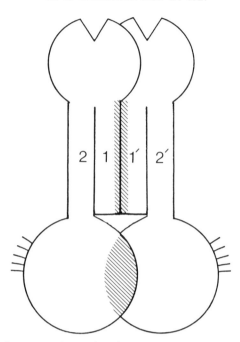

Fig. 6. A model of monomer interactions in a transducer dimer. Associations of the two monomers, indicated by cross-hatching, are thought to occur along the transmembrane helices and between the cytoplasm domains. Additional interactions may occur between the periplasmic domains.

ning helices of the two monomers of a dimer are stably packed on each other in an orientation that places the two cysteines in reactive proximity. Such a packing is consistent with the observation that this residue exhibits low reactivity to NEM. In contrast, a position five residues farther into the membrane is essentially inaccessible to NEM and exhibits almost no cross-linking, a feature it shares with a position seven residues earlier and thus on the same face of the putative helix. This would be consistent with the face defined by those two residues being packed against the second membrane-spanning region and thus being blocked for reaction with NEM and also distant from the corresponding cysteine on the other monomer. Figure 6 shows a model of dimer organization consistent with these results and those of Falke et al. (1988). Additional experiments should allow us to define in some detail the packing relationships of the four transmembrane helices in a transducer dimer and explore the possibility of structural alterations related to transmembrane signalling.

Complete information about the three-dimensional packing of transmembrane helices is available only for bacterial photosynthetic reaction centres (Deisenhofer & Michel, 1989), but a recent attempt has been made to use the patterns of packing observed for those proteins to predict exposed and buried faces of putative transmembrane helices in other membrane proteins (Rees, DeAntonio & Eisenberg, 1989). Among 35 helical segments analysed, most exhibited a distinct separation of prospective packed and exposed faces (an average angle of 128°). For the two helices of transducers, the postulated exposed and buried faces were almost contiguous on each helix. This might reflect structures designed to adopt alternative packing interactions as would be the case if the transmembrane helices rotated on each other during signal transduction.

A MODEL FOR TRANSDUCER STRUCTURE

What can be concluded about the structural organization of an individual transducer monomer? Since the predominant secondary structure in transducers is α-helix, the three-dimensional structure of these proteins should be a network of helix–helix interactions. We suggest that an important theme in transducer organization will be the association of helices along their long axes to form stacks and bundles. A speculative model based on this notion is presented in Figure 7. The organization of the periplasmic domain follows the suggestion by Moe & Koshland (1986) discussed and embellished in previous sections. The transmembrane helices are shown as packed along their length as suggested by the results of experiments using sulphydryl chemistry performed by Falke et al. (1988) and G. G. Burrows (unpublished data). The idea of helix bundles can be applied also to segments of the cytoplasmic domain, although at present there is no specific evidence for such an organization. The K1 and R1 regions would form helices with strong amphipathic character in which methyl-accepting sites, occurring every seven residues, would be in a line on the hydrophilic face. These helices could pack together and interact with helical segments of the clockwise and counterclockwise output regions of the signalling domain, making a four-helix bundle in which the switch region would be a loop connecting the output helices. Possible interactions of the linker region in connecting the transmembrane and signalling domains were discussed in an earlier section but have been left undefined in this model. As presented in Figure 6, monomers would interact along the TM1 helices (again helix packing) and also somewhere in the cytoplasmic domain. There may also be interaction at the interfaces of the periplasmic domains.

Packed helices and bundles can shift along interfaces to transmit conformational changes and to alter side chain accessibility. Such shifts could be intimately involved in transducer function. Movement at a membrane-distal, ligand-binding site (see a previous section for consideration of alternate angles of the turn between the bridging helix and helix A of the bundle) could be transmitted to the transmembrane helices by a mechanism of helix interface shear in which packed helices move relative to each other

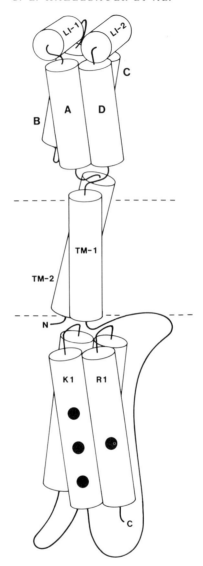

Fig. 7. A model for organization of a transducer monomer. The thin lines indicate regions of primary sequence in which no particular structure can be suggested at this time. The labels on helices correspond to those in Fig. 2 and Fig. 5. The dots on the helices in the cytoplasmic domain represent methyl-accepting sites. The diameter of the cylinders, representing α-helices, corresponds to the average size of the polypeptide backbone plus side chains.

without dissociating, the shifts being accommodated by motions of side-chain atoms arising from small changes in torsion angles (Chothia *et al.*, 1983; Lesk & Chothia, 1984). A similar mechanism could transmit a conformational change across the membrane by shifts involving either the two helices of individual monomers or the entire stack of four helices. Finally, the four-helix bundle of methyl-accepting and output regions could be affected by similarly subtle shifts as helices rolled relative to each other, making methyl-accepting sites which lie on one face of the helix more or less accessible to modification enzymes and making appropriate faces of the signalling domains more or less available for interaction with proteins of the phosphorylation cascade. Altered levels of methylation could cause the helices to roll back to their initial relative positions, poised for new inputs. The low energy nature of the conformational adjustments at each stage of the transmission and signalling process would allow fine gradations of these changes between the structural extremes. Methylation at any one of the sites lined up along the hydrophilic face of the K1 or R1 helix would have a similar effect on helix rolling along the long axis, consistent with the observation that any site can be the single methylated site on a particular protein molecule (Kehry & Dahlquist, 1982; Terwilliger *et al.*, 1986; Nowlin *et al.*, 1987) and that transducers lacking some or even most methyl-accepting glutamates are still capable of adaptation, albeit at slower rates (Nowlin, Bollinger & Hazelbauer, 1988).

We anticipate that the continued rapid accumulation of new information about transducer structure and function will lead to refinement of models such as the one presented here and thus to important insights into the mechanisms of generation and transmission of informational signals by transmembrane receptor proteins.

REFERENCES

Adler, J. (1975). Chemotaxis in bacteria. *Annual Reviews in Biochemistry*, **72**, 341–56.

Aiba, H., Nakasai F., Mizushima, S. & Mizuno, T. (1989). Phosphorylation of a bacterial activator protein, OmpR, by a protein kinase, EnvZ, results in a stimulation of its DNA-binding ability. *Journal of Biochemistry*, **106**, 5–7.

Alam, M., Lebert, M., Oesterhelt, D. & Hazelbauer, G. L. (1989). Methyl-accepting taxis proteins in *Halobacterium halobium*. *EMBO Journal*, **8**, 631–9.

Altenbach, C., Flitsch, S. L., Khorana, H. G. & Hubbell, W. L. (1989). Structural studies on transmembrane proteins. 2. Spin labeling of bacteriorhodopsin mutants at unique cysteines. *Biochemistry*, **28**, 7806–12.

Ames, P. & Parkinson, J. S. (1988). Transmembrane signaling by bacterial chemoreceptors: E. coli transducers with locked signal output. *Cell*, **55**, 817–26.

Ames, P., Chen, J., Wolff, C. & Parkinson, J. S. (1988). Structure–function studies of bacterial chemosensors. *Cold Spring Harbor Symposia on Quantitative Biology*, Vol. LIII. pp. 59–65.

Berg, H. C. (1975). Chemotaxis in bacteria. *Annual Reviews in Biophysics & Bioengineering*, **4**, 119–36.

Bollinger, J., Park, C., Harayama, S. & Hazelbauer, G. L. (1984). Structure of the Trg protein: homologies with and differences from other sensory transducers of *Escherichia coli*. *Proceedings of the National Academy of Sciences, USA*, **81**, 3287–91.

Borkovich, K. A., Kaplan, N., Hess, J. F. & Simon, M. I. (1989). Transmembrane signal transduction in bacterial chemotaxis involves ligand-dependent activation of phosphate group transfer. *Proceedings of the National Academy of Sciences, USA*, **86**, 1208–12.

Bourret, R. B., Hess, J. F. & Simon, M. I. (1990). Conserved aspartate residues and phosphorylation in signal transduction by the chemotaxis protein CheY. *Proceedings of the National Academy of Sciences, USA*, **87**, 41–5.

Boyd, A. & Simon, M. I. (1982). Bacterial chemotaxis. *Annual Reviews in Physiology*, **44**, 501–18.

Boyd, A., Kendall, L. & Simon, M. I. (1983). Structure of the serine chemoreceptor in *Escherichia coli*. *Nature, London*, **301**, 623–6.

Burrows, G. G., Newcomer, M. E. & Hazelbauer, G. L. (1989). Purification of receptor protein Trg by exploiting a property common to chemotactic transducers of *Escherichia coli*. *Journal of Biological Chemistry*, **264**, 17309–15.

Carpenter, G. (1987). Receptors for epidermal growth factor and other polypeptide mitogens. *Annual Reviews in Biochemistry*, **56**, 881–914.

Chothia, C., Lesk, A. M., Dodson, G. G. & Hodgkin, D. C. (1983). Transmission of conformational change in insulin. *Nature, London*, **302**, 500–5.

Clarke, S. & Koshland, D. E. Jr. (1979). Membrane receptors for aspartate and serine in bacterial chemotaxis. *Journal of Biological Chemistry*, **254**, 9695–702.

Craven, R. C. & Montie, T. C. (1983). Chemotaxis of *Pseudomonas aeruginosa*: involvement of methylation. *Journal of Bacteriology*, **154**, 780–6.

Dahl, M. K., Boos, W. & Manson, M. D. (1989). Evolution of chemotactic-signal transducers in enteric bacteria. *Journal of Bacteriology*, **171**, 2361–71.

Deisenhofer, J. & Michel, H. (1989). The photosynthetic reaction center from the purple bacterium *Rhodopseudomonas viridis*. *Science*, **245**, 1463–73.

Eisenbach, M. (1990). Signal transduction in bacterial chemotaxis. *Modern Cellular Biology*., in press.

Ellis, L., Morgan, D. O., Koshland, D. E. Jr., Clauser, E., Moe, G. R., Bollag, G., Roth, R. A. & Rutter, W. J. (1986). Linking functional domains of the human insulin receptor with the bacterial aspartate receptor. *Proceedings of the National Academy of Sciences, USA*, **83**, 8137–41.

Falke, J. J. & Koshland, D. E. Jr. (1987). Global flexibility in a sensory receptor: A site-directed cross-linking approach. *Science*, **237**, 1596–600.

Falke, J. J., Dernburg, A. F., Sternberg, D. A., Zalkin, N., Milligan, D. L. & Koshland, D. E. Jr. (1988). Structure of a bacterial sensory receptor. *Journal of Biological Chemistry*, **263**, 14850–8.

Foster, D. L., Mowbray, S. L., Jap, B. K. & Koshland, D. E., Jr. (1985). Purification and characterization of the aspartate chemoreceptor. *Journal of Biological Chemistry*, **260**, 11706–10.

Gomes, S. L. & Shapiro, L. (1984). Differential expression and positioning of chemotaxis methylation proteins in *Caulobacter*. *Journal of Molecular Biology*, **178**, 551–68.

Harwood, C. S. (1989). A methyl-accepting protein is involved in benzoate taxis in *Pseudomonas putida*. *Journal of Bacteriology*, **121**, 4603–8.

Hazelbauer, G. L. (1988). The bacterial chemosensory system. *Canadian Journal of Microbiology*, **34**, 466–74.

Hazelbauer, G. L. & Adler, J. (1971). Role of the galactose binding protein in chemotaxis of *Escherichia coli* toward galactose. *Nature New Biology*, **230**, 101–4.

Hazelbauer, G. L., Park, C. & Nowlin, D. M. (1989). Adaptational 'crosstalk' and the crucial role of methylation in chemotactic migration by *Escherichia coli*. *Proceedings of the National Academy of Sciences, USA*, **86**, 1448–52.

Hess, J. F., Bourret, R. B. & Simon, M. I. (1988a). Histidine phosphorylation and phosphoryl group transfer in bacterial chemotaxis. *Nature*, **336**, 139–43.

Hess, J. F., Oosawa, K., Kaplan, N. & Simon, M. I. (1988b). Phosphorylation of three proteins in the signalling pathway of bacterial chemotaxis. *Cell*, **53**, 79–87.

Hess, J. F., Oosawa, K., Matsumura, P. & Simon, M. I. (1987). Protein phosphorylation is involved in bacterial chemotaxis. *Proceedings of the National Academy of Sciences, USA*, **84**, 7609–13.

Hirota, N. (1984). Chemotaxis in thermophilic bacterium PS-3. *Journal of Biochemistry*, **96**, 645–50.

Hirota, N. (1986). Methylation and demethylation of solubilized chemoreceptors from thermophilic bacterium PS-3. *Journal of Biochemistry*, **99**, 349–56.

Igo, M. M., Ninfa, A. J. & Silhavy, T. J. (1989). A bacterial environmental sensor that functions as a protein kinase and stimulates transcriptional activation. *Genes & Development*, **3**, 598–605.

Kaplan, N. & Simon, M. (1988). Purification and characterization of the wild-type and mutant carboxy-terminal domains of the *Escherichia coli* Tar chemoreceptor. *Journal of Bacteriology*, **170**, 5134–40.

Kathariou, S. & Greenberg, E. P. (1983). Chemoattractants elicit methylation of specific polypeptides in *Spirochaeta aurantia*. *Journal of Bacteriology*, **156**, 95–100.

Kehry, M. R. & Dahlquist, F. W. (1982). The methyl-accepting chemotaxis proteins of *Escherichia coli*. Identification of the multiple methylation sites on methyl-accepting protein. *Journal of Biological Chemistry*, **257**, 10378–86.

Koshland, D. E. Jr. (1981). Biochemistry of sensing and adaptation in a simple bacterial system. *Annual Reviews in Biochemistry*, **50**, 765–82.

Kossmann, M., Wolff, C. & Manson, M. D. (1988). Maltose chemoreceptor of *Escherichia coli*: interaction of maltose-binding protein and the Tar signal transducer. *Journal of Bacteriology*, **170**, 4516–21.

Krikos, A., Mutoh, N., Boyd, A. & Simon, M. I. (1983). Sensory transducers of *Escherichia coli* are composed of discrete structural and functional domains. *Cewll*, **33**, 615–22.

Krikos, A., Conley, M. P., Boyd, A., Berg, H. C. & Simon, M. I. (1985). Chimeric chemosensory transducers of *Escherichia coli*. *Proceedings of the National Academy of Sciences, USA*, **82**, 1326–30.

Lee, L., Mizuno, R. & Imae, Y. (1988). Thermosensing properties of *Escherichia coli tsr* mutants defective in serine chemoreception. *Journal of Bacteriology*, **170**, 4769–74.

Lee, L. & Imae, Y. (1990). Role of threonine residue 154 in ligand recognition of the Tar chemoreceptor of in *Escherichia coli*. *Journal of Bacteriology*, **172**, 377–82.

Lesk, A. M. & Chothia, C. (1984). Mechanisms of domain closure in proteins. *Journal of Molecular Biology*, **174**, 175–91.

Liu, J. & Parkinson, J. S. (1989). Role of CheW protein in coupling membrane receptors to the intracellular signaling system of bacterial chemotaxis. *Proceedings of the National Academy of Sciences, USA*, **86**, 8703–7.

Lupas, A. & Stock, J. (1989). Phosphorylation of an N-terminal regulatory domain activates the CheB methylesterase in bacterial chemotaxis. *Journal of Biological Chemistry*, **264**, 17337–42.

Macnab, R. M. & DeRosier, D. J. (1988). Bacterial flagella structure and function. *Canadian Journal of Microbiology*, **34**, 442–51.

Manson, M. D. & Kossman, M. (1986). Mutations in *tar* suppress defects in maltose chemotaxis caused by specific *mal*E mutations. *Journal of Bacteriology*, **165**, 34–40.

Manson, M. D., Boos, W., Bassford, P. J. Jr, & Rasmussen, B. A. (1985). Dependence of maltose transport and chemotaxis on the amount of maltose-binding protein. *Journal of Biological Chemistry*, **260**, 9727–33.

McBride, M. J., Weinberg, R. A. & Zusman, D. R. (1989). 'Frizzy' aggregation genes of the gliding bacterium *Myxococcus xanthus* show sequence similarities to the chemotaxis genes of enteric bacteria. *Proceedings of the National Academy of Sciences, USA*, **86**, 424–8.

Milligan, D. L. & Koshland, D. E. Jr. (1988). Site-directed cross-linking. Establishing the dimeric structure of the aspartate receptor of bacterial chemotaxis. *Journal of Biological Chemistry*, **263**, 6268–75.

Moe, G. R. & Koshland, D. E. Jr. (1986). Transmembrane signalling through the aspartate receptor, in Youvan, D. C. & Daldal, F. eds. *Microbial Energy Transduction*. pp. 163–8. Cold Spring Harbor Laboratory, Cold Spring Harbor, NY.

Moe, G. R., Bollag, G. E & Koshland, D. E. Jr. (1989). Transmembrane signaling by a chimera of the *Escherichia coli* aspartate receptor and the human insulin receptor. *Proceedings of the National Academy of Sciences, USA*, **86**, 5683–87.

Mowbray, S. L., Foster, D. L. & Koshland, D. E. Jr. (1985). Proteolytic fragments identified with domains of the aspartate chemoreceptor. *Journal of Biological Chemistry*, **260**, 11711–18.

Mowbray, S. L. & Koshland, D. E. Jr. (1987). Additive and independent responses in a single receptor; aspartate and maltose stimuli on the Tar protein. *Cell*, **50**, 171–80.

Mutoh, N., Oosawa, K. & Simon, M. I. (1986). Characterization of *Escherichia coli* chemotaxis receptor mutants with null phenotypes. *Journal of Bacteriology*, **164**, 992–8.

Ninfa, A. J., Reitzer, L. J. & Magasanik, B. (1987). Initiation of transcription of the bacterial *glnAp2* promoter by purified *E. coli* components is facilitated by enhancers. *Cell*, **50**, 1039–46.

Nowlin, D. M., Bollinger, J. & Hazelbauer, G. L. (1987). Covalent modification in Trg, a sensory transducer of *Escherichia coli*. *Journal of Biological Chemistry*, **262**, 6039–45.

Nowlin, D. M., Bollinger & Hazelbauer, G. L. (1988). Site-directed mutations altering methyl-accepting residues of a sensory transducer protein. *Proteins*, **3**, 102–12.

Nowlin, D. M., Nettleton, D. O., Ordal, G. W. & Hazelbauer, G. L. (1985). Chemotactic transducer proteins of *Escherichia coli* exhibit homology with methyl-accepting proteins from distantly related bacteria. *Journal of Bacteriology*, **163**, 262–6.

Oosawa, K. & Imae, Y. (1983). Glycerol and ethylene glycol: Members of a new class of repellents of *Escherichia coli* chemotaxis. *Journal of Bacteriology*, **154**, 104–12.

Oosawa, K. & Imae, Y. (1984). Demethylation of methyl-accepting chemotaxis proteins in *Escherichia coli* induced by the repellents glycerol and ethylene glycol. *Journal of Bacteriology*, **157**, 576–81.

Oosawa, K. & Simon, M. I. (1986). Analysis of mutations in the transmembrane region of the aspartate chemoreceptor in *Escherichia coli*. *Proceedings of the National Academy of Sciences, USA*, **83**, 6930–4.

Oosawa, K., Hess, J. F. & Simon, M. I. (1988*a*). Mutants defective in bacterial chemotaxis show modified protein phosphorylation. *Cell*, **53**, 88–96.

Oosawa, K., Mutoh, N. & Simon, M. I. (1988*b*). Cloning of the C-terminal cytoplasmic fragment of the Tar protein and effects of the fragment on chemotaxis of *Escherichia coli*. *Journal of Bacteriology*, **170**, 2521–6.

Park, C. & Hazelbauer, G. L. (1986*a*). Mutations specifically affecting ligand interaction of the Trg chemosensory transducer. *Journal of Bacteriology*, **167**, 101–9.

Park, C. & Hazelbauer, G. L. (1986*b*). Mutation plus amplification of a transducer gene disrupts general chemotactic behavior in *Escherichia coli*. *Journal of Bacteriology*, **168**, 1378–83.

Rees, D. C., DeAntonio, L. & Eisenberg, D. (1989). Hydrophobic organization of membrane proteins. *Science*, **245**, 510–13.

Richardson, J. S. (1981). The anatomy and taxonomy of protein structure. *Advances in Protein Chemistry*, **34**, 167–339.

Ronson, C. W., Nixon, D. T. & Ausbel, F. M. (1987). Conserved domains in bacterial regulatory proteins that respond to environmental stimuli. *Cell*, **49**, 579–81.

Russo, A. F. & Koshland, D. E. Jr. (1983). Separation of signal transduction and adaptation functions of the aspartate receptor in bacterial sensing. *Science*, **220**, 1016–20.

Sanders, D. A., Gillece-Castro, B. L., Stock, A. M., Burlingame, A. L. & Koshland, D. E. Jr (1989). Identification of the site of phosphorylation of the chemotaxis response regulator protein, *Journal of Biological Chemistry*, **264**, 21770–8.

Schejter, E. D., Segal, D., Glazer, L. & Shilo, B. (1986). Alternative 5' exons and tissue-specific expression of the Drosophila EGF receptor homolog transcripts. *Cell*, **46**, 1091–101.

Scholle, A., Vreeman, J., Blank, V., Nold, A., Boos, W. & Manson, M. D. (1987). Sequence of the *mgl*B gene from *Escherichia coli*-K12: comparison of wild-type and mutant galactose chemoreceptors. *Mol. Gen Genet.* **208**, 247–53.

Segall, J. E., Block, S. M. & Berg, H. C. (1986). Temporal comparisons in bacterial chemotaxis. *Proceedings of the National Academy of Sciences, USA*, **83**, 8987–91.

Shaw, C. H., Ashby, A. M., Loake, G. J. & Watson, M. D. (1988). One small step: the missing link in crown gall. *Oxford Surveys Plant Molecular Cell Biology*, **5**, 177–83.

Shaw, P., Gomes, S. L., Sweeney, K., Ely, B. & Shapiro, L. (1983). Methylation involved in chemotaxis is regulated during *Caulobacter* differentiation. *Proceedings of the National Academy of Sciences*, USA, **80**, 5261–5.

Sheridan, R. P., Levy, R. M. & Salemme, F. R. (1982). α-Helix dipole model and electrostatic stabilization of 4-α-helical proteins. *Proceedings of the National Academy of Sciences, USA*, **79**, 4545–9.

Smith, J. M., Rowsell, E. H., Shioi, J. & Taylor, B. L. (1988). Identification of a site of ATP requirement for signal processing in bacterial chemotaxis. *Journal of Bacteriology*, **170**, 2698–704.

Sockett, R. E., Armitage, J. P. & Evans, M. C. W. (1987). Methylation-independent and methylation-dependent chemotaxis in *Rhodobacter sphaeroides* and *Rhodospirillum rubrum*. *Journal of Bacteriology*, **169**, 5808–74.

Sprenger, F., Stevens, L. M. & Nüsslein-Volhard, C. (1989). The *Drosophila* gene *torso* encodes a putative receptor tyrosine kinase. *Nature, London*, **338**, 478–83.

Springer, M. S., Goy, M. F. & Adler, J. (1979). Protein methylation in behavioral control mechanisms and in signal transduction. *Nature, London*, **280**, 279–84.

Stewart, R. C. & Dahlquist, F. W. (1987). Molecular components of bacterial chemotaxis. *Chemical Reviews*, **87**, 997–1025.

Stewart, R. C., Russell, C. B., Roth, A. F. & Dahlquist, F. W. (1988). Interaction of CheB with chemotaxis signal transduction components in *Escherichia coli*: Modulation of the methylesterase activity and effects on cell swimming behavior. Cold Spring Harbor Symposia on Quantitative Biology, Vol. LIII, pp. 27–40.

Stock, J. B., Clarke, S. & Koshland, D. E., Jr (1984). The protein carboxymethyltransferase involved in *Escherichia coli* and *Salmonella typhimurium* chemotaxis. *Methods in Enzymology*, **106**, 310–21.

Stock, J. & Stock, A. (1987). What is the role of receptor methylation in bacterial chemotaxis? *Trends in Biochemical Sciences*, **12**, 371–5.

Stock, J. B., Stock, A. M. & Mottonen, J. M. (1990). Signal transduction in bacteria. *Nature, London*, **344**, 395–400.

Stock, J. B., Ninfa, A. J. & Stock, A. M. (1989). Protein phosphorylation and regulation of adaptive responses in bacteria. *Microbiological Reviews*, **53**, 450–90.

Strange, P. G. & Koshland, D. E., Jr. (1976). Receptor interactions in a signalling system: Competition between ribose receptor and galactose receptor in the chemotaxis response. *Proceedings of the National Academy of Sciences, USA*, **73**, 762–6.

Terwilliger, T. C., Wang, J. Y. & Koshland, D. E. Jr. (1986). Kinetics of receptor modification. The multiply methylated aspartate receptors involved in bacterial chemotaxis. *Journal of Biological Chemistry*, **261**, 10814–20.

Ullah, A. H. J. & Ordal, G. W. (1981). *In vivo* and *in vitro* chemotactic methylation in *Bacillus subtilus*. **145**, 958–65.

Utsumi, R., Brissette, R. E., Rampersaud, A., Forst, S. A., Oosawa, K. & Inouye, M. (1989). Activation of bacterial porin gene expression by a chimeric signal transducer in response to aspartate. *Science*, **245**, 1246–9.

Vyas, N. K., Vyas, M. N. & Quiocho, F. A. (1988). Sugar and signal-transducer binding sites of the *Escherichia coli* galactose chemoreceptor protein. *Science*, **242**, 1290–5.

Weiss, R. M. & Koshland, D. E. Jr. (1988). Reversible receptor methylation is essential for normal chemotaxis of *Escherichia coli* in gradients of aspartic acid. *Proceedings of the National Academy of Sciences, USA*, **85**, 83–7.

Williams, L. T. (1989). Signal transduction by the platelet-derived growth factor receptor. *Science*, **243**, 1564–70.

Woese, C. R. (1987). Bacterial evolution. *Microbiological Reviews*, 51, 221–71.

Wolfe, A. J., Conley, M. P. & Berg, H. C. (1988). Acetyladenylate plays a role in controlling the direction of flagellar rotation. Proceedings of the National Academy of Sciences, *USA*, **85**, 6711–15.

Wolfe, A. J. & Berg, H. C. (1989). Migration of bacterial in semisolid agar. *Proceedings of the National Academy of Sciences, USA*, **86**, 6973–7.

Wolff, C. & Parkinson, J. S. (1988). Aspartate taxis mutants of the *Escherichia coli* Tar chemoreceptor. *Journal of Bacteriology*, **170**, 4509–15.

Yamamoto, K., Macnab, R., & Imae, Y. (1990). Repellent response functions of the Trg and Tap chemoreceptors of *Escherichia coli*. *Journal of Bacteriology*, **172**, 383–8.

Yarden, Y. & Ullrich, A. (1988). Growth factor receptor tyrosin kinases. *Annual Reviews in Biochemistry*, **57**, 443–78.

SIGNALLING COMPLEXES IN BACTERIAL CHEMOTAXIS

P. MATSUMURA, S. ROMAN, K. VOLZ AND D. McNALLY

Laboratory of Molecular Biology, Departments of Biological Sciences and Microbiology and Immunology, University of Illinois at Chicago, PO Box 4348 m/c 067, Chicago Illinois 60680, USA

INTRODUCTION

Bacteria respond to chemical changes in their environment by controlling their swimming behaviour to move toward higher concentrations of attractants and away from repellents. Various chemical attractants and repellents are detected by specific transmembrane receptors which transmit internal signals to control the direction of flagellar rotation (for reviews, see Stewart & Dahlquist, 1987; Parkinson & Hazelbauer, 1983; Ordal, 1985). Counterclockwise rotation causes the flagella to rotate in concert, forming a bundle which propels the bacterium forwards toward an attractant or away from a repellent. This behaviour is called smooth swimming. Clockwise rotation causes dispersal of the flagellar bundle resulting in a tumble which randomly re-orients the bacterium, thus allowing it to resume smooth swimming in a new direction. The earlier work on the bacterial chemotaxis system focused on the receptors and flagellar machinery. More recently, the nature of the intracellular signal has been under intense investigation.

There are six soluble Che proteins required for chemotactic signalling in *Escherichia coli* and *Salmonella typhimurium*. Two of these, CheR and CheB, are involved in signal adaptation. The CheR and CheB proteins reversibly methylate and demethylate the transmembrane receptors (also called methyl accepting chemotaxis proteins or MCPs), thereby modulating their function. Methylation occurs at specific glutamic acid residues on the cytoplasmic surface of the MCPs using S-adenosyl methionine as the methyl donor (Kort *et al.*, 1975, Springer *et al.*, 1979, Springer & Koshland, 1977; Stock & Koshland, 1978, Hazelbauer *et al.*, this volume).

Four other cytoplasmic Che proteins (CheA, CheW, CheY and CheZ) are necessary for integrating and transmitting the signal from the receptors to the flagellar motor. The *Che*A gene codes for two overlapping gene products with common carboxy termini, but differing in translational start sites (Smith & Parkinson, 1980). CheA$_L$ is 76kD and CheA$_S$ is 66kD. Mutations in the *cheA* and *cheW* genes both affect methylesterase activity

(Zanolari & Springer, 1984). It has been shown that CheA protein can be autophosphorylated *in vitro* and can transfer the phosphate to both CheY and CheB proteins (Hess *et al.*, 1987, 1988*a*, *b*, Borkovic *et al.*, 1989, Oosawa *et al.*, 1988, Wylie *et al.*, 1988). It is probable that phosphorylation activates CheY to increase the tumble frequency and stimulates CheB methylesterase activity. Genetic and physiological studies (Clegg & Koshland, 1984; Kuo & Koshland, 1987; Parkinson, 1978, Parkinson *et al.*, 1983, Ravid *et al.*, 1986) indicate that the CheY protein interacts with the flagellar switch to cause clockwise (CW) rotation of the flagella. CheZ inactivates CheY, thus causing a counterclockwise (CCW) signal. Addition of CheZ protein to CheY-phosphate accelerates the rate of dephosphorylation of CheY, thereby deactivating the tumble regulator (Hess *et al.*, 1988*a*).

When the CheW protein and signalling sensory receptors are added to the *in vitro* phosphotransfer reaction, the rate of phosphate transfer from CheA to CheY increases dramatically (Borkovic *et al.*, 1989). This suggested that CheW links receptor signalling to the phosphotransferase reaction. Behavioural studies have shown that, in addition to CheY, CheA and an appropriate receptor, CheW is also needed to generate a tumbly swimming behaviour (Conley *et al.*, 1989, Lui & Parkinson, 1989).

In this paper, we discuss the physical interaction between $CheA_L$, $CheA_S$, CheW, CheY and flagella motor proteins and we propose how these interactions can be correlated to the existing biochemical, physiological and genetic data.

We have used two approaches to define signalling complexes and the physical interactions of the components of the signalling system. First, we have used purified antibodies and protein affinity columns to isolate intact complexes from cells and to detect complexes formed *in vitro*. Second, we have used a combination of genetic analysis of allele-specific suppressing mutants and the three-dimensional crystal structure of the CheY protein to indirectly obtain insight into the molecular character of the interacting surfaces of CheY and a flagellar motor component, FliG.

$CheA_L/CheA_S/CheW$ COMPLEX

The interaction between CheA and CheW proteins has been seen in both biochemical and behavioural studies. *In vitro*, $CheA_L$ has been shown to have both autophosphorylation and phosphotransfer activity (Hess *et al.*, 1987), and the addition of CheW and signalling receptors has a stimulatory effect on these activities (Borkovich *et al.*, 1989). Mutations in either the *che*A or the *che*W gene have the similar effects on the activity of the methylesterase involved in sensory adaptation (Zanolari & Springer, 1984; Stewart & Dahlquist 1988), suggesting a biochemical interaction. Such mutations reduce the responsiveness of the methylesterase to chemotaxis signals. On a behavioural level, genetic and physiological data suggest that CheW couples the receptor-generated signal to CheA (Conley et al., 1989; Lui & Parkinson, 1989). In these behavioural studies, CheA and CheW were shown to be involved in the generation of a tumble signal. All of these data are consistent with either a functional or physical interaction between CheA and CheW.

We have demonstrated the existence of a complex of the CheA and CheW proteins which can be isolated from lysed cells. This complex of $CheA_L/CheA_S/CheW$ can be isolated from cells which either overproduce these proteins or express them at wild-type levels. No other flagellar or chemotaxis proteins are required for the formation of this complex, and the amount of the complex is not affected by either the absence of other chemotaxis or flagellar proteins or the signalling state of the cells (McNally & Matsumura, 1990). Figure 1 shows a Coomassie-blue stained SDS polyacrylamide gel of the $CheA_L/CheA_S/CheW$ complex isolated from cells which overexpress the *cheA* and *cheW* genes. In this experiment, purified anti-CheA antibody was covalently attached to a solid support and used as an affinity column to remove $CheA_L$ and $CheA_S$ from an S-30 sonicate of cells which did not express any flagellar or chemotaxis proteins (*flhD* mutant) but carried a plasmid (pDV4) which overexpressed the *cheA* and *cheW* genes. The level of overexpression and wild-type level of expression is shown in Table 1. Lanes 2 and 3 show the S-30 sonicate and the fraction not bound to the column. Lane 4 shows the proteins bound to the column from an S-30 fraction from cells which overexpressed *cheW* gene only. Lane 5 shows the proteins bound to the column from an S-30 sonicate of cells which overexpressed both *cheA* and *cheW* genes. CheW alone had no detectable binding to the anti-CheA antibody column, but in the presence of CheA, CheW bound to the column and was resistant to elution at high ionic strength.

While the $CheA_L/CheA_S/CheW$ complex could be isolated on the anti-CheA antibody column and was demonstrated to be stable to high ionic strength elution, the complex could only be isolated from intact cells. Attempts at *in vitro* formation of the complex were unsuccessful. Figure 2 shows the precipitation with anti-CheA antibody of S-30 sonicates from a mutant strain deleted for *cheA* (lane 1), a wild-type strain (lane 2) and a mixture of sonicates from a mutant strain deleted for *cheW* and a mutant strain deleted for *cheA* (lane 3). No detectable CheW was coprecipitated by anti-CheA antibody in lane 3. Other attempts to allow membrane and soluble fractions to interact by mixing sonicates prior to antibody exposure were also unsuccessful. The nature of the $CheA_L/CheA_S/CheW$ complex is such that its formation requires conditions not easily duplicated *in vitro*. It is possible that *cheA* and *cheW* must be cotranscribed and cotranslated for proper complex formation, but this has not been determined. Alternatively, a cofactor or small molecule might be required for the complex to form *in vivo* which is lost in the sonicates.

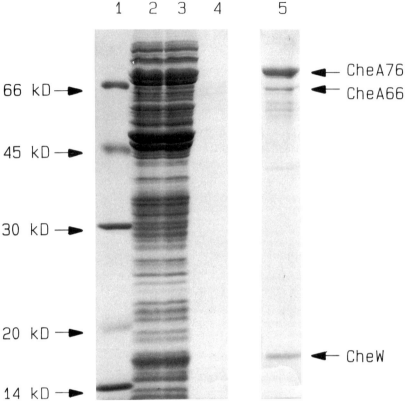

Fig. 1. Immunoadsorption of Che proteins to anti-CheA antibody column. S-30 lysates containing overexpressed levels of $CheA_L$, $CheA_S$ and CheW proteins (pDV4 in strain YK4131) or overexpressed levels of CheW protein only (PMM5 in YK4131) were circulated through columns of anti-CheA antibody. The columns were washed and bound protein was eluted and analysed by SDS PAGE and Coomassie-blue staining. Lanes 1–5 contain the following protein samples: molecular weight standards (bovine serum albumin, 66 kD; chicken ovalbumin, 45 kD; carbonic anhydrase, 30 kD; trypsin inhibitor, 20 kD; and lysozyme, 14 kD), lane 1; 100 μg of total protein from an S-30 lysate from YK4131 containing plasmid pDV4, lane 2; 100 μg of protein from a pDV4 S-30 lysate which did not bind to the anti-CheA antibody column, lane 3; protein bound to the column from a pMM5 (CheW only) S-30 lysate, lane 4; protein bound to the column from a pDV4 ($CheA_L$, $CheA_S$ and CheW) S-30 lysate, lane 5. Lanes 4 and 5 contain 10% of the total protein bound to the column.

$CheA_L/CheA_S/CheW$ COMPLEX ENHANCES $CheA_L$ AUTOPHOSPHORYLATION

As seen in Figure 1, other flagellar or chemotaxis proteins were neither required for the *in vivo* formation of the $CheA_L/CheA_S/CheW$ complex nor did they affect the amount of $CheA_L/CheA_S/CheW$ complex isolated. In addition, when $CheA_L/CheA_S/CheW$ complexes were isolated from receptor mutants which were locked in the 'tumbly' or 'smooth swimming' mode, the amount of the complex was again unchanged (McNally &

Table 1.

(a)

Plasmid	Proteins expressed	Molecules per cell $CheA_L$	$CheA_S$	CheW
pDV4	CheA, CheW	1.11×10^6	2.13×10^5	1.58×10^6
pDV4 △ EcoRV	CheA	2.75×10^5	4.08×10^4	–
pMM5	CheW	–	–	1.3×10^6

(b)

Strains	Proteins expressed	Molecules per cell $CheA_L$	$CheA_S$	CheW	CheZ
RP437	all che proteins	5.01×10^3	4.47×10^3	5.01×10^3	2.41×10^4
HCB4299	△ MCPs	4.98×10^3	4.06×10^3	5.98×10^3	2.17×10^4
RP1078	△ CheW	4.20×10^3	3.67×10^3	–	2.16×10^4
RP1788	△ CheA	–	–	3.51×10^3	1.57×10^4

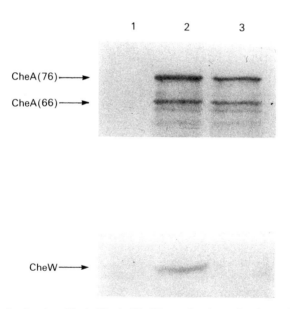

Fig. 2. Test for *in vitro* $CheA_L/CheA_S/CheW$ complex formation by anti-CheA antibody precipitation. Mixtures of various S-30 lysates were precipitated with anti-CheA antibody. The precipitates were washed and analysed by SDS PAGE and western blot. The top block shows a western blot from samples run on 7% SDS PAGE probed with anti-CheA antibody. The bottom block shows a blot from the same samples run on 15% SDS PAGE probed with anti-CheW antibody. Both blots were labelled with ^{125}I-Protein-A. Lanes 1–3 contain the following protein samples: immunoprecipitated proteins from a mixture of S-30 lysates from strains RP1788 (△ CheA) and YK4131 (*flhD*‾), lane 1; from strains RP437 (wild-type) and YK4131, lane 2 and from strains RP1788 and RP1078 (△ CheW), lane 3. All samples contain the same percentage of the total protein precipitated.

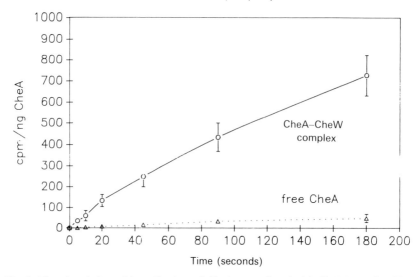

Fig. 3. Phosphorylation of free CheA$_L$ and CheA$_L$ complexed with CheW protein. CheA$_L$ and CheA$_S$ bound to CheW were immunoprecipitated from a wild-type lysate (strain RP437) using anti-CheW antibody. Free CheA$_L$ and CheA$_S$ were immunoprecipitated from a lysate devoid of CheW protein (strain RP1078) using anti-CheA antibody. Phosphorylation assays were performed on immunoprecipitated proteins resuspended in 50 mM Tris, 50 mM KCl, 5 mM MgCl$_2$, pH 7.5. Reactions were initiated by addition of [γ-^{32}P] ATP and terminated after 5, 10, 20, 45, 90 and 180 seconds by addition of Laemmli SDS sample buffer. The samples were analysed by SDS PAGE and autoradiography and the radioactivity was quantitated using a radioisotope scanning system. Identical samples were subjected to western blot analysis to determine protein levels in the precipitates. Open triangles represent phosphorylation of unbound CheA protein precipitated by anti-CheA antibody. Open circles represent phosphorylation of CheA protein bound to CheW and precipitated with anti-CheW antibody.

Matsumura, unpublished data). Therefore, the CheA$_L$/CheA$_S$/CheW complex appears to be a minimal unit which is not quantitatively altered by the signalling state or other components in the system. To determine whether complex formation has biochemical consequences, the phosphorylation rate of CheA$_L$ was assayed. Borkovich et al. (1989) have shown that the addition of CheW and membranes containing either wild-type or tumbly receptors enhanced the rate of phosphotransfer from CheA$_L$ to CheY. Figure 3 compares the autophosphorylation of CheA$_L$ when it is free and when it is complexed with CheW. In this Figure, CheA$_L$ and CheA$_S$ free of CheW was obtained by anti-CheA antibody precipitation from a mutant strain deleted for *cheW*. While fully complexed CheA$_L$ and CheA$_S$ was obtained by coprecipitation with CheW using anti-CheW antibody. It can clearly be seen that CheA$_L$ which is in a complex with CheW has a higher rate of autophosphorylation than does free CheA. In control experiments, the addition of antiserum did not effect the level

SIGNALLING COMPLEXES IN BACTERIAL CHEMOTAXIS 141

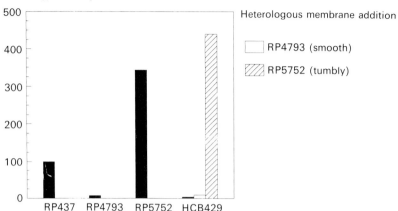

Fig. 4. Receptor effects on autophosphorylation of CheA complexed to CheW. CheA$_L$ and CheA$_S$ bound to CheW were immunoprecipitated from lysates of various mutant and wild-type strains using anti-CheW antibody. Membranes isolated from these strains were resuspended in 50 mM Tris, 50 mM KCl, 5 mM MgCl$_2$, pH 7.5. The precipitated complexes were washed with membrane resuspensions from the same strain or a heterologous strain. Phosphorylation assays were performed on immunoprecipitated proteins. Reactions were initiated by addition of [γ-^{32}P] ATP and terminated after 10 seconds by addition of Laemmli SDS sample buffer. The samples were analysed by SDS PAGE and autoradiography and the radioactivity was quantitated using a radioisotope scanning system. Identical samples were subjected to Western blot analysis to determine the level of CheA$_L$/CheA$_S$/CheW interaction in the precipitates.

of autophosphorylation. These data demonstrate that it is the CheA$_L$/CheA$_S$/CheW complex which is the active form and has the higher phosphorylation activity.

In the above experiments, wild-type MCPs were present in the reaction mixture. To determine the effect of MCPs on the autophosphorylation behaviour of preformed CheA$_L$/CheA$_S$/CheW complex, CheA$_L$/CheA$_S$/CheW complexes were isolated from wild-type strains and mutants with MCPs that only produce smooth-swimming or tumbly signals and MCP-deleted strains. Membrane preparations from these strains were presented to the isolated CheA$_L$/CheA$_S$/CheW complexes in various combinations prior to and during phosphorylation. Figure 4 shows the levels of CheA$_L$ autophosphorylation of the CheA$_L$/CheA$_S$/CheW complexes from various strains mixed with homologous and heterologous membrane fractions. The homologous mixtures of membrane fractions and isolated CheA$_L$/CheA$_S$/CheW complexes are seen in the solid bars and shows that the smooth-locked MCP mutant mixture (RP4793) had less than 10% of the level of CheA$_L$ autophosphorylation of the wild-type mixture. In the other case, the tumbly-locked MCP mutant mixture (RP5752) displayed a greater than three fold increase in $CheA_L$ autophosphorylation level. These phosphorylation levels correlated well with the observed swimming behaviour of these mutant strains: the wild-type strain tumbles approximately once per second, the tumbly mutant more frequently and the smooth swimming mutant seldom tumbles at all. $CheA/CheA_S/CheW$ complexes isolated from the strain deleted for all MCPs (HCB429) exhibited very low autophosphorylation levels and a smooth-swimming behaviour.

The $CheA_L/CheA_S/CheW$ complexes isolated from the MCP deleted strain (HCB429) were shown to respond *in vitro* to signalling and nonsignalling MCPs. $CheA_L/CheA_S/CheW$ complexes were first isolated from the MCP deleted strain (HCB429) and then mixed with the membrane fractions from the tumbly-locked MCP mutant (RP5752) and the smooth-locked MCP mutant (RP4793). In this experiment, the heterologous mixture with the tumbly mutant membranes resulted in $CheA_L$ autophosphorylation similar to the homologous $CheA_L/CheA_S/CheW$ complex isolated from the tumbly-locked MCP mutant mixed with the membrane fraction from that same strain. The heterologous combination of the smooth-locked MCP mutant membrane fraction and $CheA_L/CheA_S/CheW$ complex from the MCP deleted strain (HCB429) did not alter the $CheA_L$ autophosphorylation level. These results show that the $CheA_L/CheA_S/CheW$ complex formation is not affected by the presence of MCPs and once formed it is capable of responding appropriately *in vitro* to different signalling states of the MCPs.

$CheA_L/CheA_S/CheW$ COMPLEX HAS INCREASED AFFINITY FOR CheY

Another consequence of the formation of the $CheA_L/CheA_S/CheW$ complex is that the affinity of CheY for the complex is much greater than for either free CheA or free CheW. (McNally & Matsumura, unpublished data). Figure 5 shows the binding of $CheA_L$, $CheA_S$ and CheW to a column of immobilized CheY. Lanes 1 and 2 show that Che proteins from a mutant which is deleted for MCPs and from wild-type cells bound to the CheY column to the same degree. The MCPs have been shown not to alter the amount of $CheA_L/CheA_S/CheW$ complex formed *in vivo* and the prior exposure to MCPs did not affect the *in vitro* binding of the $CheA_L/CheA_S/CheW$ complex to the CheY column. Lanes 3 and 4 show that there was no detectable binding of $CheA_L$, $CheA_S$ or CheW to the CheY column in either *cheA*-deleted or *cheW*-deleted strains. Finally, lane 5 shows that $CheA_S$ could form a complex with CheW in the absence of $CheA_L$ and bind to the CheY column. This shows that the first 90 amino acids of $CheA_L$ are not required for either complex formation or CheY binding even though it is the $CheA_L$ protein which is autophosphorylated and subsequently transfers the phosphate to CheY. The molar ratio of the binding

SIGNALLING COMPLEXES IN BACTERIAL CHEMOTAXIS 143

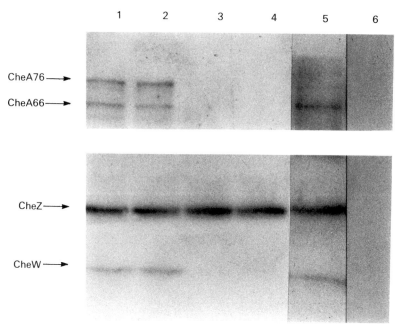

Fig. 5. CheY affinity column chromatography of lysates with wild-type levels of Che proteins. S-30 lysates containing wild-type levels of various Che proteins were circulated through columns of purified CheY protein. The columns were washed and bound protein was eluted and analysed by SDS PAGE and western blot. The top block represents a blot probed with anti-CheA antibody from protein samples run on a 7% SDS PAGE. The bottom block shows a blot probed with anti-CheZ and anti-CheW antibodies from protein samples run on 15% SDS PAGE. Both were labelled with ^{125}I-Protein-A. Lanes 1–6 contain the following protein samples: proteins bound to CheY column from an HCB429 lysate ($\triangle$ MCPs), lane 1; from an RP437 lysate (wild-type), lane 2; from an RP1078 lysate ($\triangle$ CheW), lane 3; from an RP1788 lysate ($\triangle$ CheA), lane 4; from an RP5225 lysate (CheA76Am), lane 5; proteins bound to a control column of bovine serum albumin (BSA) from an RP437 lysate (wild-type), lane 6. All lanes contain 15% of the total protein bound.

of $CheA_L$, $CheA_S$, and CheW to the CheY column was 1:1:1 as seen in Figure 6. The ratio of these proteins binding to the CheY column was the same as the molar ratio present in the cell (Table 1).

These results indicate that $CheA_L$/$CheA_S$/CheW complex formation promotes binding of CheA and CheW to the phosphorylation substrate, CheY. This increased binding of CheY, and the communication between the complex and the receptor, suggest that the $CheA_L$/$CheA_S$/CheW complex is the functional unit which connects the signalling receptor to the phosphorylation cascade.

Figures 5 and 6 also demonstrate that CheZ also binds to the CheY column. Unlike the $CheA_L$/$CheA_S$/CheW complex binding to the CheY column, this interaction did not require complex formation with any other Che protein. The amount of CheZ bound was three-fold greater than the

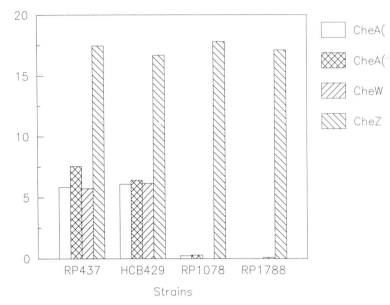

Fig. 6. Levels of Che protein interaction with CheY. The levels of $CheA_L$, $CheA_S$, CheZ and CheW proteins bound to the CheY column (Fig. 5) were quantitated from Western blots. The radioactivity from the bound protein samples was detected using a radioisotope scanning system and compared to that from known amounts of purified Che proteins. The amount of Che protein bound to CheY is graphed as picomoles bound to column. The strain name and its genotype are listed along the x axis.

$CheA_L$/$CheA_S$/CheW complex. As can be seen by comparing Table 1 with Figure 6, CheZ levels in the cell are five-fold greater than $CheA_L$, $CheA_S$ and CheW, therefore the amount of CheZ bound as a percentage of the cellular concentration is actually less than in the case of $CheA_L$, $CheA_S$ and CheW. CheZ has been shown to enhance the dephosphorylation of CheY-phosphate (Hess *et al.*, 1988c) and a separate interaction of CheZ with CheY-phosphate would therefore be expected. The phosphorylation state of the CheY on the column was not known, but since CheY-phosphate has been shown to be highly unstable (Hess *et al.*, 1988c), most, if not all, of the CheY should be in the unphosphorylated form. The binding of CheZ to the CheY column could indicate that either a small amount of CheY-phosphate was stably bound to the solid support and this would account for the CheZ/CheY interaction, or CheZ has an affinity for the unphosphorylated form of CheY itself. The latter possibility would suggest that a CheZ/CheY interaction precedes the dephosphorylation of CheY-phosphate. Since CheY-phosphate itself has an autophosphatase activity

(Hess et al., 1988c), the interaction with CheZ may enhance this CheY activity rather than CheZ itself acting as a CheY-phosphate phosphatase.

CheA$_S$/CheZ INTERACTION

Although the *cheA* gene has been shown to express CheA$_L$ and CheA$_S$ in equal stoichiometries, no specific biochemical and behavioural roles for CheA$_S$ have been established. CheA$_L$ is the form that is autophosphorylated and transfers phosphate groups to CheY and CheB (Hess *et al.*, 1988a, Wylie *et al.*, 1988). Nonsense mutations which terminate translation between the first and second start codons do not synthesize CheA$_L$ but continue to synthesize CheA$_S$. In these strains, phosphorylation of CheY and CheB can not occur and the strains display a *che*-phenotype (Smith & Parkinson, 1980). The equivalent mutation removing CheA$_S$ but not CheA$_L$ is more difficult to construct since any change in the CheA$_S$ coding region is also present in the CheA$_L$ coding region. Parkinson has, however, been able to prevent completely the expression of CheA$_S$ by site-specific mutagenesis of the start codon for CheA$_S$. A conservative substitution for the methionine start codon of CheA$_S$ results in a strain which still shows swarming on motility agar (Parkinson, personal communication). Therefore, unlike CheA$_L$, CheA$_S$ is not absolutely required for detectable chemotaxis on swarm agar. It remains to be seen whether CheA$_S$ has a subtle affect on chemotaxis which is not easily measurable on swarm plates.

We have detected a physical interaction between CheZ and CheA$_S$ which is specific for CheA$_S$ by demonstrating that antibody precipitation with anti-CheZ antibody coprecipitates only CheA$_S$ and not CheA$_L$ (McNally & Matsumura, unpublished data). A variety of strains have been assayed for the CheA$_S$/CheZ complex either by CheZ antibody or by CheA antibody precipitations from S-30 sonicates. Figure 7 and Table 2 show the results of these experiments. In Figure 7, lanes 1–5 are anti-Chez antibody precipitations of soluble proteins from wild-type (RP437) and various other mutants. Lanes 1 and 2 are a receptor deleted strain (HCB429) and wild-type (RP437) respectively. Here it can be seen that, in addition to CheZ, CheA$_S$ and CheW were also coprecipitated, but CheA$_L$ was not seen in the immune precipitate. Neither the deletion of receptors (lane 1) nor the deletion of CheW (lane 5) affected the CheA$_S$/CheZ complex. These results were quantified and summarized on Table 2. The amount of CheA$_S$ coprecipitated with CheZ was the same in wild-type (RP437), receptor deleted strain (HCB 429) and the *cheW* deleted strain (RP1078). In Figure 7, lane 4 shows that anti-CheZ precipitation of a *cheA* deletion strain resulted not only in the lack of CheA$_S$ coprecipitation, but also the absence of CheW coprecipitation. This result indicates that CheW is participating in this complex via CheA$_S$. The CheA$_S$/CheW complex seems to be independent of the CheA$_S$/CheZ complex and apparently does not affect

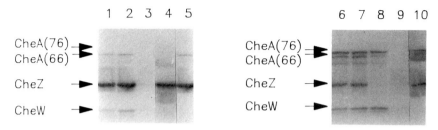

Fig. 7. Anti-CheA and anti-CheZ antibody precipitation of lysates containing wild-type levels of Che proteins. S-30 lysates containing wild-type levels of various Che proteins were incubated with anti-CheA or anti-CheZ antibody. The antigen–antibody complexes were precipitated by addition to Protein-A agarose, washed and analysed by 15% SDS PAGE and Western blot. The blot was probed with anti-CheA, anti-CheZ and anti-CheW antibodies and labelled with ^{125}I-Protein-A. Lanes 1–5 show anti-CheZ antibody precipitations and lanes 6–10 show anti-CheA antibody precipitations with the following protein samples: immunoprecipitated proteins from an HCB429 lysate ($\triangle$ MCPs), lanes 1 and 6; from an RP437 lysate (wild-type), lanes 2 and 7; from an RP5232 lysate ($\triangle$ CheZ), lanes 3 and 8; from an RP1788 lysate ($\triangle$ CheA), lanes 4 and 9; from an RP1078 lysate ($\triangle$ CheW), lanes 5 and 10. All lanes contain an equal percentage of the total precipitate.

CheA$_S$/CheZ complex formation. This could mean that CheZ and CheW bind to CheA$_S$ at nonoverlapping sites on CheA$_S$. Lanes 6–10 of Figure 7 show the Che proteins precipitated by anti-CheA antibody. In these experiments, the antibody specificity is against the carboxy terminal end of CheA so that both CheA$_L$ and CheA$_S$ are immunoreactive. These experiments show that CheZ was coprecipitated in the absence of CheW (lane 8) to the same degree as the wild-type (lane 7) or the receptor deletion strain (lane 6). In the experiments shown in Figure 7 and Table 2, the levels of CheY were not reported because levels were below the detectable range. In other experiments where CheA$_L$, CheA$_S$, CheW, CheZ and CheY were overproduced, it was shown that the coprecipitation of CheY with anti-CheZ antibody was correlated to the coprecipitation of CheA and CheW (data not shown). That is, when only CheZ and CheY were overproduced, a small amount of CheY was coprecipitated with CheZ. This result was in contrast to the CheY column binding experiments where significant amounts of CheZ were found to bind to the column.

Finally, immune precipitations with anti-CheZ and anti-CheA antibodies enabled us to establish a range of the molar ratios of the CheZ/CheA$_S$. Although using anti-CheA and anti-CheZ antibodies precipitates both complexed and free antigen, we can establish a minimum and maximum ratio for CheZ/CheA$_S$. Precipitation with anti-CheA antibody gave a minimum CheZ/CheA$_S$ ratio of 8.5 and the precipitation with anti-CheZ antibody gave a maximum CheZ/CheA$_S$ ratio of 28.3. A molar CheZ/CheA$_S$ ratio of 10:20 is very different from the CheA$_L$/CheA$_S$/CheW complex where the ratio was approximately 1:1:1. CheZ purified from *Salmonella typhi-*

Table 2. (a) Quantitation of Che proteins precipitated by anti-CheA antibody.

Strains	Phenotype	moles of Che proteins coprecipitated per mole of CheA$_S$ precipitated	
		CheW	CheZ
RP437	wild type	1.16	8.58
HCB429	△ MCPs	1.10	8.19
RP1078	△ CheW	a	9.16
RP1788	△ CheA	a	a
RP5232	△ CheZ, △ CheY	1.05	a

(b) Quantitation of Che proteins coprecipitated by anti-CheZ antibody.

Strains	Phenotype	mmoles of Che proteins coprecipitated per mole of CheZ precipitated	
		CheA$_S$	CheW
RP437	wild-type	34.7	26.0
HCB429	△ MCPs	33.5	26.9
RP1078	△ CheW	34.4	a
RP1788	△ CheA	a	a
RP5232	△ CheZ, △ CheY	a	a

a = below detectable levels.

murium has been chromatographed on a molecular sieve column and found to have an apparent native molecular mass of 115 kD (Stock & Stock, 1987). Stock & Stock (1987) also reported an aggregated form of CheZ which appeared to be a multimer of 20 or more. They also reported that no other cellular components were detected with the isolated aggregate of CheZ on Coomassie-stained SDS PAGE. Since the stoichiometry we observe is from 10–30 CheZ molecules per CheA$_S$ a CheA$_S$ band on a Coomassie-stained SDS gel is about 15-fold less intense than CheZ when precipitating with anti-CheZ antibody. Under these conditions, it is difficult to see a CheA$_S$ band above background (McNally & Matsumura, unpublished data).

An interesting possible function of the CheA$_S$/CheZ interaction would be to modulate CheZ activity by the formation of a CheA$_S$/CheZ complex. If CheZ activity is, in fact, regulated, then models of signal transduction which involve two signals could be invoked. One signal would involve the regulation of phosphorylation of CheY, and an opposite signal could involve the regulation of CheZ activity. For this type of signalling to be considered possible, the CheA$_S$/CheZ complex activity needs to be correlated to the signalling state of the cell.

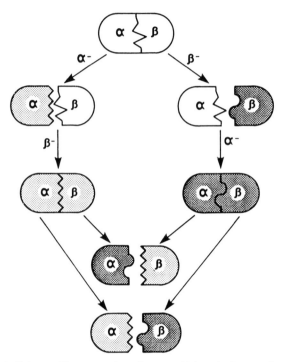

Fig. 8. Idealized allele-specific suppression of genes which code for protein α and β. Open figures represent wild-type and stippled represent mutant proteins. (Taken from S. J. Parkinson, personal communication.)

GENETIC ANALYSIS OF THE CheY-MOTOR INTERACTION: ALLELE-SPECIFIC SUPPRESSION

The use of genetic suppression to elucidate interactions between gene products has been used in a number of systems (Jarvik & Botstein, 1975; Adams et al., 1989). One of the most thorough studies of suppressor mutants has been the analysis of chemotaxis and flagellar proteins by Parkinson and his coworkers (Parkinson et al., 1983). In this pioneering study, criteria for allele-specificity were proposed and the possible ramifications of allele-specificity were explored. The most provocative prediction made from allele-specificity is that it would be a prerequisite for a conformational type of suppression involving two physically interacting proteins. A simplified view of allele-specific conformational suppression is shown in Figure 8. In this idealized case, physical interaction between proteins α and β is required for proper function to occur. If either α or β mutates in such a way that the interacting surface is disrupted, then the physical interaction and the function would be lost or reduced. However, it may be possible for a second mutation in the cognate protein to compensate for this alteration of the interacting surface. These second-site (or extragenic) suppressor mutations would be predicted to have specificity for each type of mutation which alters the interacting surface. Therefore, patterns of allele-specific suppression can be considered *prima facie* evidence for the occurrence of a conformational type of suppression between two physically interacting proteins. Parkinson has demonstrated that allele-specific patterns of suppression occur between *cheY* and *cheZ* genes and two genes *fliG* and *fliM* which are thought to be code for part of the flagella motor. In these studies, each *cheY* or *cheZ* mutant allele was tested against a number of suppressing mutations in *fliG* or *fliM*. These assays showed that patterns of suppression were unique or allele-specific for each mutant. These studies have suggested that the CheY and CheZ proteins physically interact with the FliG and FliM proteins.

The availability of the crystal structure of CheY (Stock *et al.*, 1989; Volz & Matsumura, unpublished data) has made it possible to test a primary prediction of allele-specific suppression. If allele-specific mutations affect contact points between physically interacting proteins, then the amino acid positions of these mutations should identify the interacting surfaces. A number of *cheY* suppressors of *fliG* (which are themselves suppressors of *cheY* mutants) have been generated, characterized and sequenced (Roman *et al.* unpublished data). These *cheY* mutations suppress *fliG* mutants in an allele-specific manner. The positions of these *cheY* allele-specific suppressors are dispersed on the linear amino acid sequence, but show a high degree of three-dimensional clustering when mapped on the crystal structure. Figure 9 shows the positions of these allele-specific suppressors mapped on the *Escherichia coli* CheY crystal structure. In this Figure, the β-strands ($\beta2$, $\beta1$, $\beta3$, $\beta4$, $\beta5$) are arranged front to back in the middle of the structure. The five α-helices are located around the core of β-sheets. Starting from the right front α-helix, they are arranged counter-clockwise in order of $\alpha2$, $\alpha3$, $\alpha4$, $\alpha5$, $\alpha1$. The major clustering of the suppressors are on the left surface-exposed face of the molecule. These positions are the bottom of $\alpha1$, the top of $\alpha5$, the loop connecting $\alpha5$ to $\beta5$ and the top of $\alpha4$. Collectively, these positions define a discrete surface of CheY which is consistent with it being the interacting surface between CheY and the FliG motor component. A minor cluster exists at positions 56 and 11 which are very close to each other on the three-dimensional structure but appear distinctly separate from the major cluster. These data may indicate that CheY contacts the FliG protein at two locations or, alternatively, mutations at these locations perturb the three-dimensional structure such that these changes manifest themselves as structural changes in the major cluster region.

Although further studies are required to establish the correlation between allele-specific suppressor mutants and contact points between two physically interacting proteins, the mapping of this type of mutation is, in general, consistent with the predictions made from genetic analysis.

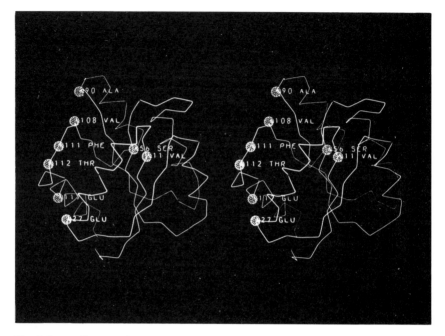

Fig. 9. Stereo diagram of the α-carbon structure of the *Escherichia coli* CheY protein. Circles represent positions of α-carbons of suppressors of *fliG* which map in *cheY*.

SUMMARY

The existence of physical interactions between complexes of the protein components of the chemotaxis system have helped to define the circuits of information flow during signal transduction. These interactions are consistent with the biochemical, behavioral and genetic data which are available. A summary of the physical interactions is shown in Figure 10. These interactions are involved in the transduction of the signal once it has been generated by the receptors (usually MCPs) and presentation of the signal to the flagellar motor.

The nature of the signal is such that a global co-ordination of all of the flagellar motors is not the result of a single intracellular event. In a seminal series of experiments, it has been shown that each individual motor is a receiver of the chemotaxis signal and each motor responds to this signal by changing the probability of reversal of the direction of rotation (Ishihara *et al.*, 1983; Macnab & Han 1983). Because the effect of signalling is the change of probabilities, the individual motors are not, in fact, co-ordinated. The probability of all motors rotating in the same direction at the same time would be extremely small if each motor was not affected by the action of other motors. Co-ordination of rotating flagella appears to be at the level of bundle formation, which has given rise to the 'voting

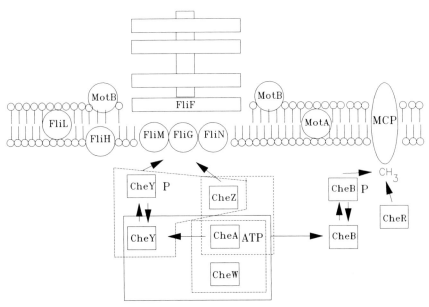

Fig. 10. Interactions of the components of the signal transduction pathway in *Escherichia coli*. Arrows represent biochemical or genetic interactions. Boxes represent known physical interactions and complexes.

hypothesis'. This states that a bundle of counter-clockwise rotating flagella is formed when a 'quorum' of flagella are rotating in a counter-clockwise direction. This 'quasi-stable' bundle is thought to hold the bundle together as the bacteria swims smoothly. When the clockwise signal level is high enough to affect the probability of clockwise rotation of a certain unspecified number of motors, then the bundle of counter-clockwise rotating flagella is disrupted and the bacterium tumbles.

In another important experiment, Segall *et al.* (1985) have shown that signal generation at one end of a cell affects motors closest to the point of stimulation and diffuses away from this point at a rate which is consistent with a protein-borne signal. It is now assumed that CheY-phosphate is this diffusible factor. In light of these studies, it may be more instructive to view the 'signal' as a ratio of two activities which alter the probability of clockwise flagellar rotation. These two activities would be CheY-phosphate and CheZ activity.

Formation of a $CheA_L/CheA_S/CheW$ complex clearly has a dramatic effect on the autophosphorylation rate of $CheA_L$ and this, in turn, would be expected to affect CheY phosphorylation. In addition to the general increase in autophosphorylation activity, the $CheA_L/CheA_S/CheW$ complex has been shown to respond to the signalling state of MCPs and therefore, couples receptor signalling to phosphorylation. The increased affinity of the $CheA_L/CheA_S/CheW$ complex for CheY probably affects the phosphorylation of CheY. The $CheA_L/CheA_S/CheW$ complex is the minimum unit of the signal transduction cascade which couples the signalling receptor to phosphorylation.

The $CheA_S/CheZ$ complex has not been shown to alter the activity of CheZ. However, the unusually high ratio of CheZ to $CheA_S$ suggests a role for the polymerization of CheZ by $CheA_S$. The $CheA_S/CheZ$ complex is similar to the $CheA_L/CheA_S/CheW$ complex with respect to formation in mutants which make no other Che proteins (*flhC* mutants). The $CheA_S/CheZ$ complex formation is not affected by the lack of other Che proteins and may represent the minimum functional unit. By analogy to the $CheA_L/CheA_S/CheW$ complex, it may be that $CheA_S/CheZ$ complex formation alters CheZ so that it is now responsive to the signalling state of the cell. One advantage of a mechanism which regulates both a positive (CheY-P) signal and a negative (CheZ activity) signal is that ratios of two regulators which are simultaneously adjusted in opposite directions are changed to a greater degree than when only one of the components is varied. For example, if a clockwise signal resulted in both the phosphorylation of CheY and the simultaneous inactivation of CheZ function, the ratio of CheY-P to CheZ activity would be perturbed more than if only CheY were phosphorylated. By this means, gain or amplification of the signal could be increased.

In this paper, we have reviewed an example of how the determination of physical interactions between proteins can contribute to the delineation of the underlying mechanism of a complex biochemical pathway. Here, in the bacterial signal transduction pathway, specific antibodies and direct protein–protein affinities have been used to create a schematic diagram (Fig. 10) of protein interactions. These interactions supply a framework which support and complement other biochemical and physiological studies.

REFERENCES

Adams, A. E. M., Botstein, D. & Drubin, D. G. (1989). A yeast actin-binding protein is encoded by *SAC6*, a gene found by suppression of an actin mutation, *Science*, **243**, 231–3.

Borkovich, K. A., Kaplan, N., Hess, J. F. & Simon, M. I. (1989). Transmembrane signal transduction in bacterial chemotaxis involves ligand-dependent activation of phosphate group transfer. *Proceedings of the National Academy of Sciences, USA*, **86**, 1208–12.

Clegg, D. O. & Koshland, D. E., Jr (1984). The role of a signal protein in bacterial sensing: behavioral effects of increased gene expression. *Proceedings of the National Academy of Sciences, USA*, **81**, 5056–60.

Conley, M. P., Wolfe, A. J., Blair, D. F. & Berg, H. C. (1989). Both CheA and CheW are required for reconstitution of signalling in bacterial chemotaxis. *Journal of Bacteriology*, **171**, 5190–3.

Hess, J. F., Oosawa, K., Matsumura, P. & Simon, M. I. (1987). Protein phosphorylation is involved in bacterial chemotaxis. *Proceedings of the National Academy of Sciences, USA*, **84**, 7609–13.

Hess, J. F., Oosawa, K., Kaplan, N. & Simon, M. I. (1988a). Phosphorylation of three proteins in the signalling pathway of bacterial chemotaxis. *Cell*, **53**, 79–87.

Hess, J. F., Bourret, R. B. & Simon, M. I. (1988b). Histidine phosphorylation and phosphoryl group transfer in bacterial chemotaxis. *Nature, London*, **336**, 139–43.

Hess, J. F., Bourret, R. B., Oosawa, K., Matsumura, P. & Simon, M. I. (1988c). Protein phosphorylation and bacterial chemotaxis. *Cold Spring Harbor Symposium on Quantitative Biology*, **LIII**, 41–7.

Ishihara, A., Segall, J. E., Block, S. M. & Berg, H. C. (1983). Coordination of flagella on filamentous cells of *Escherichia coli*. *Journal of Bacteriology*, **155**, 228–37.

Jarvik, J. & Botstein, D. (1975). Conditional lethal mutations that suppress genetic defects in morphogenesis by altering structural proteins. *Proceedings of the National Academy of Sciences, USA*, **72**, 2738–42.

Kort, E. N., Goy, M. F., Larsen, S. H. & Adler, J. (1975). Methylation of a membrane protein involved in bacterial chemotaxis. *Proceedings of the National Academy of Sciences, USA*, **72**, 3939–43.

Kuo, S. C. & Koshland, D. E. Jr. (1987). Roles of *che*Y and *che*Z gene products in controlling flagellar rotation in bacterial chemotaxis. *Journal of Bacteriology*, **169**, 51–9.

Lui, J. & Parkinson, S. J. (1989). Role of CheW protein in coupling membrane receptors to the intracellular signalling system of bacterial chemotaxis. *Proceedings of the National Academy of Sciences, USA*, **86**, 8703–7.

Macnab, R. M. & Han, D. P. (1983). Asynchronous switching of flagellar motors on a single bacterial cell. *Cell*, **32**, 109–17.

McNally, D. F. & Matsumura, P. (1990). Chemotaxis signalling complexes: physical association of CheW with CheA alters autophosphorylation rates of CheA and increases CheA binding to CheY (submitted).

Oosaway, K., Hess, J. F. & Simon, M. I. (1988). Mutants defective in bacterial chemotaxis show modified protein phosphorylation. *Cell*, **53**, 89–96.

Ordal, G. W. (1985) Bacterial Chemotaxis: Biochemistry of behavior in a single cell. *Critical Reviews of Microbiology*, **12**, 95–130.

Parkinson, J. S. (1978). Complementation analysis and deletion mapping of *Escherichia coli* mutants defective in chemotaxis. *Journal of Bacteriology*, **135**, 45–53.

Parkinson, J. S., Parker, S. R., Talbert, P. B. & Houts, S. E. (1983). Interactions between chemotaxis and flagellar gene in *Escherichia coli*. *Journal of Bacteriology*, **155**, 265–74.

Parkinson, J. S. & Hazelbauer, G. L. (1983). Molecular genetics of sensory transduction and chemotaxis gene expression. In *Gene Function in Prokaryotes*, Beckwith, J., Davies, J. & Gallant, J. A. eds, pp. 293–318. Cold Spring Harbor Laboratory, NY.

Ravid, S., Matsumura, P. & Eisenbach, M. (1986). Restoration of flagella clockwise rotation in bacterial envelopes by inserting the chemotaxis protein, CheY. *Proceedings of the National Academy of Sciences, USA*, **83**, 7157–61.

Segall, J, Ishihara, A. & Berg, H. C. (1985). Chemotactic signalling in filamentous cells of *Escherichia coli*. *Journal of Bacteriology*, **161**, 51–9.

Smith, R. A. & Parkinson, J. S. (1980). Overlapping genes at the *che*A locus of *Escherichia coli*. *Proceedings of the National Academy of Sciences, USA*, **77**, 5370–4.

Springer, W. R. & Koshland D. E., Jr (1977). Identification of a protein methyl transferase as the *che*R gene product in the bacterial sensing system, *Proceedings of the National Academy of Sciences, USA*, **74**, 533–7.

Springer, M. S., Goy, M. F. & Adler, J. (1979). Protein methylation in behavioral control mechanisms and in signal transduction. *Nature, London*, **280**, 279–84.

Stewart, R. C. & Dahlquist, F. W. (1987). Molecular components of bacterial chemotaxis. *Chemical Reviews*, **87**, 997–1025.

Stewart, R. C. & Dahlquist, F. W. (1988). N-terminal half of CheB is involved in methylesterase response to negative chemotactic stimuli in *Escherichia coli*. *Journal of Bacteriology*, **170**, 5728–38.

Stock, J. B. & Koshland, D. E., Jr (1978). A protein methylesterase in involved in bacterial sensing. *Proceedings of the National Academy of Sciences, USA*, **75**, 3659–63.

Stock, A. & Stock, J. (1987). Purification and characterization of the CheZ protein of bacterial chemotaxis. *Journal of Bacteriology*, **269**, 3301–11.

Stock, A. M., Mottonen, J. M., Stock, J. B. & Schutt, C. E. (1989). Three-dimensional structure of CheY, the response regulator of bacterial chemotaxis. *Nature, London*, **337**, 745–9.

Wylie, D., Stock, A., Wong, C. Y. & Stock, J. (1988). Sensory transduction in bacterial chemotaxis involves phosphotransfer between Che proteins. *Biochemical and Biophysical Research Communications*, **151**, 891–6.

Zanolari, B. & Springer, M. S. (1984). Sensory transduction in *Escherichia coli*: regulation of the demethylation rate by the CheA protein. *Proceedings of the National Academy of Sciences, USA*, **81**, 5061–5.

FLAGELLA BIOGENESIS IN *CAULOBACTER*

A. DINGWALL,* L. SHAPIRO* AND B. ELY†

*Beckman Center, Department of Development Biology,
Stanford University School of Medicine, Stanford,
California 94305–5427, USA
†Department of Biology, University of South Carolina,
Columbia SC 29208, USA

INTRODUCTION

The bacterial flagellum is a complex structure that has been conserved, with some minor variations, among all motile species that have been studied. With the exception of the innermost ring of the basal body (rotor), which anchors the flagellum to the cell, the entire structure is outside of the cell. The flagellum is built by assembling the most cell proximal subassembly (the basal body) first, followed by the hook, and finally, the filament (Shapiro, 1985). This ordered assembly process poses important questions about the mechanisms used by the cell to build a structure at the cell surface. What kinds of instructions are used to ensure that the ground floor is built before the penthouse? Are the protein components synthesized in the order in which they are assembled? How is the expression of large numbers of structural genes co-ordinated? In addition to the problems faced by the cell in translating an architectural design to a finished product at the cell surface, *Caulobacter* provides a positional constraint. The single flagellum is assembled at only one site on the cell, at the pole opposite the stalk. The opportunity to investigate the translation of a linear code to three-dimensional space makes *Caulobacter* an attractive addition to the elegant work being done on flagellar biogenesis in *Salmonella* and *Escherichia coli* (see Macnab, this volume). There are also temporal constraints. The biogenesis of the flagellum is known to be tightly coupled to the *Caulobacter* cell cycle, and this has recently been shown to be true for *Salmonella* (S-I Aizawa, personal communication) and *E. coli* (Nishimura & Hirota, 1989). Because it is relatively easy to synchronize *Caulobacter* cultures, a significant body of work has accumulated that is directed towards understanding the mechanisms that control the time of flagellar gene expression as a function of the cell cycle. In this review, we present current information about the structure of the *Caulobacter* flagellum, and we explore the mechanisms that are used by the cell to control the temporal and spatial expression of flagellar regulatory and structural genes.

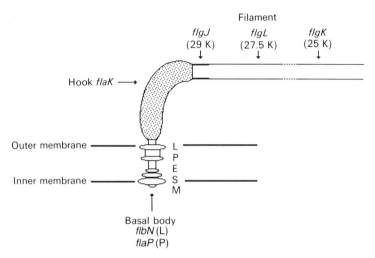

Fig. 1. Diagram of the *C. crescentus* flagellum, which is composed of a basal body anchored in the membrane, a hook, and a filament. The basal body contains five rings threaded on a rod. The P-ring is encoded by the *flaP* gene (Khambaty & Ely, unpublished data), and the L-ring is encoded by the *flbN* gene (Dingwall *et al.*, 1990). The hook is attached to the basal body rod and is composed of a 70K protein encoded by the gene *flaK* (Sheffery & Newton, 1979). The filament is composed of 25K and 27.5K flagellins encoded by *flgK* and *flgL*, respectively. A 29K flagellin, encoded by *flgJ*, resides at the junction of the hook and filament (Driks *et al.*, 1989).

THE STRUCTURE OF THE *CAULOBACTER* FLAGELLUM

The *Caulobacter* flagellum has been analysed by electron microscopy, immuno-microscopy, and image processing, and three-dimensional reconstructions of the flagellar components have been generated. As is the case for all bacterial flagella studied thus far, the *Caulobacter* flagellum is made up of three sub-assemblies: the basal body, the hook, and the filament (Fig. 1). The proximal end of the basal body anchors the flagellum to the inner membrane of the cell and the hook and filament reside outside of the cell.

Basal body

The basal body, a structure that functions as the flagellar motor, is composed of five rings (M, S, E, P and L) threaded on a rod that spans the inner membrane, periplasm, and the outer membrane. Low resolution three-dimensional structures of the basal body were generated from single-particle averaging of negatively stained basal bodies (Stallmeyer *et al.*, 1989). This analysis revealed a new feature, an axial button at the cytoplasmic end of the rod, on the inner face of the M-ring (Fig. 1). The M-ring, which appears to be within the inner membrane, does not contact the rod but is in contact with the S-ring. The membrane-spanning rod appears to have two domains. The E-ring, a distinct structure that lies at a junction between two domains of the rod, may be involved in the ejection of the flagellum during the swarmer to stalked cell transition. An E-ring is never observed in basal body preparations from *Salmonella typhimurium*, an organism that does not shed its flagella as a normal function of the cell cycle. The released *Caulobacter* flagellum consists of the filament, hook, and the hook-proximal portion of the rod. It is not known if the rings of the basal body are released or remain on the cell surface. The site of new stalk formation is the site vacated by the released flagellum.

Although many of the protein components of the *Salmonella* basal body, and the genes that encode them, have been identified by Macnab and his associates (Macnab & DeRosier, 1988), only two *Caulobacter* genes encoding basal body proteins have been definitely identified: The gene encoding the P-ring protein, *flaP* (Khambaty & Ely, unpublished data), and the gene encoding the L-ring protein, *flbN* (Dingwall *et al.*, 1990). The *flbN* gene resides within a large cluster of genes required for basal body formation (Hahnenberger & Shapiro, 1987). Three of the genes in this seven-gene cluster were found by DNA sequence analysis to encode proteins with potential cleavable signal peptides (Hahnenberger & Shapiro, 1988; Dingwall *et al.*, 1990). Mutations in one of these genes, *flaD*, results in the formation of a partial basal body made up of the cell proximal portion of the rod and the M-, S- and E-rings, suggesting that the *flaD* product is involved in basal body assembly. The *flaP* gene resides elsewhere on the chromosome (Ely *et al.*, 1984).

Hook

The filament is coupled to the flagellar motor by a curved hook that appears to function as a universal joint. The *Caulobacter* hook is composed of a single 70K protein (Lagenaur & Agabian, 1978; Johnson *et al.*, 1979; Sheffery & Newton, 1979). Three-dimensional reconstruction from electron micrographs suggests that the hook is arranged in a right-handed helix (Wagenknecht *et al.*, 1981). The model predicts that the protein monomers are finger-like subunits oriented with their long axis 45° to the axis of the hook. The hooks from both *Caulobacter* and *Salmonella* have a notch at the end that joins the hook and the filament, and a cone at the other end that joins the hook to the rod (Wagenknecht *et al.*, 1981, 1982). Macnab (1987) has suggested that the hook is assembled from monomers that pass down the centre of the basal body rod, traverse the growing hook and add on at the cell-distal tip of the elongating structure.

The *Caulobacter* gene encoding the hook monomer, *flaK*, (Ohta *et al.*, 1985), has been sequenced (A. Newton, unpublished). Alignment of the amino acid sequence of the 42K hook protein from *Salmonella* (Homma *et al.*, 1990) and the predicted amino acid sequence of the larger *Caulobacter* hook monomer shows strong homology at the amino terminus and at the carboxy terminus. There are two large internal regions present in the *Caulobacter* protein that are missing in the *Salmonella* protein (Homma et al., 1990; DeRosier, personal communication). Comparison of the three-dimensional reconstructions of the *Salmonella* and *Caulobacter* hook (Wagenknect et al., 1982) showed that the *Caulobacter* hook monomer projects farther out from the body of the hook. DeRosier predicts that this bulge represents the extra region of internal sequence within the hook monomer.

Filament

The filament of the *Caulobacter* flagellum is unusual in that it is composed of three distinct flagellin monomers. All of the flagellins are closely related in amino acid sequence and are immunologically cross-reactive. The flagellin gene family probably arose as a result of gene duplication and, with time, the individual genes acquired base changes that led to proteins with separate functions (Weissborn et al., 1982; Gill & Agabian, 1982). Two clusters of flagellin genes encode the three flagellins. The α-cluster (*flgJ*, *flgL*, and *flgK*) encodes a 29K, a 27.5K and a 25K flagellin, respectively (Gill & Agabian, 1982; Minnich & Newton, 1987). The β-cluster, which is not linked to the α cluster, contains three additional 25 K flagellin genes that appear to be redundant with the α *flaK* gene (N. Agabian and B. Ely, personal communication). Peptide analysis and immunomicroscopy of flagellin filaments derived from wild-type and mutant strains revealed that the filament is a highly ordered structure (Fukuda et al., 1978; Lagenauer & Agabian, 1978; Koyasu et al., 1981; Weissborn et al., 1982; Driks et al., 1989). A 60 nm segment of the filament adjacent to the hook appears to be composed of the 29K flagellin. This segment also contains a short (10 nm) hook-proximal region that differs structurally from the rest of the 60 nm region and may contain hook-associated proteins (HAPs) that have been found in the *Salmonella* hook-filament junction. The next segment of the filament contains the 27.5K flagellin (1 to 2μm), the product of the *flgL* gene. The boundary between the 29K segment and the 27.5K segments is sharp. However, at the distal end of the 27.5K segment there is a mixture of the 27.5K flagellin and the 25K flagellin, the product of the *flgK* gene. The 25K flagellin then increases in amount and is the sole monomer for the rest of the $5.8\mu m \pm 0.3\mu m$ filament. The synthesis of the three types of flagellin monomers occurs in the same temporal order as their appearance in the filament (Lagenauer & Agabian, 1978; Loewy et al., 1987). All three flagellin monomers are required for normal swimming behaviour (Minnich et al., 1988; Driks et al., 1989). Strains with deletions of either the *flgJ* or the *flgK* genes are still able to assemble a functional filament, but have a slower swarm rate in motility agar. A mutant strain that lacks a 27.5K flagellin is unable to assemble the 29K flagellin and produces a filament composed solely of the 25K flagellin. A strain that contains a deletion of the 29K flagellin assembles both the 27.5K and the 25K monomers (Driks et al., 1989). This suggests that the 29K and the 25K flagellin are unable to make the contact normally made by the 27.5K and the 25K monomers. Therefore, the order of assembly appears to be dictated by allowed structural interactions among the flagellar monomers.

Three-dimensional maps of the *Caulobacter* filament were generated and compared to the *Salmonella* filament (Trachtenberg & DeRosier, 1988). Both filaments are composed of 11 protofilaments of flagellin monomers with protein fingers projecting towards the axis of the filament. Each finger of the *Salmonella* monomer has a surface knob. The *Caulobacter* monomer, which is half the molecular weight of the *Salmonella* monomer, is missing this knob and therefore yields a filament with a smoother surface. The knob in *Salmonella* arises from an additional internal peptide segment (Trachtenberg & DeRosier, 1988).

GENETIC ANALYSIS OF FLAGELLAR AND CHEMOTAXIS GENES

Mutant isolation

Over 200 mutants that are unable to form chemotactic swarms in semi-solid medium have been characterized with respect to their ability to swim and to form a flagellum. Mutations in genes designated *fla* or *flb* result in the inability to form a flagellum. Approximately half of these mutations result in cells that assemble the flagellar hook and basal body but fail to assemble the filament (Johnson et al., 1983). Most of these mutants exhibit normal hook protein synthesis, but they have reduced synthesis of the flagellin proteins, suggesting that filament assembly is required for normal flagellin synthesis. The remainder of the *fla* and *flb* mutations result in cells that fail to form hook structures and in some cases no basal body is formed. These mutants exhibit an altered synthesis of the hook protein and a reduced synthesis of the flagellin proteins. Thus, hook assembly appears to be required for normal levels of hook protein synthesis.

Mutations in genes designated *mot* yield cells with a structurally normal but paralysed flagellum. These cells have normal flagellin and hook protein synthesis. Mutations in genes designated *che* result in motile cells which are unable to form a chemotactic swarm. In addition, mutations in genes designated *ple* or *pod* result in pleotropic defects that involve several aspects of polar organelle development including stalk formation, flagellum release, resistance to bacteriophage which use polar receptors, and the ability to form swarms.

The six genes encoding the flagellin monomers have been designated *flg*, and they occur in two unlinked clusters designated region α and region β. Mutations or deletions in the three α region genes result in the production of a filament with an altered subunit composition, but they have only minor effects on the size of chemotactic swarms (Minnich *et al.*, 1988; Driks *et al.*, 1989; Schoenlein & Ely, unpublished data). Deletions of the three *flg* genes in the β region have not been obtained; however, they appear to be redundant copies of the 25 K gene (*flgK*) found in the α cluster.

Genetic mapping

The flagellar mutants were initially separated into linkage groups by cotransductional analysis (Johnson & Ely, 1979). Despite the crude nature of these analyses, the interpretation of the data proved to be accurate (Ely *et al.*, 1984) and provided the basis for all subsequent studies of motility in *C. crescentus*. Recently, a genomic map has been constructed for *C. crescentus* using pulsed field gel electrophoresis (PFGE) that correlates the genetic and physical maps of the chromosome (Fig. 2) (Ely & Ely, 1989). This map made it possible to rapidly determine the locations of many mutations affecting motility using PFGE instead of the more cumbersome methods of traditional bacterial genetics. The results of these studies indicated that at least 48 genes are involved in motility and chemotaxis in *C. crescentus* (Ely *et al.*, 1986; Ely & Ely, 1989). Nearly half of these genes are grouped in three large clusters, and the remainder are scattered throughout the chromosome.

REGULATION OF *fla* AND *che* GENE EXPRESSION

The assembly of the polar flagellum, as described above, requires the co-ordinated expression of a large number of genes. These genes must be expressed in a defined temporal order as well as in appropriate amounts to assure the biogenesis of a functional structure. How does the cell co-ordinate the time of expression of these genes relative to each other and to other cell cycle events?

As is true in many developmental systems that must co-ordinate the ordered expression of large numbers of genes (Losick *et al.*, 1989; Herskowitz, 1989), the flagellar biogenesis pathways in *E. coli* and *Salmonella* (Komeda, 1982, 1986; Kutsukake *et al.*, 1990), and in *C. crescentus* (Champer *et al.*, 1985, 1987; Ohta *et al.*, 1985; Chen *et al.*, 1986; Newton *et al.*, 1989; Xu *et al.*, 1989; Bryan *et al.*, 1990) utilize a *trans*-acting regulatory hierarchy of positive genetic interactions to control the ordered expression of the flagellar genes. The position of the flagellar genes in the cascade reflects their temporal order of transcription, which, in turn, reflects the order of assembly of their protein products into the growing flagellum. Negative genetic interactions appear to be involved in the shut-off of flagel-

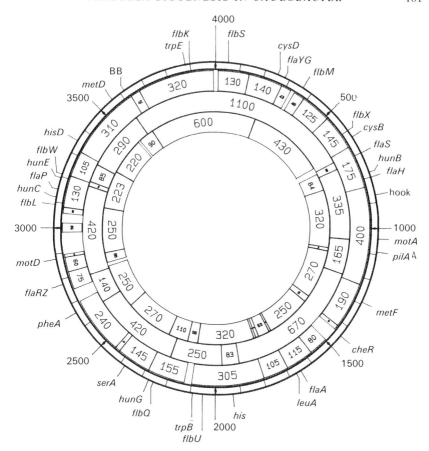

Fig. 2. Correlation between the physical and genetic maps of the *C. crescentus* chromosome. The physical size of the chromosome was determined to be approximately 4000 kilobases based on the sum of restriction fragment lengths using DraI, AseI and SpeI (Ely & Ely, 1989) as shown inside the genetic map. The SpeI map is shown as the innermost ring, then the AseI map (middle ring) and the DraI map (outer ring), respectively. The sizes of the restriction fragment lengths are indicated in kilobases. Gene locations were determined by hybridization to pulsed-field gels. The *flaYG* gene cluster contains the three α flagellin genes; whereas hook refers to the hook operon gene cluster and BB refers to a cluster of genes (including *flbN*) involved in basal body biogenesis.

lar gene transcription as each portion of the structure is completed (Xu et al., 1989; Newton et al., 1989; Schoenlein et al., 1989).

Temporal control of flagellar biogenesis

Flagellar biogenesis occurs within a defined period of the *C. crescentus* cell cycle (Fig. 3). The time of expression of the major flagellin component was initially determined in pulse-labelled synchronized cultures (Shapiro

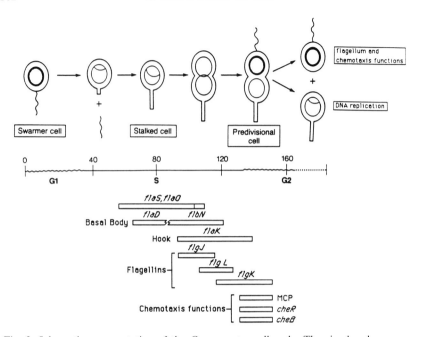

Fig. 3. Schematic representation of the *C. crescentus* cell cycle. The circular chromosome is indicated within the cells, where a heavy circle represents the non-replicating chromosome (G1-phase). After ejection of the flagellum and initiation of stalk synthesis, chromosome replication begins (S-phase). Prior to cell division the newly replicated chromosomes are segregated into the incipient swarmer and stalked cells (G2-phase). Below the cell cycle diagram is a schematic representation of the temporal pattern of synthesis of several flagellar genes. Included are regulatory genes near the top of the hierarchy (*flaS, flaO*), early structural genes (basal body and hook) and late assembly genes (flagellins and chemotaxis proteins).

& Maizel, 1973). The order of synthesis of each of the flagellum components, including the hook and the other filament proteins, was determined using immunoprecipitation of *in vivo* labelled cell extracts with antibodies raised against specific flagellar proteins. These studies (Lagenaur & Agabian, 1978; Osley *et al.*, 1977; Sheffrey & Newton, 1981; Loewy *et al.*, 1987) showed that the flagellum components are synthesized in the order in which they are assembled. The time of synthesis was found to be controlled at the level of transcription by direct measurement of the accumulation of mRNA in nuclease S1 protection assays and by the use of reporter gene fusions to the flagellar gene promoters (Champer *et al.*, 1985, 1987; Ohta *et al.*, 1985; Chen *et al.*, 1986; Loewy *et al.*, 1987; Minnich & Newton, 1987; Hahnenberger & Shapiro, 1988; Kaplan *et al.*, 1989; Mullin & Newton, 1989; Dingwall *et al.*, 1990).

Recent work has shown that the earliest transcribed genes, those at, or near, the top of the regulatory hierarchy (Fig. 4), are required for the transcription of the structural genes. For example, transcription of the *flaS*

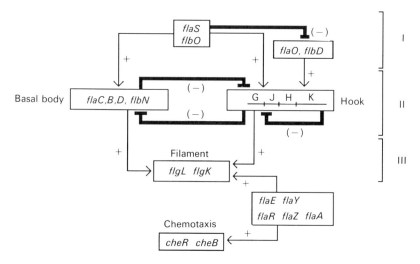

Fig. 4. Schematic of the genetic regulatory network of *C. crescentus* flagellar and chemotaxis genes showing positive (+) and negative (−) interactions among the genes. The negative interactions are indicated by bold lines and positive control is indicated by arrows. The hierarchy is presented as a network of interactions on three levels: Level I is composed of the early genes whose synthesis is required for transcription of genes in Level II. The genes in Level II encode the proteins for the basal body and hook structures. Level III genes include the flagellins, which are assembled last. The genes shown outside the hierarchy are required for filament assembly and chemotaxis functions and their expression is independent of the genes shown within the cascade.

gene precedes the transcription of *flbN*, a gene which appears to encode the L-ring of the basal body (Dingwall *et al.*, 1990). Transcription of the hook structural gene (*flaK*) and the 29 K flagellin gene (*flgJ*) follows the transcription of the basal body genes, and they encode the next components to be assembled (Osley *et al.*, 1977; Lagenaur & Agabian, 1978; Ohta *et al.*, 1985; Loewy *et al.*, 1987). The structural genes transcribed last are those encoding the 27.5 K (*flgL*) and 25 K (*flgK*) flagellins. Expression of the structural component genes and the early regulatory genes terminates prior to cell division, with the exception of the *flgK* gene encoding the 25 K flagellin, whose synthesis continues in the progeny swarmer cell (Agabian *et al.*, 1979). The proteins that carry out chemotaxis functions are synthesized only in the pre-divisional cell and they are segregated to the progeny swarmer cell upon division (Shaw *et al.*, 1983; Gomes & Shapiro, 1984).

The order of synthesis of the flagellar structural components reflects their order of assembly into the nascent flagellum. For example, the basal body must be assembled prior to the addition of hook monomers which, in turn, must be assembled prior to addition of the flagellins. A mutation in a basal body gene (*flbN*) results in a dramatic decrease in flagellin synthesis and an accumulation of the 68K hook precursor protein (Johnson *et al.*, 1983). Hook monomers are not assembled in these mutants. A similar decrease in flagellin synthesis occurs in strains that contain a null mutation in the hook gene, although basal body proteins are made. In contrast, the hook protein is synthesized at normal wild-type levels and is assembled correctly in strains that contain mutations in the flagellin genes, which suggests that the basal body genes are also unaffected in these mutant strains. Therefore, once a subassembly (basal body or hook) is complete, the cell begins to make the next components.

The ordered assembly model predicts that specific signals act at the gene promoters to determine the order of transcription, based on their order of assembly. However, the structural genes identified thus far all contain promoters with a σ^{54} consensus sequence (Mullin *et al.*, 1987; Kaplan *et al.*, 1989; Mullin & Newton, 1989; Ninfa *et al.*, 1989; Dingwall *et al.*, 1990). Therefore, ordered temporal synthesis is not merely the result of a cascade of alternate sigma factors as has been demonstrated in *B. subtilis*. Temporal control of flagellar genes in *C. crescentus* requires not only a cascade of developmental events, but also the co-ordination of unique combinations of *trans*-acting factors capable of discriminating among these promoters. The framework through which these mechanisms exert their control over gene expression is the genetic regulatory hierarchy.

REGULATORY HIERARCHY

A positive *trans*-acting cascade modulates the level of transcription of the flagellar genes (Fig. 4). Within this hierarchy, there are at least three levels of *trans*-acting control. Level one is composed of the early whose genes transcription is essential for the expression of the structural genes in level two and indirectly for the expression of genes below them in the hierarchy. Proper assembly of the structural components encoded by the genes in level two, such as the basal body (e.g. *flbN*) and hook (*flaK*), is required for the synthesis and assembly of the level three structural genes, the flagellins. How is the cascade of positive control initiated, and what are the mechanisms that determine the temporal order of transcription? The transcription of genes at all levels of the hierarchy is being examined and a molecular dissection of their promoter regions has been initiated to identify the *cis*- and *trans*-acting elements that contribute to timed gene expression. Although we do not know what triggers the positive cascade, mutations in regulatory genes near the top of the hierarchy result in complex phenotypes. For example, *flaS* mutants fail to produce any of the components of the flagellum, and are also unable to divide normally. This dual phenotype suggests that early regulatory genes may be involved in the co-ordination of the time of flagellar biogenesis with other cell cycle events, such as DNA replication and cell division. Newton *et al.* (1989) have postulated that a master timing gene begins the process, although no genes have yet been found that are required for the transcription of the genes near the top of the hierarchy, such as *flaS* and *flaO*.

The transcription start site and time of synthesis have been determined for *flaS* (A. Dingwall, W.-Y. Zhuang & L. Shapiro, unpublished data). Transcription of this gene initiates earlier than either the basal body *flbN* gene (Dingwall *et al.*, 1990), or the hook protein gene, *flaK* (Ohta *et al.*, 1985). The predicted promoter sequences for early genes, *flaS* and *flaO*, are similar to each other in both the -10 and -35 region; however, they do not share any significant similarity to the σ^{70} promoter, to the σ^{54} promoter (Mullin & Newton, 1989; Ninfa *et al.*, 1989; Dingwall *et al.*, 1990), or to the σ^{28} promoter that appears to be used by the chemotaxis genes (Frederikse & Shapiro, 1989; M. R. K. Alley & L. Shapiro, unpublished data). These promoter sequence comparisons suggest that the early flagellar regulatory genes are subject to common regulatory controls and might be recognized by an alternate sigma factor.

At least five *C. crescentus* flagellar structural genes contain promoters that are similar to the recognition sequence associated with the *E. coli* alternate sigma factor σ^{54} (for review, see Newton & Ohta, 1990). In a number of gram-negative bacteria, the σ^{54} conserved promoter sequence is utilized for regulated expression of a wide variety of genes (for review, see Kustu *et al.*, 1989; Santero *et al.*, 1989). The identified σ^{54} promoters all contain conserved nucleotides at -24 and -12 with respect to the transcript initiation site. The earliest transcribed *C. crescentus* flagellar gene that contains a σ^{54} promoter is the basal body gene *flbN*. This promoter is similar in sequence and spacing to the promoter of the hook operon and flagellin genes *flgK* and *flgL*, that were shown by Ninfa *et al.* (1989) to be transcribed *in vitro* using purified *E. coli* σ^{54}. Most known promoters of this type require a *trans*-acting protein that binds at an upstream enhancer and activates transcription (for review, see Kustu *et al.*, 1989). The *C. crescentus* σ^{54} promoters behave in a similar fashion. The *flbN* promoter region contains an upstream activator sequence at -110bp from the transcription start site that has sequence similarity to the NifA binding site found upstream of nitrogenase (*nif*) promoters. When these sequences are mutated, timed expression of the *flbN* promoter still occurs, but at a significantly lower level. Therefore, this upstream binding site, while necessary for full transcription activation, is not solely responsible for temporal control of *flbN* (Dingwall *et al.*, 1990). Mutation analysis has shown that the *flbN* gene requires sequences both 5' and 3' to the transcription start site for temporal control (Dingwall *et al.*, 1990). A 70K protein, RF-2, which binds to the upstream region of the *flbN* promoter has been purified (J. Gober, unpublished). The RF-2 protein binding site, analogous to the NifA binding site associated with genes containing σ^{54} promoters in a variety of organisms, can functionally serve as a NifA site for transcriptional activation of *flbN* in *E. coli* when the NifA protein is provided in *trans* from a plasmid-borne *Rhizobium meliloti nifA* gene (Dingwall et al., 1990). Alteration or deletion of the RF-2 site abolishes activation by NifA.

At least four different proteins have been identified that are directly involved in transcriptional control of the hook operon. In addition to the sigma factor, three *trans*-acting proteins have been observed that bind to the upstream region of the hook promoter. Specific binding was determined by a combination of methods, including gel mobility shift and Southwestern analyses, as well as *in vitro* DNAse footprinting (J. Gober, H. Xu, A. Dingwall & L. Shapiro, submitted for publication; Gober & Shapiro 1990). One of these binding proteins, designated RF-1, has been purified from *Caulobacter* cell extracts and was shown by Southwestern analysis to be approximately 95 K (Gober et al., submitted for publication). Gel mobility shift assays have demonstrated that RF-1 binds to a region that resides about 100 bases upstream from the transcription start site and that the binding activity is temporally controlled. Recently, the *flbD* gene was shown by Ramakrishnan & Newton (1990) to encode a protein of approximately 52 K that has homology to the response class of proteins in bacterial two component regulatory systems. This class includes several proteins that activate transcription in response to environmental cues. The predicted *flbD* amino acid sequence is similar to NifA and NtrC, two known transcription activators of σ^{54} promoters. *In vivo* experiments have shown that the product of *flbD* activates transcription of the hook operon promoter and also the *E. coli* glutamine synthetase promoter. The similarity between the *flbD* amino acid sequence and other response regulators indicates that its ability to activate transcription is probably modulated by other proteins in response to some unknown cell cycle cues (Newton & Ohta, 1990). Although the exact recognition sequences of the RF-1 and FlbD proteins have not been identified, Southwestern analyses using *Caulobacter* cell extracts have shown that two proteins of approximately 95 K and 50 K are present that can bind to a region upstream of the hook promoter designated *ftr* (Gober et al., submitted for publication). The 18 bp conserved *ftr* sequence, identified by Mullin & Newton (1989), is also present in the upstream region of other *Caulobacter* flagellar genes that have σ^{54} promoters, including *flgK* and *flgL*, and downstream of the *flaN* promoter that is transcribed divergently from the hook promoter. Mutational analysis has shown that the *ftr* element is required for transcription of the hook operon (Mullin & Newton, 1989; Gober et al., submitted for publication).

A third *trans*-acting protein involved in hook transcription has been observed to bind in a relatively A+T-rich region between the *ftr* element and the promoter that contains a consensus binding sequence for the *E. coli* protein integration host factor (IHF). Consensus IHF sites are also located between the *ftr* element and the promoter of the *flgK* and *flgL* genes. IHF is a small, basic, sequence specific DNA binding protein that was originally identified as a host factor required for bacteriophage lambda integration. Upon binding to DNA, IHF introduces a sharp bend in the DNA helix (Robertson & Nash, 1988). Purified *E. coli* IHF has been shown by *in vitro* DNAse footprinting to bind to the IHF site in the *flbG* promoter region and mutational analysis of this site indicated that it is necessary for transcription *in vivo* (Gober & Shapiro, 1990). A similar requirement for IHF has recently been demonstrated with the *Klebsiella pneumoniae nifH* promoter in an *in vitro* transcription system (Hoover, *et al.*, 1990). In these promoters, the IHF induced bend is presumably required to bring a protein bound at the upstream activator site in close proximity to RNA polymerase bound at the promoter. Transcriptional activation of the hook operon promoter is clearly a multicomponent process, with RF-1, FlbD and IHF all contributing to the precise regulation of gene expression.

The genes involved in the chemotaxis machinery are transcribed late in the cell cycle. These include the methyl accepting chemotaxis proteins (MCP), the carboxymethylesterase (*cheB*) and the carboxymethyltransferase (*cheR*). In *C. crescentus*, several of these genes appear to be organized into an operon (M. R. K. Alley, W. Alexander, S. L. Gomes, & L. Shapiro, unpublished data). The promoter sequence for this operon appears to be similar to the σ^{28} consensus sequence (M. R. K. Alley & L. Shapiro, unpublished data) associated with flagellar genes from *B. subtilis* and *E. coli* (Helmann & Chamberlin, 1987; Helmann *et al.*, 1988; Frederikse & Shapiro, 1990). Mutations in the chemotaxis genes have no effect on the assembly of the flagellum and the cells remain motile, although they are unable to reverse direction of swimming. The expression of the chemotaxis genes is dependent on two flagellar genes, *flaE* and *flaY* (Bryan *et al.*, 1987), that appear to be involved in assembly of the filament.

The use of multiple *trans*-acting proteins and sigma factors for the activation of flagellar gene transcription provides a large number of combinations to assure that the flagellar genes are expressed at the correct time and in the appropriate amounts to build a functional flagellum. However, positive control of transcriptional activation is insufficient to assure proper assembly of the flagellum. In addition to the positive cascade of *trans*-acting control, the *C. crescentus* flagellar genes also utilize a network of negative regulatory interactions that function to modulate the level and duration of gene expression (Newton *et al.*, 1989; Xu *et al.*, 1989). Negative control has also been implicated in the flagellar gene cascade in *E. coli* (Komeda, 1986). In the *C. crescentus* flagellar hierarchy, negative interactions are defined by the increased transcription of a given *fla* gene in the presence of another *fla* mutation. These effects have been observed using both nuclease S1 protection assays and transcription fusions to reporter genes encoding neomycin phosphotransferase (Champer *et al.*, 1987; Xu *et al.*, 1989), β-galactosidase (Newton *et al.*, 1989) and luciferase (Xu *et al.*, 1989). Negative regulation is observed in autoregulated genes, such as the hook operon

(Chen et al., 1986); Mullin et al., 1987), and between unlinked genes at the same level of the hierarchy (Newton et al., 1989; Xu et al., 1989).

In the context of the *C. crescentus* flagellar hierarchy, positive control is epistatic to negative regulation. Evidence for this comes from an analysis of double mutants in *flaS* and the basal body or hook genes (Champer et al., 1987). The *flbN* and hook operon genes are dependent on expression of *flaS* and this positive control is exerted at the level of transcription. In double mutant strains, there is no transcription from either the hook operon or *flbN* promoters, thereby preventing negative regulation of these genes.

How does the cell know when to turn on flagellar genes and when to turn them off? A cell cycle cue must signal the start of the cascade, resulting in the expression of the early regulatory genes. The next step is the coordinate expression and assembly of the basal body and hook proteins. Assembly of these structural proteins then yields a signal that, directly or indirectly, exerts a negative control that functions to turn off their own synthesis. Failure to complete assembly of the basal body and hook in mutant strains would therefore result in overexpression of those early components, which in some manner prevents the transcription of the filament genes.

POSITIONAL INFORMATION

Cell division in *Caulobacter* yields progeny cells that have distinct functional and structural characteristics. The swarmer cell, but not the stalked cell, has a polar flagellum, is motile and can carry out chemotaxis. Many of the flagellar and chemotaxis proteins are synthesized prior to cell division and are segregated to the swarmer cell upon division. The progeny stalked cell, but not the swarmer cell, is able to initiate chromosome replication. Several heat shock proteins, including the product of the *dnaK* gene, are synthesized in the predivisional cell and then specifically sequestered to the stalked cell upon division. Thus, subsets of proteins are targeted to one of the two progeny at each division. Are these proteins localized prior to cell division? What are the mechanisms that regulate protein segregation? To provide answers to these questions, the genes that encode these sequestered proteins are being analysed.

Chemotaxis apparatus

Bacteria with a single polar flagellum swim in either a forward or reverse direction. In *Caulobacter*, clockwise rotation of the flagellar rotor results in a long forward swim (Koyasu & Shirakihara, 1984). Counterclockwise rotation results in a short backwards swim. The swimming pattern is biased by the presence of chemoattractants or repellents in the media. The chemoattractants identified for *C. crescentus* include glucose, galactose, xylose, ribose, alanine, proline, and glutamine (Ely *et al.*, 1986). Several *Caulobacter* general non-chemotactic mutants were identified and the genes were mapped. Two strains carrying mutations in the *cheR* and *cheB* genes were analysed by a computerized cell-tracking system (Ely *et al.*, 1986), which showed that *cheR* mutants swam only in the forward direction (clockwise flagellar rotation) and *cheB* mutants swam only in the reverse direction (counterclockwise flagellar rotation). Biochemical assays showed that the *cheR* mutants lacked carboxymethyltransferase activity and *cheB* mutants lacked carboxymethylesterase activity (Gomes & Shapiro, 1984; Ely *et al.*, 1986). These soluble enzymes facilitate a change in the pattern of MCP methylation in response to chemical stimuli. A third *che* gene was identified by a Tn5 insertion mutation in strain SC1130. This mutant lacked over 80% of the cellular MCP (methyl accepting chemotaxis protein) activity (Ely *et al.*, 1986). The MCPs are integral membrane receptor proteins that can bind attractants and repellents. The major MCP from *C. crescentus* is transiently carboxymethylated at glutamate residues (Gomes & Shapiro, 1984). In *Caulobacter*, all three of these protein components of the chemotaxis machinery are developmentally regulated (Shaw *et al.*, 1983; Gomes & Shapiro, 1984; Shapiro *et al.*, 1985; Shapiro & Gober, 1989). The methyltransferase, methylesterase and the MCPs are synthesized only in the predivisional cell and then all three proteins are sequestered to the progeny swarmer cell (Gomes & Shapiro, 1984). When the swarmer cell ejects its flagellum and differentiates into a stalked cell, there is a turnover of all three chemotaxis proteins which are not synthesized again until just prior to cell division. Thus, the genes encoding these proteins are both temporally and spatially regulated and a specific proteolytic event must occur during the swarmer to stalked cell transition. There are several major questions arising from these observations. How do the chemotaxis proteins (which include both soluble and membrane proteins) segregate to only one of the two progeny cells? What are the signals that control the timed 'turn-on' and 'turn-off' of gene expression? What controls the programmed turnover of these proteins at a specific time in the cell cycle?

To address the mechanism of protein segregation to the swarmer cell, assays were done to determine if the MCPs were localized to the incipient swarmer cell portion of the predivisional cell (Nathan *et al.*, 1986). Membrane vesicles were prepared from predivisional cells. Those originating from the 'swarmer' portion of the cells contained a flagellum while those arising from other portions of the cell did not. The flagellated and non-flagellated vesicles were separated by affinity chromatography (Nathan *et al.*, 1986) using anti-flagellin antibody (Huguenel & Newton, 1984). The two groups of vesicles were assayed for the presence of newly synthesized MCPs and it was found that the MCPs are preferentially localized to the swarmer pole of the predivisional cell (Nathan et al., 1986). Thus, localization occurs prior to division. What is the mechanism that allows the newly synthesized MCP receptor proteins to be targeted to a specific location in the predivisional cell? One possibility is that there are regions within the MCP protein that interact with localized membrane sites at the swarmer pole of the predivisional cell. This explanation requires the asymmetric distribution of these 'localized' sites and the mechanism for site distribution is unknown. Another possibility is that the MCP gene is transcribed from only one of the two newly replicated chromosomes in the predivisional cell. Coupled transcription and translation from the chromosome localized in the swarmer cell portion of the predivisional cell, followed by the immediate incorporation of the MCP translation product into the membrane would result in the localization of the MCPs to the new swarmer cell. We cannot rule out the possibility that a compartment is formed in the predivisional cell but electron microscopy has not revealed a membrane separating the 'swarmer' and 'stalked' cell portions of the predivisional cell. How, then, are the soluble chemotaxis proteins segregated to the swarmer cell? The mechanisms of protein targeting in the predivisional cell are not known, but, as described below, there is evidence that differential transcription of a set of flagellar genes may contribute to their localization to the progeny swarmer cell.

Localization of the flagellum

The components of the flagellum are synthesized in the order in which they are assembled. The fact that this assembly occurs at a specific site on the cell raises several fundamental questions about cellular morphogenesis. How do the flagellar components get to the site of assembly? Where is the genetic information for localizing gene products in a three-dimensional field? Are the proteins to be localized synthesized throughout the cell or is flagellar protein synthesis localized to specific domains of the cell? In an attempt to answer these questions, the localization of several of the cloned flagellar gene products has been analysed.

The gene encoding the 29K flagellin (*flgJ*) is transcribed, translated and its product is assembled prior to cell division. Consequently, the 29 K flagellin that is synthesized in the predivisional cell appears in the progeny cell that carries the new flagellum. Thus, it is reasonable to assume that segregation is a function of assembly. Surprisingly, we found that the swarmer cell segregation of the 29 K flagellin occurred even in mutants that were unable to assemble a hook and filament (Loewy et al., 1987). This would argue that the protein might bind to a specific site at the cell pole, independent of assembly. To determine if the *flgJ* protein carries the information for its localization to the cell pole, a gene fusion was constructed between the 5' regulatory region of the *flgJ* gene and a promoterless *neo* reporter gene. This construction encodes a chimeric mRNA with a separate and intact *neo* gene that encodes neomycin phosphotransferase II (NPTII). A protein fusion between the 29 K flagellin and NPTII is precluded by the presence of translation stop codons at the junction of the genes. Using this construction, it was found that NPTII synthesis occurred at the correct time in the cell cycle, but that the reporter NPTII protein was not segregated to the swarmer cell. Therefore, the 29 K flagellin carries the signal for its correct localization (Loewy et al., 1987). What might the 29 K flagellin be 'seeing' at the cell pole? The pole at which the new flagellum is assembled is the site of the previous cell division; therefore, the 'receptor' for the new flagellar components might be laid down at the site of cell division.

A subset of flagella genes appear to have an additional mechanism for protein localization. One of these genes, *flgK* encodes the 25K flagellin that is incorporated into the major portion of the distal filament. Although *flgK* transcription and translation is initiated in the pre-divisional cell, the expression of this gene continues in the swarmer, but not the stalked progeny cell. Milhausen & Agabian (1983) presented evidence that the flagellin mRNA was segregated to the progeny swarmer cell and that this might contribute to the localization process. Recently, experiments supporting this apparent *flgK* mRNA localization were done with *flgK–neo* and *flgK–lux* transcription reporter gene fusions (J. Gober, R. Champer, S. Reuter & L. Shapiro, submitted for publication). It was found that the reporter gene product was segregated to the swarmer cell and that this segregation occurred in the presence of rifampicin. Since rifampicin inhibits RNA synthesis, the chimeric *flgK*-reporter gene mRNA synthesized in the predivisional cell prior to the addition of rifampicin must, in some manner, be sequestered to the swarmer cell. Differential transcription from the newly replicated chromosomes would provide a mechanism for this mRNA segregation. In this context, another example of asymmetry is provided by the fact that the newly replicated chromosomes have different potentials for the initiation of DNA replication. Only the chromosome that has segregated to the progeny stalked cell is able to initiate the replication of the chromosome. The chromosome in the progeny swarmer cell is unable to initiate replication until later in the cell cycle. Because the newly replicated DNA strands sort randomly during semi-conservative replication (Osley & Newton, 1974; Marczynski et al., 1990), differential marking of the two chromosomes must occur after replication is completed. The sedimentation coefficients of the chromosome isolated from swarmer and stalked cell progeny is different (Evinger & Agabian, 1977; Swoboda et al., 1982; M. Rizzo & L. Shapiro, unpublished data). When the swarmer cell differentiates into a stalked cell, the chromosomal sedimentation coefficient reverts to that observed in the progeny stalked cell. These cyclical differences in sedimentation coefficient might reflect the differential marking of the two chromosomes.

Segregation of heat shock proteins

In the two previous sections, we discussed flagellar and chemotaxis proteins that are segregated to the progeny swarmer cell. Another subset of proteins, those involved in the heat shock response, were found to be synthesized in the predivisional cell and segregated to the progeny stalked cell upon cell division (Reuter & Shapiro, 1987). Thus, protein targeting in *Caulobacter* occurs with both progeny cells. The targeting of the flagellar and chemotaxis proteins to the swarmer cell clearly reflects the functions of the swarmer cell. Of the known heat shock proteins that are targeted (DnaK, Lon, and the heat shock sigma factor) to the stalked cell, only the product of the *dnaK* gene reflects a known stalked cell-specific function, the initiation of DNA replication. In *E. coli* the DnaK protein appears to be involved in replisome function, apparently by mediating the availability of active DnaA protein (J. Kaguni, unpublished data). The segregation of DnaK to the stalked cell may reflect an interaction of this protein with the DNA initiation complex in the portion of the predivisional cell that will become the stalked cell. This explanation would require a strict stoichiometric relationship between the amount of DnaK synthesized and protein–substrate complex formation. Because segregation of DnaK to the stalked cell still occurred when the DnaK concentration was increased ten-fold after exposure to heat shock, it is unlikely that this is the explanation for DnaK targeting to the stalked cell. The *Caulobacter dnaK* gene has recently been cloned and sequenced (Gomes *et al.*, 1990). The gene is transcribed early in the cell cycle just prior to the initiation of DNA replication, suggesting that there are both temporal and spatial restrictions on the expression of this gene. Experiments are in progress to define the regulatory sequences that control the timing of *dnaK* expression and the portions of the protein that might contribute to localization.

REFERENCES

Agabian, N., Evinger, M. & Parker, G. (1979). Generation of asymmetry during development. *Journal of Cell Biology*, **81**, 123–36.

Bryan, R., Champer, R., Gomes, S., Ely, B. & Shapiro, L. (1987). Separation of temporal control and *trans*-acting modulation of flagellin and chemotaxis genes in *Caulobacter*. *Molecular & General Genetics*, **206**, 300–6.

Bryan, R., Glaser, D. & Shapiro, L. (1990). A genetic regulatory hierarchy in *Caulobacter* development. In *Advances in Genetics*, J. G. Scandalios & T. R. F. Wright, eds, vol. 27, London: Academic Press, pp 1–31.

Champer, R., Bryan, R., Gomes, S. L., Purucker, M. & Shapiro, L. (1985) Temporal and spatial control of flagellar and chemotaxis gene expression during *Caulobacter* cell differentiation. *Cold Spring Harbor Symposia in Quantitative Biology*, **50**, 831–40.

Champer, R., Dingwall, A. & Shapiro, L. (1987). Cascade regulation of *Caulobacter* flagellar and chemotaxis genes. *Journal of Molecular Biology*, **194**, 71–80.

Chen, L.-S., Mullin, D. & Newton, A. (1986). Identification, nucleotide sequence, and control of developmentally regulated promoters in the hook operon region of *Caulobacter crescentus. Proceedings of the National Academy of Sciences, USA*, **83**, 2860–4.

Dingwall, A., Gober, J. W. & Shapiro, L. (1990). Identification of a *Caulobacter* basal body structural gene and a *cis*-acting site required for the activation of transcription. *Journal of Bacteriology*, in press.

Driks, A., Bryan, R., Shapiro, L. & DeRosier, D. J. (1989). The organization of the *Caulobacter crescentus* flagellar filament. *Journal of Molecular Biology*, **206**, 627–36.

Driks, A., Schoenlein, P. V., DeRosier, D. J., Shapiro, L. & Ely, B. (1990). A *Caulobacter* gene involved in polar morphogenesis. *Journal of Bacteriology*, **172**, 2113–23.

Ely, B., Croft, R. J. & Gerardot, C. J. (1984). Genetic mapping of genes required for motility in *C. crescentus. Genetics*, **108**, 523–32.

Ely, B. & Ely, T. (1989). Use of pulsed field gel electrophoresis and transposon mutagenesis to estimate the minimal number of genes involved in motility in *Caulobacter crescentus. Genetics*, **123**, 649–54.

Ely, B., Gerardot, C. J., Fleming, D. L., Gomes, S. L., Frederikse, P. & Shapiro, L. (1986). General nonchemotactic mutants of *Caulobacter crescentus. Genetics*, **114**, 717–30.

Evinger, M. & Agabian, N. (1977). Envelope-associated nucleoid from *Caulobacter crescentus* stalked and swarmer cells. *Journal of Bacteriology*, **132**, 294–301.

Frederikse, P. H. & Shapiro, L. (1989). An *Escheridia coli* chemoreceptor gene is temporally controlled in *Caulobacter. Proceedings of the National Academy of Sciences, USA*, **86**, 4061–5.

Fukuda, A., Koyasu, S. & Okada, Y. (1978). Characterization of two flagella-related proteins from *Caulobacter crescentus. FEBS Letters*, **95**, 70–5.

Gill, P. R. & Agabian, N. (1982). A comparative structural analysis of the flagellin monomers of *Caulobacter crescentus* indicates that these proteins are encoded by two genes. *Journal of Bacteriology*, **150**, 925–33.

Gober, J. W. & Shapiro, L. (1990). Integration host factor is required for the activation of developmentally regulated genes in *Caulobacter. Genes and Development*, **4**, 1494–504.

Gomes, S. L., Gober, J. W. & Shapiro, L. (1990). Expression of the *Caulobacter* heat shock gene *dnaK* is developmentally controlled during growth at normal temperatures. *Journal of Bacteriology*, **172**, 3051–59.

Gomes, S. L. & Shapiro, L. (1984). Differential expression and positioning of chemotaxis methylation proteins in *Caulobacter. Journal of Molecular Biology*, **177**, 551–68.

Hahnenberger, K. & Shapiro, L. (1987). Identification and isolation of a gene cluster involved in flagellar basal body biogenesis in *Caulobacter crescentus. Journal of Molecular Biology*, **194**, 91–103.

Hahnenberger, K. & Shapiro, L. (1988). Organization and temporal expression of a flagellar basal body gene in *Caulobacter crescentus. Journal of Bacteriology*, **170**, 4119–24.

Helmann, J. D. & Chamberlin, M. J. (1987). DNA sequence analysis suggests that expression of flagellar and chemotaxis genes in *Escherichia coli* and *Salmonella typhimurium* is controlled by an alternate sigma factor. *Proceedings of the National Academy of Sciences, USA*, **84**, 6422–4.

Helmann, J. D., Marquez, L. M. & Chamberlin, M. J. (1988). Cloning, sequencing, and distribution of the *Bacillus subtilis* σ^{28} gene. *Journal of Bacteriology*, **170**, 1568–74.

Herskowitz, I. (1989). A regulatory hierarchy responsible for cell specialization in yeast. *Nature, London*, **342**, 749–57.

Homma, M., DeRosier, D. J. & Macnab, R. M. (1990). Flagellar hook and hook-associated proteins of *Salmonella typhimurium* and their relationship to other components of the flagellum. *Journal of Molecular Biology*, **213**, 819–32.

Hoover, T. R., Santero, E., Porter, S. and Kustu, S. (1990). The integration host factor (IHF) stimulates interaction of RNA polymerase with NifA, the transcriptional activator for nitrogen fixation operons. *Cell*, in press.

Huguenel, E. & Newton, A. (1984). Isolation of flagellated membrane vesicles from *Caulobacter crescentus* cells. *Proceedings of the National Academy of Sciences, USA*, **81**, 3409–3413.

Johnson, R. C. & Ely, B. (1979). Analysis of non-motile mutants of the dimorphic bacterium *Caulobacter crescentus*. *Journal of Bacteriology*, **137**, 627–35.

Johnson, R. C., Walsh, J. P., Ely, B. & Shapiro, L. (1979). Flagellar hook and basal complex of *Caulobacter crescentus*. *Journal of Bacteriology*, **138**, 984–9.

Johnson, R. C., Ferber, D. M. & Ely, B. (1983). Synthesis and assembly of flagellar components by *Caulobacter crescentus* motility mutants. *Journal of Bacteriology*, **154**, 1137–44.

Kaplan, J. B., Dingwall, A., Bryan, R., Champer, R. & Shapiro, L. (1989). Temporal regulation and overlap organization of two *Caulobacter* flagellar genes. *Journal of Molecular Biology*, **205**, 71–83.

Komeda, Y. (1982). Fusions of flagellar operons to lactose genes on a Mu *lac* bacteriophage. *Journal of Bacteriology*, **150**, 16–26.

Komeda, Y. (1986). Transcriptional control of flagellar genes in *Escherichia coli* K-12. *Journal of Bacteriology*, **168**, 1315–18.

Koyasu, S., Asada, M., Fukuda, A. & Okada, Y. (1981). Sequential polymerization of flagellin A and flagellin B into *Caulobacter* flagella. *Journal of Molecular Biology*, **153**, 471–5.

Koyasu, S. & Shirakihara, Y. (1984). *Caulobacter crescentus* flagellar filament has a right-handed helical form. *Journal of Molecular Biology*, **173**, 125–30.

Kustu, S., Santero, E., Keener, J., Popham D. & Weiss, D. (1989). Expression of σ^{54} (ntrA)-dependent genes is probably united by a common mechanism. *Microbiological Review*, **53**, 367–76.

Kutsukake, K., Ohya, Y. & Iino, T. (1990). Transcriptional analysis of the flagellar regulon of *Salmonella typhimurium*. *Journal of Bacteriology*, **172**, 741–7.

Lagenaur, C. & Agabian, N. (1978). *Caulobacter* flagella organelle: Synthesis, compartmentation, and assembly. *Journal of Bacteriology*, **135**, 1062–9.

Loewy, Z. G., Bryan, R. A., Reuter, S. H. & Shapiro, L. (1987). Control of synthesis and positioning of a *Caulobacter crescentus* flagellar protein. *Genes and Development*, **1**, 626–35.

Losick, R., Kroos, L., Errington, J. & Youngman, P. (1989). Pathways of developmentally regulated gene expression in the spore-forming bacterium *Bacillus subtilis*. In *Genetics of Bacterial Diversity*, K. P. Chater & D. A. Hopwood, eds. London: Academic Press.

Macnab, R. (1987). Flagella. In *Escherichia coli and Salmonella typhimurium Cellular and Molecular Biology*, F. C. Neidhardt, ed., vol. 1, pp. 70–83. Washington, DC: American Society for Microbiology Press.

Mcnab, R. M. & DeRosier, D. J. (1988). Bacterial flagellar structure and function. *Canadian Journal of Microbiology*, **34**, 442–51.

Marczynski, G. T., Dingwall, A. & Shapiro, L. (1990). Plasmid and chromosomal

DNA replication and partitioning during the *Caulobacter crescentus* cell cycle. *Journal of Molecular Biology*, **212**, 709–22.

Milhausen, M. & Agabian, N. (1983). *Caulobacter* flagellin mRNA segregates asymmetrically at cell division. *Nature, London*, **302**, 630–2.

Minnich, S. A. & Newton, A. (1987). Promoter mapping and cell cycle regulation of flagellin gene transcription in *Caulobacter crescentus*. *Proceedings of the National Academy of Sciences, U.S.A.*, **84**, 1142–6.

Minnich, S. A., Ohta, N., Taylor, N. & Newton, A. (1988). Role of the 25-, 27- and 29-Kd flagellins in *Caulobacter crescentus* cell motility: method for construction of deletion and Tn5 insertion mutants by gene replacement. *Journal of Bacteriology*, **170**, 3953–60.

Mullin, D., Minnich, S. A., Chen, L.-S. & Newton, A. (1987). A set of positively regulated flagellar gene promoters in *Caulobacter crescentus* with sequence homology to the *nif* gene promoters of *Klebsiella pneumoniae*. *Journal of Molecular Biology*, **195**, 939–43.

Mullin, D. A. & Newton, A. (1989). Ntr-like promoters and upstream regulatory sequence *ftr* are required for transcription of a developmentally regulated *Caulobacter crescentus* flagellar gene. *Journal of Bacteriology*, **171**, 3218–27.

Nathan, P., Gomes, S. L., Hahnenberger, K., Newton, A. & Shapiro, L. (1986). Differential localization of membrane receptor chemotaxis proteins in *Caulobacter* predivisional cells. *Journal of Molecular Biology*, **191**, 433–40.

Newton, A. & Ohta, N. (1990). Regulation of the cell division cycle and differentiation in bacteria. *Annual Reviews in Microbiology* (in press).

Newton, A., Ohta, N., Ramakrishnan, G., Mullin, D. & Raymond, G. (1989). Genetic switching in the flagellar gene hierarchy of *Caulobacter* requires negative as well as positive regulation of transcription. *Proceedings of the National Academy of Sciences, USA*, **86**, 6651–5.

Ninfa, A. J., Mullin, D. A., Ramakrishnan, G. & Newton, A. (1989). *Escherichia coli* σ^{54} RNA polymerase recognizes *Caulobacter crescentus flbG* and *flaN* flagellar gene promoters *in vitro*. *Journal of Bacteriology*, **171**, 383–91.

Nishimura, A. & Hirota, Y. (1989). A cell division regulatory mechanism controls the flagellar regulon in *Escherichia coli*. *Molecular & General Genetics*, **216**, 340–6.

Ohta, N., Chen, L. S., Swanson, E. & Newton, A. (1985). Transcriptional regulation of a periodically controlled flagellar gene operon in *Caulobacter crescentus*. *Journal of Molecular Biology*, **186**, 107–15.

Osley, M. A. & Newton, A. (1974). Chromosome segregation and development in *Caulobacter crescentus*. *Journal of Molecular Biology*, **90**, 359–70.

Osley, M. A., Sheffery, M. & Newton, A. (1977). Regulation of flagellin synthesis in the cell cycle of *Caulobacter*: Dependence on DNA replication. *Cell*, **12**, 393–400.

Ramakrishnan, G. & Newton, A. (1990). FlbD of *Caulobacter crescentus* is a homologue of the NtrC (NR$_1$) protein and activates σ^{54}-dependent flagellar gene promoters. *Proceedings of the National Academy of Sciences, USA*, **87**, 2369–73.

Reuter, S. & Shapiro, L. (1987). Asymmetric segregation of heat-shock proteins upon cell division in *Caulobacter crescentus*. *Journal of Molecular Biology*, **194**, 653–62.

Robertson, C. A. & Nash, H. A. (1988). Bending of the bacteriophage lambda attachment site by *Escherichia coli* integration host factor. *Journal of Biological Chemistry*, **263**, 3554–7.

Santero, E., Hoover, T., Keener, J. & Kustu, S. (1989). *In vitro* activity of the nitrogen fixation regulatory protein NIFA. *Proceedings of the National Academy of Sciences, USA*, **86**, 7346–50.

Schoenlein, P. V., Gallman, L. S. & Ely, B. (1989). Organization of the *fla*FG gene cluster and identification of two additional genes involved in flagellum biogenesis in *Caulobacter crescentus*. *Journal of Bacteriology*, **171**, 1544–53.

Shapiro, L. (1985). Generation of polarity during *Caulobacter* cell differentiation. *Annual Reviews in Cell Biology*, **1**, 173–207.

Shapiro, L., Alexander, W., Bryan, R., Champer, R., Frederikse, P., Gomes, S. L., Hahnenberger, K. & Ely, B. (1985). Biogenesis of a polar flagellum and a chemosensory system during *Caulobacter* cell differentiation. In *Sensing and Response in Microorganisms*, M. Eisenbach & M. Balaban, eds., pp. 93–106. New York: Elsevier Science Publishers.

Shapiro, L. & Gober, J. W. (1989). Positioning of gene products during *Caulobacter* cell differentiation. *Journal of Cell Science, Suppl.*, **11**, 85–95.

Shapiro, L. & Maizel, J. V. (1973). Synthesis and structure of the *Caulobacter crescentus* flagella. *Journal of Bacteriology*, **113**, 478–85.

Shaw, P., Gomes, S. L., Sweeney, K., Ely, B. & Shapiro, L. (1983). Methylation involved in chemotaxis is regulated during *Caulobacter* differentiation. *Proceedings of the National Academy of Sciences, USA*, **80**, 5261–5.

Sheffery, M. & Newton, A. (1979). Purification and characterization of a polyhook protein from *Caulobacter crescentus*. *Journal of Bacteriology*, **138**, 575–83.

Sheffery, M. & Newton, A. (1981) Regulation of periodic protein synthesis in the cell cycle. Control of initiation and termination of flagellar gene expression. *Cell*, **24**, 49–57.

Stallmeyer, M. J. B., Hahnenberger, K. M., Sosinsky, G. E., Shapiro, L. & DeRosier, D. J. (1989). Image reconstruction of the flagellar basal body of *Caulobacter crescentus*. *Journal of General Microbiology*, **205**, 511–18.

Swoboda, U. K., Dow, C. S. & Vitkovic, L. (1982). Nucleoids of *Caulobacter crescentus* CB15. *Journal of General Microbiology*, **128**, 279–89.

Trachtenberg, S. & DeRosier, D. J. (1988). Three-dimensional reconstruction of the flagellar filament of *Caulobacter crescentus*: A flagellin lacking the outer domain and its amino acid sequence lacking an internal segment. *Journal of Molecular Biology*, **202**, 787–808.

Wagenknecht, T., DeRosier, D., Aizawa, S. I. & Macnab, R. M. (1982). Flagellar hook structures of *Caulobacter* and *Salmonella* and their relationship to filament structure. *Journal of Molecular Biology*, **162**, 69–87.

Wagenknecht, T., DeRosier, D., Shapiro, L. & Weissborn, A. (1981). Three-dimensional reconstruction of the flagellar hook from *Caulobacter crescentus*. *Journal of Molecular Biology*, **151**, 439–65.

Weissborn, A., Steinman, H. M. & Shapiro, L. (1982). Characterization of the proteins of the *Caulobacter crescentus* flagellar filament: Peptide analysis and filament organization. *Journal of Biological Chemistry*, **257**, 2066–74.

Xu, H., Dingwall, A. & Shapiro, L. (1989). Negative transcriptional regulation in the *Caulobacter* flagellar hierarchy. *Proceedings of the National Academy of Sciences, USA*, **86**, 6656–60.

METHYLATION-INDEPENDENT TAXIS IN BACTERIA

JUDITH P. ARMITAGE, WENDY A. HAVELKA and R. ELIZABETH SOCKETT

Microbiology Unit, Biochemistry Department, University of Oxford, Oxford OX1 3QU, UK

INTRODUCTION

Bacterial motility and its chemotactic control is arguably the best understood behavioural system in nature. The elegant explanation of sensory transduction at a molecular and biochemical level from methyl accepting chemotaxis protein (MCP) through the phosphorylation pathway to the motor switch (see chapters 5 and 6, this volume) is, however, only a part of the total integrative system that results in a bacterium reaching its optimum environment for survival and growth. In addition to responding to the chemicals sensed through the MCPs, bacteria also respond to a range of chemicals that do not interact with these dedicated sensory proteins. They may respond either positively or negatively to different electron acceptors such as oxygen, nitrate or dimethyl sulphoxide and swim towards, or away from, different wavelengths and intensities of light. The signals caused by the different non-MCP systems and their mechanism of interaction with the other pathways to control the frequency of motor switching have not been identified. The sensory signals from these different stimuli must however integrate with those from the MCP system to produce a net response.

The basic metabolic nature of the MCP-independent systems suggests that they probably represent very early sensory mechanisms, in evolutionary terms. In natural environments, these systems may be the dominant sensory mechanisms as they are often involved in responding to primary growth limiting factors such as oxygen, light and carbohydrates. The dedicated receptor-dependent chemotaxis systems may be more important for directing 'well-fed' bacteria towards environments where the growth rate may be increased. This chapter will try to bring together data on methylation-independent tactic responses from several different bacterial species. It is probable, given the evidence from methylation-dependent chemotaxis, that the mechanisms involved in methylation-independent taxis will prove to be common across the bacterial world.

METHYLATION-INDEPENDENT CHEMOTAXIS

Chemotaxis towards phosphotransferase carbohydrates

The early experiments by Adler and others on *Escherichia coli* showed that, in addition to the range of chemicals sensed via the non-transporting MCP system, there are a large number of carbohydrates that require transport through the phosphotransferase (PTS) system to cause a behavioural response (Adler & Epstein, 1974; Lengeler *et al.*, 1981; Niwano & Taylor, 1982). It now seems probable that the intracellular signalling pathway for chemotaxis to PTS sugars is closely allied to the MCP system. However, the primary chemotactic response depends on the carbohydrates being transported and phosphorylated through a dual transport/chemotaxis pathway, rather than on the binding to a dedicated periplasmic-facing sensory receptor. The mechanisms involved in the interaction with the intracellular sensory transduction pathway may provide insight into the integration of methylation-dependent and independent pathways in general.

In many bacterial species, the phosphotransferase system is a major uptake system for a wide range of carbohydrates (Postma & Lengeler, 1985; Saier, 1989). In enteric bacteria, carbohydrates are translocated through the membrane via specific membrane-spanning proteins, enzymes II (EII). In combination with another specific protein, enzyme III, which in some cases is an integral part of a larger enzyme II, the substrate is phosphorylated as it is transported into the cell. The phosphorylation of the substrate depends on the different enzymes II/III being themselves phosphorylated by a non-substrate specific pathway involving the transfer of a phosphoryl group from phosphoenolpyruvate (PEP) via a non-substrate specific soluble enzyme I to a hexose protein (HPr).

Mutants in the chemotactic response towards phosphotransferase carbohydrates of *E. coli* fall into three groups: one group has lost the ability to respond to single carbohydrates, the second group shows no chemotaxis towards any PTS carbohydrate but still responds to MCP-dependent chemoeffectors, and the third group has lost chemotaxis to both PTS and MCP-dependent chemoeffectors (Lengeler & Vogler, 1989).

In all the EII–PTS systems analysed, there is always a correlation between transport/phosphorylation and chemotaxis to that specific carbohydrate. The use of analogues that are either non-transportable, or are transported and phosphorylated but not metabolized further, shows that a chemotactic response depends on the compound being transported and phosphorylated but that no further metabolism is required for a signal (Adler & Epstein, 1974; Lengeler *et al.*, 1981). The response is not dependent on an increase in cellular energy as phosphorylation of transported, non-metabolizable analogues is an energy drain on the cells, but still causes a chemoattractant response (Lengeler *et al.*, 1981). To show that the phosphorylated form of the PTS carbohydrate is not itself an intracellular signal, substrates that can be transported in the phosphorylated form have been tested. Transport of the phosphorylated carbohydrates does not result in a chemotactic response, although the intracellular form of the carbohydrate is the same as the EII-dependent chemotactic substrate (Pecher, Renner & Lengeler, 1983). These results all suggest the essential role of both transport and phosphorylation in causing the chemotactic signal. Maximum chemotactic responses are only seen if the EII activities for the specific carbohydrate are fully induced. Measurements of K_m values for many PTS substrates show very good correlations with the chemotaxis response curves for those compounds (Lengeler et al., 1981). Hybrid PTS proteins have been made in which the EIII type part of a large EII protein, e.g. that for carrying n-acetyl glucosamine, has been deleted and complemented with a soluble EIII, such as that for glucose. As long as the EII for glucose has been deleted these cells transport and respond to n-acetyl glucosamine, but not glucose (Lengeler et al., 1981). Together these data provide extremely good evidence that the chemoreception through EII complexes of carbohydrates requires substrate binding and subsequent phosphorylation of the substrate.

The transport and phosphorylation of the carbohydrate must then be signalled to the flagellar motor. The membrane-spanning structure of the EII protein has some resemblance to the MCPs, but the mechanism of signalling is very different. Methylation is not involved in either sensory signalling or adaptation to PTS substrates. Mutants of *E. coli* and *Salmonella typhimurium* that are unable to methylate MCPs because of s-adenosyl methionine depletion or mutations in the methyltransferase or esterase genes still show normal chemotaxis and adaptation to PTS substrates, although methylation-dependent chemotaxis is lost (Taylor, Johnson & Smith, 1988).

Early work suggested that the cytoplasmic signal might be a cyclic nucleotide, but subsequent examination of the mutants involved has shown that the results were caused by a secondary effect on the synthesis of the motor apparatus of the cell, which in *E. coli* is under catabolite control (Black, Hobson & Adler, 1983; Tribhuwan, Johnson & Taylor, 1986). The mechanism of intracellular signalling is still therefore speculative. Mutants have been isolated that lack the EI kinase and HPr but still possess active EII enzymes. These proteins will bind the substrate, allowing some limited diffusion of the substrate into the cell but without phosphorylation. Under these conditions the bacteria show no chemotactic response to any PTS substrate (Lengeler, 1975; Lengeler et al., 1981). This, coupled with the lack of chemotactic response to transportable, phosphorylated, carbohydrates, strongly suggests that it is the process of phosphorylation that causes the sensory signal. It is possible that a conformational change in the EII protein accompanying the transport and phosphorylation of the substrate is sensed by one of the chemotaxis proteins involved in the cytoplasmic pathway from the MCP system. This, however, seems unlikely as the

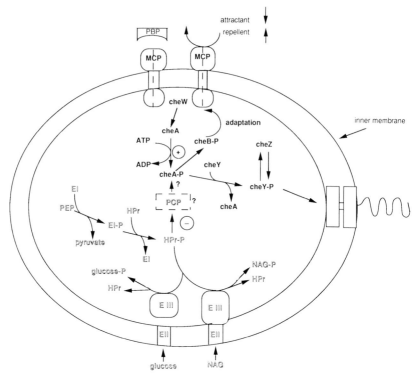

Fig. 1. Possible mechanisms involved in integrating methylation-dependent chemotaxis with the sensory transduction pathway from the phosphotransferase carbohydrate transport system. The removal of an attractant results in an increase in CheA-P and in consequence CheY-P and increased clockwise rotation (tumble); the increase in a PTS carbohydrate decreases CheA-P and in consequence causes counterclockwise rotation (smooth swimming). (Adapted from Lengeler & Vogler (1989) with permission). MCP: two representative methyl-accepting chemotaxis proteins; PBP: periplasmic binding protein; PCP: phosphotransferase chemotaxis protein; EII and EIII two representative phosphotransferase proteins; NAG: n-acetyl glucosamine. + = increase in CheA-P, − = decrease in CheA-P. For explanations of other abbreviations see text.

chemotaxis protein would have to interact directly with the membrane bound EII protein. The EII protein for sucrose transport from a naturally non-motile enteric bacterium, *Klebsiella pneumoniae*, has been expressed in *E. coli* and causes a chemotactic response to sucrose in *E. coli* (Sprenger & Lengeler, 1984). Any structural change essential for chemotaxis would be expected to have been lost during evolution in a non-motile species. It is likely, therefore, that the signalling mechanism depends on a biochemical change in the cell which occurs as a consequence of the phosphorylation reaction occurring through the HPr-EI cascade (Fig. 1).

Mutants that have lost their response to all PTS substrates but still show normal responses to MCP-dependent chemoeffectors have mutations in the Hpr or EI proteins involved in phosphorylation of the EII proteins. It seems possible, therefore, that it is the change in these proteins that is sensed by the chemotactic apparatus. This is supported by the finding that phosphorylation and transport can be restored in a mutant lacking hexose-protein (Hpr) by the addition of fructose-protein (Fpr) without the restoration of chemotaxis (Lengeler & Vogler, 1989). HPr may therefore be involved in the integration of the signal from the EII proteins to the motor. HPr mutants have been isolated with a reduced ability to carry out phosphorylation. The limited level of EII phosphorylation is just enough to allow some PTS transport, but no chemotaxis. If expression of the Hpr is increased, the level of phosphorylation increases until chemotaxis to the PTS sugar again occurs (Lengeler, personal communication). Lengeler has proposed that there may be a protein, the phosphoryl-chemotaxis-protein (PCP), which interacts with the CheY–CheZ complex involved in altering the switching frequency of the motor (Lengeler & Vogler, 1989). His hypothesis is that the transport of a specific carbohydrate results in the reduction of the phosphorylated state of the PCP via HPr. This might directly or indirectly reduce the phosphorylation level of CheY either by altering the autophosphorylation state of CheA or by directly removing phosphoryl groups from CheY (Fig. 1). There is some evidence from experiments with mutants with different components of the sensory transduction pathway deleted that both CheA and CheW may be required for a response to PTS sugars, making the indirect effect on CheY-P more probable (E. Rowsell, J. Smit & B. L. Taylor, unpublished observations). As a net excess of CheY-phosphate probably causes a tumble (flagellar reversal), the removal or reduction of CheY-P would result in a period of smooth swimming, or a positive chemotactic response to the increase in the carbohydrate concentration. Adaptation to PTS chemoeffectors could be the result of restoration of the equilibrium within the phosphorylation cascade of enzymes involved in carbohydrate transport and phosphorylation and the glycolytic supply of PEP.

The interaction of the phosphorylation pathway of the PTS carbohydrate system with the phosphorylation pathway from the MCP-receptor system has led to the suggestion that the PTS-dependent control of switch phosphorylation may be the phylogenetic ancestor of the MCP-dependent system, the dedicated chemosensory pathway evolving later (Lengeler & Vogler, 1989; Taylor, Johnson & Smith, 1988).

Chemotaxis in R. sphaeroides

In addition to PTS carbohydrates, there are many other compounds to which bacteria show behavioural responses that are independent of the MCP system. Unlike the PTS chemotaxis system, many of these depend on at least some limited metabolism of the effectors. In enteric bacteria, for example, experiments have shown that chemotactic responses to proline depend on the metabolism of that compound, although the extent and result of metabolism are not known (Clancy, Madill & Wood, 1981).

The majority of research into MCP-independent chemotaxis has been carried out on the photosynthetic bacterium *Rhodobacter sphaeroides*, and a large section of this review will be centred on this species. This is the only bacterial species in which biochemical, behavioural and genetic studies have failed to identify any methylation-dependent sensory transducers (Sockett, Armitage & Evans, 1987). The behavioural responses of *R. sphaeroides* show several differences from those described in other chemotactic species: 1. unlike other chemotactic organisms there is not an all-or-none response when chemotactically stimulated, 2. no repellent response has been seen to any chemical (including glycerol which apparently causes a repellent response in other species by causing general structural changes to MCPs), 3. only metabolites cause behavioural responses, 4. methionine auxotrophs still show chemotaxis and, 5. two different, separable responses can be measured, depending on whether the cells are given a temporal or spatial gradient. It seems likely that MCP-type sensors have been lost from this species during evolution. Under photosynthetic conditions, the bacteria have an extensive intracytoplasmic membrane system housing the photosynthetic apparatus (Kaplan, 1978). The concentration of light harvesting pigments varies with the growth conditions and, as s-adenosyl methionine is required for protoporphorin synthesis, its role in chemotaxis may have been sacrificed (Jones, 1978). *Rhodospirillum rubrum*, another photosynthetic bacterium, does, however, have MCPs (Sockett, Armitage & Evans, 1987).

Although no MCPs have been identified by s-adenosyl methionine labelling, anti-Trg and anti-Tar immunoprecipitation, or methanol release, *R. sphaeroides* still shows strong behavioural responses to a wide range of stimuli. As yet it is not known whether any part of the sensory transduction pathway identified in enteric bacteria is present in *R. sphaeroides*. Chromosomal DNA from *R. sphaeroides* has been probed with genes from *E. coli* coding for the different sensory components and cellular extracts tested with antisera raised against those proteins, all with negative results (W. A. Havelka & J. P. Armitage, unpublished). The difference between the G + C content of *R. sphaeroides* and *E. coli*, and the weak immunoreactivity of the antisera with control proteins makes the results, however, of limited significance.

Rhodobacter sphaeroides *motility*

R. sphaeroides swims differently from other flagellate bacteria, possibly reflecting the simplified chemosensory system used in this species. It is

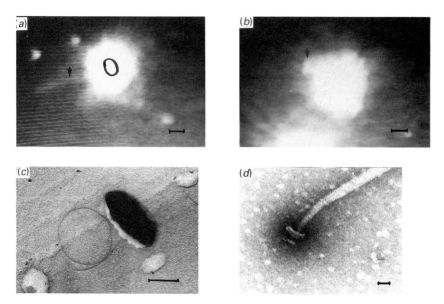

Fig. 2. Flagellum and swimming behaviour of R. sphaeroides: (a) in vivo active flagellum rotating clockwise and (b) relaxed flagellum using high intensity dark field microscopy. The arrows indicate the rotating and relaxed flagellum, the actual size of the cell body (masked by flare) is indicated on (a); bars 1.5 μm. (c) Electron micrograph of whole cell showing relaxed flagellum, bar 1.0 μm and (d) negative stained isolated basal body from R. sphaeroides. Note that the hook is straight and more narrow than the filament. Bar 20 nm.

therefore worth describing the structure and function of the R. sphaeroides flagellum before describing the chemotactic behaviour of this species.

It has only one flagellum, made up of 55 kD flagellin subunits, which only rotates in a clockwise direction (Armitage & Macnab, 1987). The cells have a cocco-bacillus shape and the flagellum arises from the middle of the long side of the cell (Fig. 2(a),(c)). The new flagellum on the daughter cell invariably arises from the opposite side of the cell with the result that, before two cells separate after division, pairs of cells can be seen spinning if both flagella are operating simultaneously. R. sphaeroides is motile under both aerobic and anaerobic conditions, with the flagellum located in the same position on the cell. The structure of the cytoplasmic membrane is, however, very different in the two growth condition. Aerobically grown cells have a membrane structure similar to most Gram-negative bacteria, but, when photosynthetic growth is induced by a reduction in the oxygen tension, there is a massive increase in the cytoplasmic membrane with the result that the cytoplasm appears packed with invaginated membrane housing the photosynthetic apparatus (Kaplan, 1978). This membrane has been shown to be continuous with the cytoplasmic membrane. The mechanism by which the flagellum is targeted to the same, non-polar site under

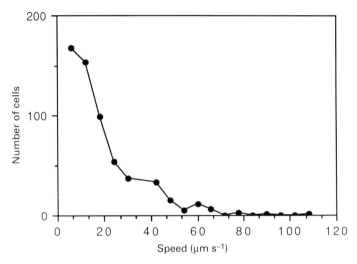

Fig. 3. Polygon showing mean speed variation within a population of swimming R. sphaeroides. Individual cells were followed for 400 ms and the results averaged as described in Poole, Sinclair & Armitage (1988).

all growth conditions, and its fate during the switch from one growth state to the other, are currently being investigated. Analysis of isolated basal bodies shows an additional ring when compared to enteric bacteria (Fig. 2(d)). This extra ring might be comparable with that identified in *Caulobacter crescentus* (Stallmeyer *et al.*, 1989), and is perhaps related to membrane targeting of the single flagellum.

The change of direction of swimming cells is brought about by periodic stops in flagellar rotation. During these stops, the flagellar structure relaxes from a functional helix to a short wavelength, large amplitude structure and Brownian motion appears to cause reorientation of the cell body (Armitage & Macnab, 1987). When the motor starts to rotate again, a functional helix reforms and the cell swims in a different direction. Unlike the rather uniform behaviour of a population of enteric bacteria, a culture of *R. sphaeroides* shows an enormous variation in both the behaviour of individual cells with time and between individuals (Fig. 3). Swimming speeds may vary from a few μm s^{-1} to 100 μm s^{-1}, with stopping frequencies varying from every few seconds to no stops over several minutes. Stopping times also vary from a fraction of a second to several minutes, even under conditions where all the cells in a population may be expected to have a full proton motive force (pmf). The variation in behaviour, and the very fact of stopping flagellar rotation, suggests that the motor can either control the proton flow through the motor proteins, switching it off periodically (disengaging the motor), or changes in the motor structure can alter the extent to which proton flow is coupled to the mechanical work of rotation ('change of gear'). Analysis of the changes in swimming behaviour in response to stimuli is therefore complex, and must rely on the averaging of a large number of cells (Poole, Sinclair & Armitage, 1988). It is possible that both this stop–start mechanism of swimming, and the variation in population behaviour is a consequence of the lack of the MCP-dependent sensory transduction machinery acting as a damper on the cytoplasmic signal.

R. sphaeroides exhibits two distinct behavioural responses to environmental stimuli, one to spatial gradients of chemicals and one to 'step-up' or temporal changes in concentration (Poole, Williams & Armitage, 1990). Both of these responses are shown primarily to metabolites such as weak acids, nitrogen sources and sugars and in addition to cations such as potassium. *R. sphaeroides* is a photoheterotroph and therefore also responds to light, oxygen and alternative electron acceptors; the mechanism involved in these systems will be discussed later.

Temporal responses

If the concentration of a metabolite is increased, the average swimming speed of the cells in the population increases (Poole, Williams & Armitage, 1990) (Fig. 4). This increase in population swimming speed, or excitation, as measured by motion analysis, is the result of both a decrease in stopping frequency and an increase in the swimming speed of individual cells. Unlike the responses seen to gradients in enteric bacteria, there is no apparent adaptation to the step-up response (Poole & Armitage, 1988). It might be expected that, as the chemoeffectors are metabolites, the increase in swimming speed might be the direct result of an increase in pmf caused by an increased metabolic rate. Experiments have shown, however, that the increase in swimming speed is independent of any change in the membrane potential across the cytoplasmic membrane and is also independent of electron transport (Poole, Brown & Armitage, 1990*a*; Poole, Brown & Armitage, 1990*b*). The excitation response occurs both under conditions where the excitant causes the membrane potential to increase and also under conditions where it decreases. The change in swimming speed is therefore independent of the driving force for flagellar rotation, the pmf, and may represent a change in the interaction of the different motor proteins coupling the membrane potential to rotation of the flagellum.

The change in swimming speed continues apparently for as long as the increased concentration is present and has been shown to be additive, the speed can be increased in a step-wise manner up to maximum swimming speed by increasing the concentration of the excitant (Poole, Williams, & Armitage, 1990). Not all metabolites cause the excitant response, and not all excitants cause the increase in speed under all conditions. The behaviour shown by populations of cells to different metabolites under different

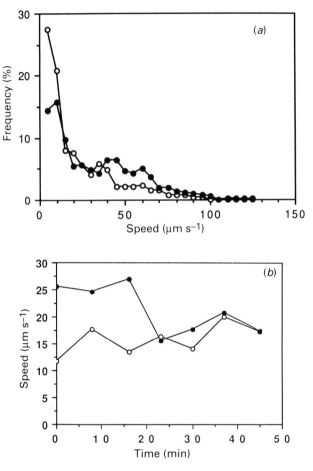

Fig. 4. (a) Increase in mean swimming speed of a population of R. sphaeroides after the addition of 1 mM potassium. (○) control, (●) plus potassium. (From Poole, Brown & Armitage, 1990.) (b) Lack of short-term adaptation to the addition of 1 mM propionate. 200 cells were tracked for 400 ms each at each time point and results plotted as the population average. (○) no addition (●) 1 mM propionate (Poole & Armitage, 1988).

conditions suggests that this response is essentially a metabolic response to limiting nutrients, and serves to disperse the bacteria within their environment, rather than cause their accumulation.

The bacteria only show a strong response to the nutrient which is currently growth limiting. Therefore, for example, if the bacteria have been grown in a nitrogen-rich medium, they will not show a strong excitation response when presented with an increased concentration of ammonia or glutamate, but they will respond to a carbon source if this is now the limiting nutrient. The response is not specific for particular chemicals, but rather for nutrient

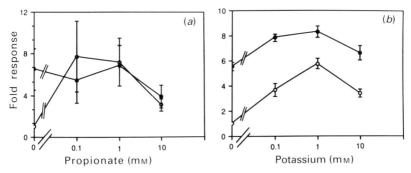

Fig. 5. (a) Response of *R. sphaeroides* to a gradient of a carbon source, propionate, with (●) and without (○) a background of another carbon source, 5 mM pyruvate. (b) Response of *R. sphaeroides* to a gradient of potassium, with (●) and without (○) a background of 5 mM pyruvate. (See text for details, adapted from Poole, Williams & Armitage, (1990) with permission.)

groups, therefore the carbon response could be to pyruvate or to acetate, and the nitrogen response could be to ammonia or to glutamate.

The fact that this response causes the dispersal rather than accumulation of the bacteria has been demonstrated experimentally (Poole, Williams & Armitage, 1990). *R. sphaeroides* will respond to a spatial gradient of a chemical by swimming from one compartment of a chamber to a second compartment containing an attractant. If, however, the attractant is also an excitant and the bacteria are resuspended in the attractant an increased number will still swim from the compartment, now containing the attractant/excitant, into the second, attractant free, compartment. An increased response can be seen if a second attractant is added to the top chamber, particularly if it belongs to a second nutrient group (Fig. 5).

It seems probable that this gradient-independent, membrane potential-independent response serves to disperse the bacteria in their normally nutrient limited environment. If there is an increase in the concentration of the currently limiting nutrient, then there is an increased swimming speed within the population, spreading the bacteria within their environment. When saturated, e.g. for carbon, the bacteria then remain at that elevated level of swimming, but show no further response to an increase in concentration of any carbon source. If an increased concentration of the now limiting nutrient, e.g. nitrogen is encountered, there may again be an increase in swimming speed until the speed reaches a maximum level. *R. sphaeroides* can swim at speeds of up to 120 μm s^{-1}.

Transport and at least some limited metabolism are required for the excitation response. Experiments using transportable amino acid analogues (P. S. Poole & J. P. Armitage, unpublished observations) and inhibitors of ammonia metabolism (Poole & Armitage, 1989) have shown that transport, and therefore receptor binding, is not enough for excitation; at least some metabolism of the effector is required. Not all nutrients cause this excitation response. *R. sphaeroides* shows a spatial response to both PTS and non-PTS carbohydrates, but neither of these groups has been shown to cause a chemokinetic response to an increase in their background concentration (Jeziorska & Armitage, unpublished observations). This suggests that a specific metabolic change, common to a wide range of excitants, does not occur during sugar metabolism. This is currently being investigated.

Accumulation response

When presented with a nutrient gradient, either in a soft agar plate or in liquid culture, *R. sphaeroides* accumulates in a region of increased concentration. The precise cellular behaviour that brings about this accumulation is unknown because, unlike enteric bacteria, not all cells in a population show a response, possibly because the response depends on the metabolic state of the individual cell. This has made the analysis of the behaviour of individual tethered cells very difficult. In addition, the same metabolites can cause a chemokinetic response or an accumulation response, depending on whether they are presented as a temporal or spatial gradient, and this has meant that the increase in swimming speed has tended to mask any behavioural changes that would bring about the gradient response.

We have now isolated mutants that have lost the gradient response to a limited range of metabolites but retain the chemokinetic response (W. A. Havelka & J. P. Armitage, unpublished observations). In particular, one mutant, isolated by its failure to show normal swarms on soft agar plates when grown on acetate or propionate, shows an excitation response when given a step-up increase in either of these chemicals, but shows no accumulation response to either. Unstimulated cells grown on either of these organic acids show an increase in stopping frequency. The mutation is pleomorphic, with abnormal accumulation of photosynthetic pigments and storage granules. The mutant shows normal accumulation and excitation responses to many other chemoeffectors, and appears normal when grown on other carbon sources. These results suggest an alteration in metabolic control; a change in intracellular concentration of a metabolite affecting flagellar switching frequency as well as the synthesis of cellular components. The gene complementing this mutation maps close to a cluster of flagellar genes and its DNA sequence is currently being determined. These results show that the gradient response, like the chemokinetic response, is dependent on at least limited metabolism. However, although both responses affect the operation of the flagellar motor the signals are probably not the same. The temporal response may be a result of the metabolic state directly affecting the flagellar motor. The metabolic signal causing the response to a spatial chemical gradient may be related to the PTS signalling pathway described above, where the chemotactic response depends on a change in the phosphorylation potential of the cell altering the phosphorylation state of CheA or CheY. Alternatively, it may be related to the acetyl adenylate requirement for clockwise flagellar rotation recently described in *E. coli*. The addition of acetate to an *E. coli* strain deleted for all MCPs and sensory transduction proteins except CheY resulted in an enhanced clockwise rotation of the motor (Wolfe, Conley & Berg, 1988). Biochemical analysis of these mutants, and mutants with a normal sensory pathway but deficient in acetate metabolism, suggested that either adenylation or acetylation may be needed for sensory signalling in *E. coli*. Although this pathway is not present in *R. sphaeroides*, it provides evidence that an intermediary metabolite can alter flagellar behaviour in a species with a well-defined sensory pathway. It is possible that an equivalent change in *R. sphaeroides* in response to an alteration in metabolic activity may control flagellar stopping frequencies.

ELECTRON ACCEPTOR TAXIS

A group of compounds to which most bacterial species respond through a receptor-independent mechanism are the electron acceptors of aerobic and anaerobic respiration. The aerotactic response was one of the earliest to be identified. Suspensions of bacteria will segregate in different regions around an air bubble or the coverslip edge of a microscope slide. Some species will swim very close to the bubble interface, others localize as far from the air bubble as possible, and yet others take an intermediate position. The location of different species at different distances from the air interface reflects their current metabolic requirement for oxygen. It is probable that these are amongst the strongest chemoeffectors as they are responsible for maintaining the proton gradient at a level required to sustain, e.g. nutrient transport at a maximum level for growth. The precise response to a particular terminal acceptor will depend on the current mode of growth of a bacterium. For example, *R. sphaeroides* can grow as a photoheterotroph, using photosynthetic electron transport to maintain its pmf, or as a heterotroph, using respiratory electron transport. Obviously, under the former growth conditions, oxygen is not required, and therefore an aerotactic response is unnecessary. However, under the second condition, oxygen is the terminal electron acceptor and it would therefore be an advantage to the bacteria to swim into regions of increased oxygen.

Results from *R. sphaeroides*, *E. coli*, *S. typhimurium* and *Bacillus subtilis* have all shown that active electron transport to a particular electron acceptor is necessary for an attractant response to that electron acceptor (Laszlo & Taylor, 1981; Laszlo *et al.*, 1984*b*; Armitage, Ingham & Evans, 1985; Dang *et al.*, 1986). Many bacterial species can induce different terminal acceptors under different growth conditions, to maintain electron transport under both high and low aerobic conditions and anaerobic conditions. This has been used to investigate the roles of different acceptors and pathways of electron transport in behavioural responses. For example, *E. coli* can carry out electron transport anaerobically using nitrate as a terminal electron acceptor under conditions where nitrate reductase is induced. *E. coli* only shows a chemotactic response to nitrate under these conditions (Taylor *et al.*, 1979). In the presence of oxygen, the nitrate reductase is inhibited and chemotaxis towards nitrate is lost, but the cells now show an aerotactic response. Mutants have been isolated in both *E. coli* and *S. typhimurium* which lack one or both of the terminal oxidase enzymes of the branched electron transport chain. There was no aerotaxis in the mutants lacking both the terminal oxidases, cytochrome o and cytochrome d, but normal aerotaxis in mutants with just one of these oxidases present (Laszlo *et al.*, 1984a; Shioi *et al.*, 1988). Electron transport down either branch can therefore result in a positive aerotactic response. Binding of oxygen to a terminal oxidase is not, however, sufficient to cause a chemotactic signal. There must be metabolism of the oxygen, i.e. active electron transport, to produce a behavioural response. In the past, it has been suggested that HPr of the PTS system might be involved in integrating the aerotactic signal to the motor but Taylor examined a range of mutants in the PTS system and showed that this was not the case. Mutants lacking different components of the PTS system, including EII and EI, all responded normally to oxygen gradients (Shioi *et al.*, 1988). There is also no measurable competition between oxygen and PTS sugars in chemotaxis experiments.

These, and earlier results, show that active electron transport is essential for an aerotactic response. The signal sent from the electron transport chain to control the switching frequency of the flagellar motor has not been identified. Several studies suggest that changes in the proton gradient accompany the response, but it is impossible to unequivocally separate changes in pmf from changes in the redox potential of one of the electron transport components. Under some conditions, uncouplers of respiratory electron transport have been shown to act as repellents, suggesting that a perturbation of the pmf can cause a negative signal to be sent to the flagellar motor (Ordal & Villani, 1980). However, these inhibitors have very wide-ranging effects on many cellular functions.

Experiments with *R. sphaeroides* have also shown that active electron transport is required for an aerotactic response (Armitage, Ingham & Evans, 1985). It is possible to grow this species aerobically with a branched electron transport chain, where one branch is to an alternative oxidase, while the main pathway leads to a coupled terminal aa_3-type cytochrome. The advantage of this system is that electron transport through the cytochrome aa_3 results in pmf generation at that oxidase but not at the other terminal cytochrome. By using combinations of inhibitors and electron donors, it is possible to allow electron transport to either terminal oxidase under conditions that only allow one pathway to generate an increase in pmf. Only the pathway allowing both active electron transport and a pmf increase results in a response to a spatial gradient of oxygen. Again, however, it is impossible to separate the increase in pmf from a change in the redox poise of one of the intermediates.

R. sphaeroides can also use dimethylsulphoxide as a terminal electron acceptor when grown anaerobically in the dark (Ferguson, Jackson & McEwan, 1987). Under these conditions, DMSO acts as a strong attractant for the bacteria, as would be expected from the results with nitrate and oxygen. Respiratory electron transport to oxygen inhibits electron transport to DMSO and therefore the attractant response to DMSO is inhibited in the presence of oxygen. Photosynthetic electron transport has also been shown to inhibit both respiratory and DMSO-dependent electron transport in *R. sphaeroides*, presumably by altering the redox state of the shared electron transport components (Ferguson, Jackson & McEwan, 1987). Light would, therefore, be expected to inhibit both the aerotactic and DMSO dependent responses. Surprisingly, although light does reduce the aerotactic response, it has no effect on the attractant response to DMSO (Armitage, unpublished observations). Therefore, although photosynthetic and respiratory electron transport have both been shown to inhibit DMSO reductase, only respiratory electron transport inhibits the chemoattractant response. These results suggest that electron transport-dependent taxis may be more complex than current models suggest.

Under some conditions oxygen can act as a repellent. This is obviously the case for strict anaerobes, but also applies to facultative bacteria under some conditions. For example, *R. sphaeroides* is attracted to oxygen when grown aerobically, but repelled from oxygen when grown photosynthetically. In the case of *R. sphaeroides*, it seems probable that the repellent response seen to oxygen in light-incubated photosynthetically grown cells is the result of oxygen binding to a terminal cytochrome oxidase and altering the electron flow through the electron transport components shared with the photosynthetic electron transport chain (Jackson, 1988). This will alter both the pmf generated and the redox poise of the components of the chain, one of these probably causing the repellent signal. The cause of the repellent response in obligate anaerobes is unknown, but is probably caused by the oxidation of a metabolic component.

The aerotactic response must be integrated with the signals from the other sensory pathways. The point at which the interaction occurs is still uncertain. Mutants of enteric bacteria lacking the methylation-dependent sensory transducers respond normally to oxygen, but a mutant with an altered methyl esterase, which shows an inverted response to attractants, also shows an inverted response to oxygen. It seems likely that, in enteric bacteria, the aerotactic signal interacts with the CheA–CheY phosphorylation sequence, and it is at this point that integration of the chemotactic and aerotactic pathway occurs.

PHOTOTAXIS

As with electron acceptor taxis described above, the phototactic response shown by photosynthetic eubacteria depends on electron transport (for details of photoresponses in halobacteria see Oesterhelt & Marwan, this volume). The classical description of the photoresponses in eubacteria is a 'photophobic' response where the bacterial cell reverses its direction of swimming when entering a region of either lower light intensity or a less photosynthetically useful wavelength of light (Hader, 1987; Armitage, 1988). As a result, bacteria become trapped in regions of maximum light intensity and optimum wavelength as they reverse back into that region when passing over the light edge. The swimming speed probably also changes transiently as the rate of photosynthetic electron transport changes, before feedback control on the rate of electron transport brings the pmf level back to equilibrium. This transient decrease in speed on lowering of the light intensity, and increase on a step-up, can be seen with *R. sphaeroides*.

Most of the studies on phototactic behaviour have centred on large rod-shaped or spiral photosynthetic species with polar tufts of flagella, that reverse to change their direction of swimming (Pfennig, 1968; Clayton, 1957; Hader, 1987). Reversal every time they enter a darker region will therefore hold these species in the light. Clayton showed that *R. rubrum* could respond to reductions in light intensity as small as 1%, the response being caused by the transient change in the rate of photosynthesis caused by the change in intensity rather than the overall change in the steady state (Clayton, 1953b). This suggests that bacteria adapt to the phototactic response. When exposed to a step-down in light the cells continue rhythmic reversals for up to 30 s after the stimulus, a response reminiscent of the tumbling response to repellents shown by enteric bacteria (Clayton, 1953c). Similarly, the phototactic response is additive, the response to two repetitive sub-threshold step-down stimuli resulting in a behavioural response, provided the time between the two step-down stimuli is less than about 3 s and greater than 250 ms. The strength of the phototactic response is dependent on the change in intensity and the duration. Therefore a change in intensity close to the threshold needs to last about 1.5 s to cause a behavioural response, whereas a change of twice the strength will cause a response in about 0.5 s (Clayton, 1953c). The continued reversal of cells after a single step-down in light intensity suggests a long-term signal followed by adaptation; reversal back into high light results in a return to normal swimming.

For *R. sphaeroides*, which stops rather than reverses to change swimming direction, an increase in stopping frequency on entering a dark area can only result in the observed accumulation back in the light if there is a 'memory' in the system and if the time between stops allows the cells to swim the distance back into the illuminated region.

Adaptation may be caused, as suggested for PTS chemotaxis, by the restoration of equilibrium in a critical metabolic pathway, but the mechanism involved in cells remembering the previous photosynthetic state is unknown. All investigations carried out so far have centred on temporal gradients of light, or step-down responses. The behaviour of individual photosynthetic bacteria in spatial light gradients has not been examined.

Early experiments with photosynthetic bacteria showed that different species accumulated in regions of light of a wavelength more or less coincident with maxima in the absorption spectrum of the photosynthetic pigments, leading to the suggestion that the pigments, specifically bacteriochlorophyll, were acting as the receptors (Clayton, 1953a). From his experiments, Clayton suggested that the signal was a change in photosynthetic activity rather than a response to sensory pigments. The observation of the phototactic behaviour of different photosynthetic species with mutations in different components of the photosynthetic apparatus, and the use of inhibitors of photosynthetic electron transport, has shown that active electron transport is essential for a phototactic response (Harayama & Iino, 1976; Harayama, 1977; Armitage, Ingham & Evans, 1985). Indeed, the isolation of phototactically incompetent mutants is a reliable means of isolating mutants defective in photosynthetic electron transport (Armitage & Evans, 1981). As with behavioural responses to other MCP-independent systems, the mechanisms involved in transmitting changes in photosynthetic electron transport to the flagellar motor, and integrating the signals with other sensory pathways are currently unknown. The time taken for *R. rubrum* to accommodate to a step-down response is longer than would be expected for photosynthetic electron transport to reach a new steady state. It seems likely, therefore, that some metabolic intermediate is involved in signalling the change in photosynthetic activity to the flagellar motor.

SUMMARY

Motile bacteria respond to a wide range of stimuli within their environment, and under *in vivo* conditions they must integrate many different signals to produce an overall net response (Fig. 6). The best-understood sensory pathway is currently that through the dedicated methylation-dependent pathway. However, as discussed above, many important environmental changes are sensed by methylation-independent pathways. The sensory signals from these diverse pathways must eventually integrate with the signals from other methylation-independent and from methylation-dependent

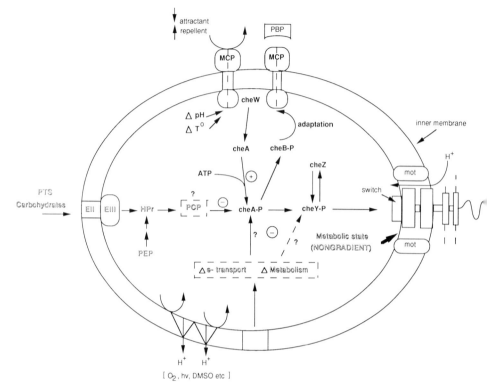

Fig. 6. Possible mechanisms involved in integrating all sensory stimuli via the phosphorylation level of the cell to produce a net response. The increase in metabolic state (excitation) would operate outside the integrative pathway. Drawn to show the effect of a decrease in attractant and an increase in PTS or metabolic rate. For abbreviations see Fig. 1.

pathways. Whether or not CheA and/or CheY turn out to be universal sensory proteins, responding perhaps to the phosphorylation state of the cell and consequently controlling behaviour of the flagellar motor, remains to be seen. Further investigation of the rather neglected area of methylation-independent chemotaxis will provide information on both environmental sensing and the critical intracellular metabolic changes that are involved in sensory responses in general.

ACKNOWLEDGEMENTS

The work on *R. sphaeroides* described here was supported by the SERC, the Wellcome Trust and the Leverhulme Foundation.

REFERENCES

Adler, J. & Epstein, W. (1974). Phosphotransferase-system enzymes as chemoreceptors for certain sugars in *Escherichia coli* chemotaxis. *Proceedings of the National Academy of Sciences, USA*, **71**, 2895–9.

Armitage, J. P. & Evans, M. C. W. (1981). The reaction centre in the phototactic and chemotactic responses of *Rhodopseudomonas sphaeroides*. *FEMS Microbiology Letters*, **11**, 89–92.

Armitage, J. P., Ingham, C. & Evans, M. C. W. (1985). Role of the proton motive force in phototactic and aerotactic responses of *Rhodopseudomonas sphaeroides*. *Journal of Bacteriology*, **163**, 967–72.

Armitage, J. P. & Macnab, R. M. (1987). Unidirectional intermittent rotation of the flagellum of *Rhodobacter sphaeroides*. *Journal of Bacteriology*, **169**, 514–18.

Armitage, J. P. (1988). Tactic responses in photosynthetic bacteria. *Canadian Journal of Microbiology*, **34**, 475–81.

Black, R. A., Hobson, A. C. & Adler, J. (1983). Adenylate cyclase is required for chemotaxis to phosphotransferase system sugars by *Escherichia coli*. *Journal of Bacteriology*, **153**, 1187–95.

Clancy, M., Madill, K. A. & Wood, J. M. (1981). Genetic and biochemical requirements for chemotaxis to L-proline in *Escherichia coli*. *Journal of Bacteriology*, **146**, 902–6.

Clayton, R. K. (1953*a*). Studies in the phototaxis of *Rhodospirillum rubrum*. I. Action spectrum, growth in green light and Weber-law adherence. *Archiv für Mikrobiologie*, **19**, 107–24.

Clayton, R. K. (1953*b*). Studies in the phototaxis of *Rhodospirillum rubrum*. II. The relation between phototaxis and photosynthesis. *Archiv für Mikrobiologie*, **19**, 125–40.

Clayton, R. K. (1953*c*). Studies in the phototaxis of *Rhodospirillum rubrum*. III. Quantitative relationship between stimulus and response. *Archiv für Mikrobiologie*, **19**, 141–65.

Clayton, R. K. (1957). Patterns of accumulation resulting from taxes and changes in motility of microorganisms. *Archiv für Mikrobiologie*, **27**, 311–19.

Dang, C. V., Niwano, M., Ryu, J.-I. & Taylor, B. L. (1986). Inversion of aerotactic response in *Escherichia coli* deficient in *cheB* protein methylesterase. *Journal of Bacteriology*, **166**, 275–80.

Ferguson, S. J., Jackson, J. B. & McEwan, A. G. (1987). Anaerobic respiration in the Rhodospirillaceae: characterisation of pathways and evaluation of roles in redox balancing during photosynthesis. *FEMS Microbiology Reviews*, **46**, 117–43.

Hader, D.-P. (1987). Photosensory behavior in prokaryotes. *Microbiological Reviews*, **51**, 1–21.

Harayama, S. (1977). Phototaxis and membrane potential in the photosynthetic bacterium *Rhodospirillum rubrum*. *Journal of Bacteriology*, **131**, 34–41.

Harayama, S. & Iino, T. (1976). Phototactic responses of aerobically cultivated *Rhodospirillum rubrum*. *Journal of General Microbiology*, **94**, 173–9.

Jackson, J. B. (1988). Bacterial photosynthesis. In *Bacterial energy transduction*, Anthony, C. (ed.), pp. 317–75. London: Academic Press.

Jones, O. T. G. (1978). Biosynthesis of porphyrins, hemes and chlorophylls. In *The photosynthetic bacteria*, Clayton, R. K. and Sistrom, W. R. (eds), pp. 751–77. New York: Plenum Press.

Kaplan, S. (1978). Control and kinetics of photosynthetic membrane development. In *The photosynthetic bacteria*, Clayton, R. K. and Sistrom, W. R. (eds), pp. 809–39. New York: Plenum Press.

Laszlo, D. J. & Taylor, B. L. (1981). Aerotaxis in *Salmonella typhimurium*: role of electron transport. *Journal of Bacteriology*, **145**, 990–1001.

Laszlo, D. J., Fandrich, B. L., Sivaram, A., Chance, B. & Taylor, B. L. (1984a). Cytochrome o as a terminal oxidase and receptor for aerotaxis in *Salmonella typhimurium*. *Journal of Bacteriology*, **159**, 663–7.

Laszlo, D. J., Niwano, M., Goral, W. W. & Taylor, B. L. (1984b). *Bacillus cereus* electron transport and proton motive force during aerotaxis. *Journal of Bacteriology*, **159**, 820–4.

Lengeler, J. (1975). Mutations affecting transport of the hexitols D-mannitol, D-glucitol, and galactitol in *Escherichia coli* K-12: isolation and mapping. *Journal of Bacteriology*, **124**, 26–38.

Lengeler, J., Auburger, A. -M., Mayer, R. & Pecher, A. (1981). The phosphoenolpyruvate-dependent carbohydrate: phosphotransferase system enzyme II as chemoreceptors in chemotaxis of *Escherichia coli* K-12. *Molecular and General Genetics*, **183**, 163–70.

Lengeler, J. W. & Vogler, A. P. (1989). Molecular mechanisms of bacterial chemotaxis towards PTS-carbohydrates. *FEMS Microbiology Reviews*, **63**, 81–92.

Niwano, M. & Taylor, B. L. (1982). Novel sensory adaptation mechanism in bacterial chemotaxis to oxygen and phosphotransferase substrates. *Proceedings of the National Academy of Sciences, USA*, **79**, 11–15.

Ordal, G. W. & Villani, D. P. (1980). Action of uncouplers of oxidative phosphorylation as chemotactic repellents of *Bacillus subtilis*. *Journal of General Microbiology*, **118**, 471–8.

Pecher, A., Renner, I. & Lengeler, J. (1983). The phosphoenolpyruvate-dependent carbohydrate: phosphotransferase system enzymes II, a new class of chemosensors in bacterial chemotaxis. In *Mobility and recognition in cell biology*, Sund, H. and Veeger, C. (eds), pp. 517–31. Berlin: Walter de Gruyter and Co.

Pfennig, N. (1968). *Chromatium akenii* (Thiorhodaceae). pp. 3–9. Göttingen: Institut für den Wissenschaftlichen Film.

Poole, P. S. & Armitage, J. P. (1988). Motility response of *Rhodobacter sphaeroides* to chemotactic stimulation. *Journal of Bacteriology*, **170**, 5673–9.

Poole, P. S. & Armitage, J. P. (1989). Role of metabolism in the chemotactic response of *Rhodobacter sphaeroides* to ammonia. *Journal of Bacteriology*, **171**, 2900–2.

Poole, P. S., Brown, S. & Armitage, J. P. (1990a). Swimming changes and chemotactic responses in *Rhodobacter sphaeroides* do not involve changes in the steady state membrane potential or respiratory electron transport. *Archiv für Mikrobiologie*, **153**, 614–18.

Poole, P. S., Brown, S. & Armitage, J. P. (1990b). Potassium chemotaxis in *Rhodobacter sphaeroides*. *FEBS Letters*, **260**, 88–90.

Poole, P. S., Sinclair, D. R. & Armitage, J. P. (1988). Real time computer tracking of free-swimming and tethered rotating cells. *Analytical Biochemistry*, **175**, 52–8.

Poole, P. S., Williams, R. L. & Armitage, J. P. (1990). Chemotactic responses of *Rhodobacter sphaeroides* in the absence of apparent adaptation. *Archiv für Mikrobiologie*, **153**, 368–72.

Postma, P. W. & Lengeler, J. (1985). Phosphoenolpyruvate: carbohydrate phosphotransferase system of bacteria. *Microbiological Reviews*, **49**, 232–69.

Saier, Jr M. H. (1989). Protein phosphorylation and allosteric control of inducer exclusion and catabolite repression by the bacterial phosphoenolpyruvate: sugar phosphotransferase system. *Microbiological Reviews*, **53**, 109–20.

Shioi, J., Tribhuwan, R. C., Berg, S. T. & Taylor, B. L. (1988). Signal transduction in chemotaxis to oxygen in *Escherichia coli* and *Salmonella typhimurium*. *Journal of Bacteriology*, **170**, 5507–11.

Sockett, R. E., Armitage, J. P. & Evans, M. C. W. (1987). Methylation-independent and methylation-dependent chemotaxis in *Rhodobacter sphaeroides* and *Rhodospirillum rubrum*. *Journal of Bacteriology*, **169**, 5808–14.

Sprenger, G. A. & Lengeler, J. (1984). L-sorbose metabolism in *Klebsiella pneumoniae* and Sor+ derivatives of *Escherichia coli* K-12 and chemotaxis towards sorbose. *Journal of Bacteriology*, **157**, 39–45.

Stallmeyer, M. J. B., Hahnenberger, K. M., Sosinsky, G. E., Shapiro, L. & DeRosier, D. J. (1989). Image reconstruction of the flagellar basal body of *Caulobacter crescentus*. *Journal of Molecular Biology*, **205**, 511–18.

Taylor, B. L., Johnson, M. S. & Smith, J. M. (1988). Signaling pathways in bacterial chemotaxis. *Botanica Acta*, **101**, 101–4.

Taylor, B. L., Miller, J. B., Warrick, H. M. & Koshland, D. E. Jr (1979). Electron acceptor taxis and blue light effect on bacterial chemotaxis. *Journal of Bacteriology*, **140**, 567–73.

Tribhuwan, R. C., Johnson, M. S. & Taylor, B. L. (1986). Evidence against the direct involvement of cyclic GMP or cyclic AMP in bacterial chemotactic signalling. *Journal of Bacteriology*, **168**, 624–30.

Wolfe, A. J., Conley, M. P. & Berg, H. C. (1988). Acetyladenylate plays a role in controlling the direction of flagellar rotation. *Proceedings of the National Academy of Sciences, USA*, **85**, 6711–15.

CONTROL OF DIRECTED MOTILITY IN
MYXOCOCCUS XANTHUS

D. R. ZUSMAN, M. J. McBRIDE, W. R. McCLEARY AND K.A. O'CONNOR

Department of Molecular and Cell Biology, University of California, Berkeley, CA 94720, USA

INTRODUCTION

Myxococcus xanthus is a Gram-negative rod-shaped bacterium, typically about 5 µm in length and 0.5 µm in diameter. Cells of *M. xanthus* move by gliding motility and exhibit multicellular interactions, the most noted of which is the formation of raised mounds called fruiting bodies (for reviews, see Zusman, 1984; Kaiser, 1986; Shimkets, 1987). This article will discuss the motility patterns of *M. xanthus* and relate these patterns to the multicellular life-cycle of the bacterium.

CELL MOTILITY AND VEGETATIVE GROWTH OF *M. XANTHUS*

M. xanthus, in nature, is typically found in damp fertile soils, rich in decaying organic matter. Like the cellular slime moulds, which also inhabit this ecological niche, *M. xanthus* cells are predators which kill and digest both Gram-positive and Gram-negative bacteria. They can be grown in the laboratory, however, on defined or semi-defined media. The myxobacteria do not engulf other microorganisms but rely instead on extracellular antibiotics that kill or immobilize their prey, and on extracellular proteases, nucleases, lipases, phosphatases, and various polysaccharide degrading enzymes. The soil habitat of *M. xanthus* and the reliance on so many extracellular bacteriolytic and degradative activities are probably responsible for the unusual motility patterns of these organisms:

1. *multicellular associations*. The bacteria maintain close cell-to-cell associations (social interactions) in all phases of their life cycle. They are normally found in large 'hunting groups' (called swarms) that may contain several million individuals. Presumably, the swarms ensure that a high enough concentration of antibiotics and lytic activities are present at one location to successfully digest their prey. Time-lapse films by Kühlwein & Reichenbach (1968) show individual cells moving out from the group but most return a short time later.
2. *movement in slime trails*. The bacteria cannot move in liquid culture

but require a solid surface for movement. Cells usually glide in slime trails of their own species or of another unrelated gliding bacteria such as a cyanobacterium (Burchard, 1984), although they are capable of much slower movement in the absence of these trails. The slime trails may provide a matrix for cell-to-cell communication by binding potential food substrates, degradative activities or intercellular signals. In addition, this behaviour may allow myxobacteria that prey on other gliders to track their prey.

3. *rate of cell movement.* The bacteria move at about 2–4 μm per minute with some variations in rate depending on substrate,nutrition, and temperature. This is very slow in comparison to about 1500 μm/min for enterobacteria (which are flagellated) and 60–600 μm/min for some other gliding bacteria such as *Cytophaga johnsonae* (Pate, 1988). The slow rate of movement for *M. xanthus* may be important for keeping the 'hunting group' together. A more critical function, however, may be to keep the bacteria in close proximity to their degradative enzymes and potential food sources.

4. *substratum interactions.* Irregularities in the substratum (for example, scratch marks) cause cells to reorient themselves within the grooves. This occurs because *M. xanthus* shows elasticotaxis, a phenomenon in which bacteria tend to follow stress lines in the surface (Stanier, 1942). For example, when glass beads are placed on an agar surface, the surface is stressed. The bacteria follow the stress lines and appear to move towards the nearest bead. They surround the bead and then move away. This phenomenon occurs even in *frz* mutants and probably involves the interaction of the gliding mechanism with the surface.

GLIDING MOTILITY

The mechanism of gliding motility is very poorly understood. It is usually described as movement in the absence of flagella at a solid–liquid or an air–liquid interface. Pate (1988) has summarized nearly everything known about gliding motility: cells must be in contact with a surface, extracellular slime is involved, no organelles of motility have been identified and no obvious morphological changes in the cells occur during their movements. In addition, gliding has been shown, at least in some species, to require a membrane potential rather than ATP (Glagoleva *et al*, 1980; Pate, 1988), and to require calcium ions (Burchard, 1984). Continuous protein synthesis is not required for gliding since short term movement occurs in the presence of chloramphenicol. However, motility of *M. xanthus* is inhibited by treatment with proteolytic enzymes indicating that cell surface proteins are required for gliding. In addition, osmotic shock causes a reversible inhibition of gliding (Burchard, 1974).

Current theories of the mechanism of gliding motility include: (i) use of contractile elements (Burchard, 1981); (ii) directional propagation of waves (Humphrey et al., 1979); (iii) directional extrusion of slime (Ridgway et al., 1975) (iv) use of rotary motors similar to those found in flagellated bacteria (Pate & Chang, 1979); (v) outer membrane components that are moved in tracks associated with the underlying peptidoglycan structure (Lapidus & Berg, 1982); and (vi) controlled release of surfactants from poles of cells (Dworkin et al., 1983). Despite this abundance of potential mechanisms for cell movement, clear-cut data on the gliding mechanism have not been found. Fortunately, Kaiser and his collaborators have isolated a large number of motility mutants, several of which may be involved in the mechanism of motility in *M. xanthus* (Hodgkin & Kaiser, 1979a,b). Some of these motility genes have been cloned and sequenced (Stephens & Kaiser, 1987; Stephens et al., 1989). Identification of the gene products with the aid of antibodies may help to identify possible motility structures in *M. xanthus*.

'A' AND 'S' MOTILITY IN *M. XANTHUS*

Hodgkin & Kaiser (1977, 1979a,b) isolated hundreds of motility mutants of *M. xanthus*. From their genetic and morphological analysis of the mutants, they deduced that *M. xanthus* contains two parallel multigene systems that control motility: (a) system A (for 'Adventurous') is required for the movement of single cells or small groups of cells and has at least 23 genetic loci; (b) system S (for 'Social') is required for the movement of cells in groups only and also contains numerous genetic loci. Except for one locus that is shared by both systems (*mgl*), all non-motile mutants were found to be double mutants having at least one mutation in system A and one in system S. Figure 1 shows the morphology of edges of colonies with defects in A or S motility. Recently, Fink & Zissler (1989) showed that several mutants which are defective in lipopolysaccharide O-antigen represent a new class of A-mutants, unaltered in S motility.

One particularly interesting observation made by Hodgkin & Kaiser was that many of the nonmotile mutants, although unable to move on their own, became transiently motile after contact with wild-type cells. In addition, certain pairs of mutants, each individually nonmotile, were able to move when mixed together. Hodgkin & Kaiser were able to identify five 'stimulation groups' which exhibit extracellular complementation among the system A mutants and at least one among the system S mutants. Each mutant in a particular 'stimulation group' was able to move when mixed with a member of another group or with wild-type cells, but it could not move when mixed with a different mutant from the same group. The existence of so many extracellular complementation groups for gliding motility suggests that cell–cell signalling is an important regulatory mechanism for the movement of cells.

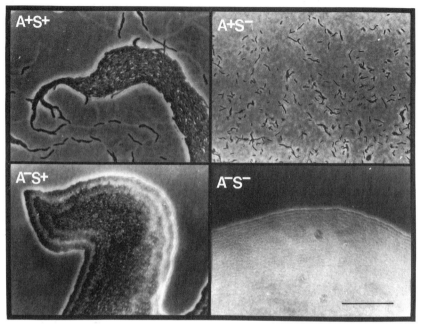

Fig. 1. Effect of motility mutations on cell behaviour. *M. xanthus* cells were inoculated onto CTT agar and incubated at 32 °C for 3 days and the colony edge was photographed under phase contrast. Bar, 50 μm. (Courtesy of L. Shimkets, Shimkets, 1986.)

Kaiser (1979) observed that mutants in S motility are defective in pili production. Wild-type cells contain 4 to 10 pili, about 8 nm in diameter and 3–10 μm in length. The pili are usually located at one pole but occasionally are found at both poles. One of the S motility mutants which can become transiently motile after contact with wild-type cells, can also transiently produce pili after being stimulated. Thus expression of the S motility system and pili production are correlated. This suggests that the pili may be important in the cell–cell recognition and signalling process.

FRUITING BODY FORMATION IN *M. XANTHUS*

In nature, when *M. xanthus* cells are unable to find sufficient nutrients, they aggregate into fruiting bodies where they sporulate. The spores are resting cells which can withstand prolonged periods of starvation. In the laboratory, fruiting body formation in *M. xanthus* is triggered by plating cells under starvation conditions, at high cell density on a solid surface (Shimkets, 1987). While the large majority of cells (80 to 90%) aggregate to form raised mounds in which all of the cells sporulate (see Fig. 2), the remaining cells follow a different developmental fate. These cells, called

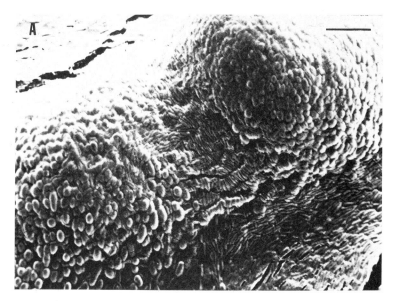

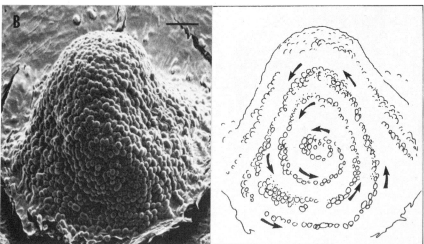

Fig. 2. Fruiting bodies of *M. xanthus* retain evidence of spiral patterns of cell movements. Scanning electron micrographs of cells developing on a Nuclepore filter at 28 °C for 48 hours. An artist's sketch accompanies (*b*) Bars, 10 μm (O'Connor & Zusman, 1989).

peripheral rods, remain as rod-shaped cells around and between fruiting bodies. These cells do not normally aggregate or sporulate unless they are harvested and resuspended at high concentration on a fresh substrate (O'Connor & Zusman, unpublished data). The peripheral rods express many biochemical markers which distinguish them from both vegetative cells and sporulating cells. If a low-level or transient nutrient source becomes available, these cells have been shown to respond immediately, unlike spores which require several hours and a relatively high threshold of nutrients to germinate.

The aggregating cells have been followed morphologically by scanning electron microscopy and time-lapse video microscopy (O'Connor & Zusman, 1989). The first sign of aggregation is the formation of large spiral patterns of cells in stacked monolayers parallel to the substratum. The cells in the spirals are tightly coherent, remaining associated even after being scraped from the agar surface. Terraces form through the stacking of aggregation centres in adjacent monolayers. The formation and maintenance of terraced monolayers suggest that cells participate in side-to-side and pole-to-pole interactions which differ from dorsal-to-ventral interactions. As cells continue to accumulate, the terraces become mounds. The mounds become hemispherical in shape as development proceeds. Sporulation can be concurrent with the morphogenesis of mounds. Eventually, at least 98% of the cells in the aggregates become spores.

Kroos et al. (1988) have described a link between cell motility and developmental gene expression in M. xanthus. Non-motile mutants not only failed to execute the morphogenetic movements required to shape a fruiting body, but they also produced very few spores when plated at cell densities appropriate for wild-type cells to form fruiting bodies. Even when cells are plated at ten times normal cell density, only 1% of the non-motile cells sporulated. These cells showed an altered pattern of gene expression as monitored by transcriptional fusions of the sporulation genes to Tn5-lac. The transposon Tn5-lac creates transcriptional fusions between upstream promoters and the lacZ gene of Escherichia coli which functions as a reporter of gene expression. The lac fusions which are normally expressed during the first six hours of development were expressed normally in the non-motile mutant. However, Tn5-lac fusions which are expressed at a later stage showed reduced or no expression in the non-motile mutant. The stage of development at which the non-motile mutants are arrested is similar if not identical to that of the csgA mutants. The csgA mutants are blocked in development at a relatively early stage. These mutants are of interest because they can be rescued for sporulation by codevelopment with wild-type cells or with cells of another mutant class (Hagen et al., 1978). The C-factor was recently purified and found to be a 17 kd protein which at approximately 1 to 2 nM restores normal development to csgA mutants (Kim & Kaiser, 1990). Kroos et al. (1988) hypothesized that if C-signalling caused cells to move, then a positive feedback loop would be created (movement, C-signalling, movement, etc) which could produce 'ripples', ridge-like accumulations of cells which move co-ordinately and rhythmically over the substrate like ripples on a water surface (Shimkets & Kaiser, 1982). Rippling is often observed during developmental aggregation in M. xanthus wild-type but is not observed in csgA mutants.

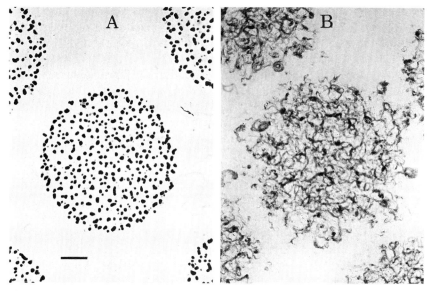

Fig. 3. Morphology of frizzy mutants. Liquid cultures of *M. xanthus* DZF1, the fruiting parental strain, and DZF 1227, a FrzE mutant were spotted on fruiting agar. After 7 days of incubation at 28 °C, the spots were photographed. Bar, 1 mm (A) strain DZF1; (B) strain DZF 1227. (Zusman, 1982.)

DIRECTIONAL CONTROL OF MOTILITY: THE FRIZZY GENES

Isolation and genetic analysis of the frz genes

During the course of isolating and characterizing fruiting body defective mutants of *M. xanthus*, Morrison & Zusman (1979) found a class of mutants which made swirls at low cell density and 'frizzy filaments' at high cell density (Fig. 3). The *frizzy* (*Frz*) mutants were unable to aggregate into discrete mounds on fruiting agar, but were still able to sporulate. Thirty-six mutants exhibiting this phenotype were mapped by transductional analysis and all were found to be genetically linked to the same Tn5 insertion site (Zusman, 1982). Using the kanamycin resistance drug marker in Tn5 as a selectable marker, the *frz* genes were cloned and mapped on a 7.5 kbp region of DNA (Blackhart & Zusman, 1985a). The cloned DNA was analysed by isolating and characterizing new Tn5 insertions at short intervals within the *M. xanthus* DNA and by constructing deletions *in vitro*. The mutated DNA was then transduced into *M. xanthus* where the cloned DNA became integrated into the bacterial chromosome, replacing the wild-type gene or becoming merodiploids. The merodiploid strains were used for complementation analysis: five *frz* complementation groups were identified (*frzA*, *frzB*, *frzC*, *frzE*, and *frzF*). Strains with mutations in any one of these genes had the Frz phenotype: (i) The mutants were defective in aggregation but not sporulation, and (ii) the mutants caused cells to move primarily in one direction and were recessive to the wild-type alleles. The *frzD* mutants, which were defined by two Tn5 mutations, were dominant. These mutants formed non-spreading colonies of motile cells. The mutations caused cells to very frequently reverse their direction of movement. Tn5 insertions in the one kbp region between *frzE* and *frzF* did not prevent fruiting body formation and were thought to define a gap in the *frz* region. However, recent work has shown that this DNA does indeed encode a relevant *frz* gene product and that mutations in this region cause defective aggregation and fruiting body irregularities (McCleary *et al.*, 1990). This gene is now called *frzG*.

Motility pattern of the frz mutants

The next clue to the function of the *frz* genes came from observations of the motility pattern of the *frz* mutants (Blackhart & Zusman, 1985b). Since *M. xanthus* cells move very slowly, 2–4 μm/min, the movements of the bacteria were followed by time-lapse video microscopy. When wild-type cells were plated at low cell density on motility agar (0.5% Casitone + 8 mM MgSO$_4$), the cells were found to reverse their direction of gliding approximately every 7–8 minutes; net movement occurred since the interval between reversals was found to vary widely. Mutants which showed the Frz phenotype (*frzA*, *frzB*, *frzC*, *frzE*, and *frzF*) were found to rarely reverse direction. Cells averaged one to two reversals per hour but some cells did not reverse their direction of movement over many hours. In contrast, *frzD* mutants reversed their direction of movement very frequently, about once every 2 minutes and did not vary the interval between reversals; individual cells showed little net movement and formed smooth-edged 'non-motile' type colonies. *frzG* mutants reversed direction more frequently than wild-type cells, about every 4–5 minutes, but did show net directional movement. These behaviour patterns are reminiscent of chemotaxis mutants of enteric bacteria (Fig. 4). *E. coli* and *Salmonella typhimurium* regulate their chemotaxis activity by changing the direction of flagellar rotation from counterclockwise, which causes directed movement (running), to clockwise, which causes tumbling (Stock & Koshland, 1984). The net directional movement of the bacterium is then determined by the frequency of tumbling and running in response to a chemotactic gradient. In *Caulobacter crescentus*, which contains only a single polar flagellum, the reversal of flagellar rotation results in the bacteria swimming in reverse, rather than tumbling (Ely & Shapiro, 1984). *M. xanthus*, unlike flagellated organisms that resume running in a random direction after an episode of tumbling, usually moves within the confines of a pre-existing slime trail in an essentially two-dimensional plane. The irregularities in the substrate, however, cause frequent reorientation of the bacteria because of elasticotaxis. In addition, *M. xanthus* cells are extremely flexible and

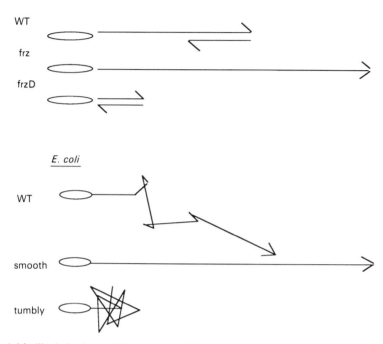

Fig. 4. Motility behaviours of *M. xanthus* and *E. coli* are compared.

it is not uncommon to see individual cells bending to make a right turn or even a U-turn. Thus the regulation of cell reversal in *M. xanthus* is extremely important for the directed movement of this organism.

Sequence similarities between the frz *genes and the chemotaxis genes of enteric bacteria*

The nucleotide sequence of about 8 kbp of the *frz* region of *M. xanthus* DNA has recently been completed (McBride *et al.*, 1989; McCleary *et al.*, 1990; McCleary & Zusman, 1990). The deduced gene products, with the exception of FrzB, show strong homology to the chemotaxis proteins of *S. typhimurium* and *E. coli* (Fig. 5; for a detailed description of these chemotaxis proteins, see chapters 5 and 6, this volume):

(a) The putative FrzA protein (17 kD) shares considerable sequence similarity with CheW of *S. typhimurium*. The FrzA sequence contains 45 identical residues with CheW (28% amino acid identity on the entire FrzA sequence). The striking similarity between FrzA and CheW suggests that they are of common evolutionary origin and may perform similar functions in directing cell movement.

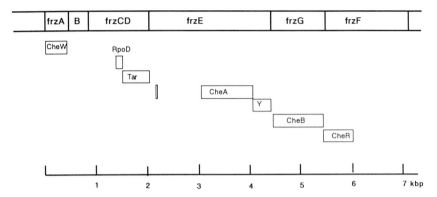

Fig. 5. Sequence similarities between the 'frizzy' genes of M. xanthus and the enteric chemotaxis genes.

(b) The *frzCD* open-reading frame was found to encode a single protein of 417 amino acids (43,571 Da). The FrzCD protein contains a large internal region of similarity to the methyl accepting chemotaxis receptor proteins (MCPs) of the enteric bacteria (Russo & Koshland, 1983; Krikos et al., 1983). The regions of strongest similarity are those that are also highly conserved among the enteric MCPs. For example, FrzCD and *S. typhimurium* Tar exhibit 40% amino acid identity over 141 residues. The degree of similarity is even more striking if conservative amino acid substitutions are allowed.

The *frzD* locus which was defined by two Tn5 insertions, was found to be located within the C-terminus of FrzCD. These Tn5 insertions result in the formation of stable truncated proteins which are poor substrates for methylation, see later (McCleary et al., 1990). Since *frzD* mutations are dominant, the truncated proteins must interfere with the functioning of the wild-type product. For example, enteric cells containing certain truncated MCPs continually tumble or continually swim (Krikos et al., 1985). These phenotypes are thought to result from the truncated MCPs being locked in tumbly or smooth swimming signalling states. Similarly, truncated FrzCD might continually transmit a signal which results in a change in direction of movement.

The N-terminal region of the FrzCD protein shows little similarity to known chemotaxis proteins. The two extended hydrophobic regions that anchor the MCPs of enteric bacteria to the cytoplasmic membrane do not appear to be present in FrzCD (Fig. 6). The FrzCD protein appears to be a freely soluble cytoplasmic protein although it may be loosely associated with the membrane or a membrane protein (unpublished data). Curiously, the FrzCD protein does contain a region which shows 31% amino acid identity over 79 residues to *Bacillus subtilis* RpoD (σ^{43}). This region of RpoD has been implicated in the binding of the σ factor to DNA (Gitt

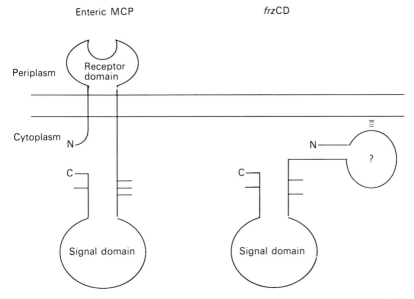

Fig. 6. Proposed similarities between enteric methylated chemotaxis proteins (MCPs) and *M. xanthus* frzCD. The bars are proposed sites of methylation.

et al., 1985). Interaction of FrzCD with a specific DNA sequence might regulate gene expression or might be important for signal transduction.

(c) The *frzE* gene encodes a 777 amino acid protein with a molecular weight of 83,095 Da (McCleary & Zusman, 1990). Analysis of the protein sequence data showed that FrzE is homologous to two enteric chemotaxis proteins: CheA and CheY (Fig. 5). There is a short conserved stretch of 20 amino acids near the N-terminus of FrzE which is homologous to the putative histidine phosphorylation site of CheA as well as a large central region of 360 amino acids which shows 33% amino acid identity to CheA. Immediately adjacent to this region is a 124 amino acid domain which shows 31% amino acid identity to CheY; this homology spans the entire CheY protein. FrzE contains an unusual 68 amino acid domain (between the small and large regions of homology to CheA) which is 38% alanine and 34% proline. Similar (alanine + proline) regions have been identified in the pyruvate dehydrogenase multienzyme complex of *E. coli* (Radford et al., 1989). In this complex, substrates are shuttled between physically separate active sites by swinging lipoyl arms. The (alanine + proline) region is believed to facilitate this movement. As represented in Figure 7, the FrzE (alanine + proline) domain might constitute an interdomain hinge to allow phosphotransfer between spatially separate domains. The homologies between FrzE and CheA and CheY suggest that FrzE is a multidomain/multifunctional protein that functions as a second messenger linking FrzCD

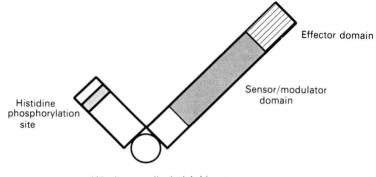

Fig. 7. Multifunctional domains of FrzE. FrzE has two domains which are homologous to CheA (shaded region); these are proposed to contain the sensor/modulator region. It also contains one domain which is homologous to CheY (striped region). This domain is hypothesized to be an effector domain which interacts with the mechanism of gliding. The centre region (circle) consists of an (alanine + proline) region which would be expected to be very flexible. This region is hypothesized to act as a hinge which facilitates the intramolecular transfer of a phosphate group from the histidine phosphorylation site to an aspartate phosphorylation site within the effector domain.

to the gliding motor. The conservation of functional domains between FrzE and CheA suggests common mechanisms of action.

(d) The 1 kbp region between *frzE* and *frzF* was originally designated as a 'gap' region since strains with Tn*5* mutations in that region did not show the Frz phenotype and were able to form fruiting bodies (Blackhart & Zusman, 1985*a*). The DNA sequence of the 'gap' region showed an open reading frame (which we now designate *frzG*) with significant sequence homology to CheB of *E. coli* (McCleary *et al.*, 1990). CheB is the chemotactic methylesterase that acts on the MCPs during the adaptive response. In addition to demethylating MCPs during adaptation, this enzyme also deamidates several glutaminyl residues of the MCPs which subsequently become available for methylation. FrzG shows 31% amino acid identity (105 residues out of 334) with CheB. It has a calculated molecular mass of 35 kD.

The similarity of FrzG with CheB prompted a reevaluation of the effect of *frzG* mutations on gliding behaviour of individual cells and on fruiting body formation. Using time-lapse video microscopy, the *frzG* mutants were found to reverse direction about twice as often as wild-type cells, about 14 times per hour. This phenotype is somewhat similar to the tumbly behaviour of *cheB* mutants in the absence of stimuli. When the *frzG* mutants were placed on fruiting agar, they formed fruiting bodies which showed subtle yet significant differences in morphology from the wild type. The fruiting bodies formed by *frzG* cells were, on average, smaller in size and appeared irregular in shape. These characteristics together with the known homology to CheB suggest that adaptation to aggregation signals during fruiting body morphogenesis was impaired.

(e) There are two possible initiation codons for *frzF* (McCleary et al., 1990). The calculated molecular mass of the peptide initiated from the upstream GTG is 64 kD while the 554 amino acid peptide initiated by the downstream ATG has a predicted molecular weight of 60 kD. The *frzF* gene encodes a protein whose N-terminal region is homologous to the entire enteric methyltransferase, CheR. The C-terminal half of the protein is not similar to any known chemotaxis proteins and we do not yet know its function.

Protein methylation and phosphorylation are catalysed by the frz *gene products*

The sequence similarities between FrzCD and the enteric MCPs suggested that the FrzCD protein might be methylated and demethylated *in vivo* in response to certain stimuli. Polyclonal antisera against a ProteinA–FrzCD fusion protein was therefore prepared (McCleary et al., 1990). The antisera was able to recognize purified Tar protein (the aspartate receptor) from *S. typhimurium*. In *M. xanthus*, the antisera recognized a ladder of bands migrating between 44.5 and 41kD on a Western immunoblot of extracts of wild type cells (Fig. 8). These bands are of the approximate molecular weight predicted for FrzCD by sequence analysis (43.6 kD) and were not present in extracts of a *frzC* mutant. This ladder of bands is consistent with the pattern seen for methylated MCPs: MCPs contain multiple sites for reversible methylation of specific glutamate residues. The addition of each methyl group causes the MCP to migrate faster in the gel. Like the enteric MCPs, treatment of wild-type extracts of *M. xanthus* with 0.1M NaOH, chased the ladder of FrzCD bands into two slower migrating bands of 44.5 kD and 41 kD. In a *frzF* mutant the ladder of FrzCD material was absent; only two bands were observed and these comigrated with the bands produced by 0.1M NaOH treatment of wild-type. These data suggested that the multiple bands observed for FrzCD were indeed the result of methylation and that FrzF was required for methylation. These results were confirmed by direct labelling of *M. xanthus* cells with S-adenosyl-L[methyl-^{3}H] methionine.

In order to follow the methylation of FrzCD during the life cycle of *M. xanthus*, extracts from vegetative and developmental cells were analysed by Western immunoblots using antisera to FrzCD (McCleary et al., 1990). As shown in Figure 8, FrzCD is highly methylated during vegetative growth on rich media (first lane). When cells are starved in buffer, it becomes demethylated within 30 minutes (lane B). On fruiting agar, FrzCD remains unmethylated for several hours after which it is again methylated. The changes in methylation pattern suggest that FrzCD is sensing, responding

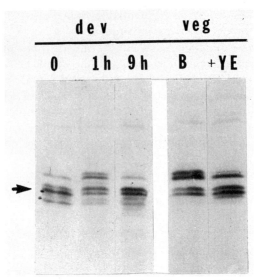

Fig. 8. *In vivo* methylation of FrzCD. Western blots of extracts of *M. xanthus* were analysed for FrzCD using polyclonal rabbit antisera to a proteinA–FrzCD fusion protein. Developmental extracts were prepared at 0 (direct from CYE broth), 1 hour, and 9 hours of development on fruiting agar. Vegetative extracts were made from cells starved in buffer (MOPS, pH7) for 1 hour or cells which were incubated in yeast extract (0.5%) for 1 hour following starvation in buffer. The arrow indicates the position of a 43 kD molecular weight marker.

and adapting to one or more factors in rich medium. Since FrzCD is a soluble cytoplasmic protein, it seems likely that it functions in cooperation with one or more accessory proteins which probably are localized in the membrane and act as primary signal receptors. Figure 8 shows that starved cells respond to yeast extract (they do not respond to Casitone or amino acids, however). During development, FrzCD may respond to one or more factors which may be communicated between cells. Thus the methylation state of FrzCD may comprise a biochemical assay for signals which are generated during cellular aggregation.

In our current working model, the *frz* sensory transduction pathway consists of a signalling protein, FrzCD, which is covalently modified by methylation to mediate adaption. FrzF and FrzG are the enzymes responsible for the methylation and demethylation, respectively. In analogy to the chemotaxis system, FrzCD probably interacts with FrzE by means of FrzA. Genetic evidence shows that FrzE interacts with FrzCD. *frzD* mutants (which form truncated FrzCD protein) are dominant to wild-type and give the non-Frz phenotype of hyperreversible motility. *frzD frzE* double mutants, however, show the Frz phenotype and rarely reverse direction (Blackhart & Zusman, 1985*b*). This shows that *frzD* is dependent on *frzE* for its phenotypic expression.

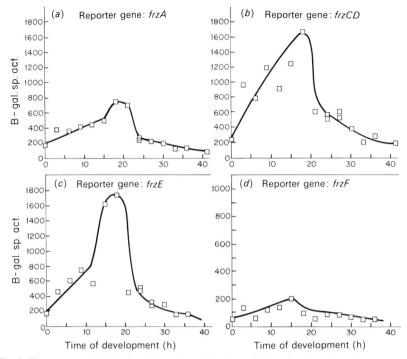

Fig. 9. Time course of induction of β-galactosidase (β-gal) from *frz*:: Tn*5*-*lac* during development of CF agar (Weinberg & Zusman, 1989).

The sequence similarities between FrzE and CheA and CheY suggest that FrzE should have autokinase activity, since CheA autophosphorylates and then transfers its phosphate to CheY (Stock *et al.*, 1989). In order to test for kinase activity, the FrzE protein was overproduced in a T7 expression vector and the purified protein incubated with ^{32}P labelled ATP. The protein was then run on a gel and shown to be phosphorylated (McCleary and Zusman, manuscript in preparation). The role of phospho-FrzE in activating/deactivating other Frz proteins has not yet been determined.

Expression of the frz genes

The expression of the first four genes of the *frz* gene cluster (*frzA*, *frzB*, *frzCD*, and *frzE*) was studied by assaying the level of β-galactosidase produced from strains containing a transcriptional fusion resulting from the insertion of the transposon Tn*5*-*lac* into this cluster (Fig. 9; Weinberg & Zusman, 1989). These *frz* genes were all found to be expressed vegetatively when cells were grown on rich media. However, when cells were placed on fruiting agar, there was a significant induction of activity that peaked at about 12–18 hours into development. The time of peak expression corresponded to the period when discrete mounds of cells became evident if wild-type cells were plated on the same medium. The level of β-galactosidase activity measured at the peak of expression was as much as ten-fold greater than vegetative levels. Northern blot hybridization analysis suggested that the *frz* genes were arranged as an operon. To test this hypothesis, double mutants were constructed which contained Tn5–*132* (a modified form of Tn5 which has a tetracycline resistance element rather than a kanamycin resistance element) either upstream or downstream of the reporter Tn5-*lac*. The expression of the *frz* genes in the double mutants was consistent with the hypothesis that the first four genes are organized as an operon with additional internal promoters. Insertion mutations in *frzCD* and *frzE* lowered expression of both upstream and downstream reporter Tn5-*lac* insertions, suggesting that the FrzCD and FrzE proteins regulate transcription of the entire operon from a promoter upstream of *frzA*. Since the FrzCD protein has a potential DNA binding domain as part of its structure (McBride *et al.*, 1989), it is quite possible that this protein is involved in the autoregulation of the *frz* operon. A model for the autoregulation of the *frz* genes is presented in Fig. 10.

Contrasts between the motility of the enteric bacteria and M. xanthus

There are many differences between enteric bacteria and the myxobacteria in both the means of movement and its directional control. The enteric bacteria are motile in liquid media and control their movements in three dimensions by random redirection during tumbling. *M. xanthus* can only move on a solid substrate in two dimensions; redirection occurs principally because of inconsistencies on the surface on which the cells glide. Another striking difference is the time scales over which excitation and adaptation must take place in each system. *E. coli* moves at a rate of approximately 1500 μm/min. Adapted cells punctuate their runs approximately every 1 to 2 s with a 0.1 to 0.2 s tumble. Excitation occurs in less than a second following exposure to a chemoeffector and adaptation follows this over a period of seconds to minutes. In contrast, *M. xanthus* cells move at approximately 2 μm/min and change direction approximately every 7–8 min. If a proportional reduction occurs in the speed of excitation and adaptation, some aspects of the molecular details of signal transduction may actually be more easily studied in the myxobacteria.

Since the *frz* genes appear to be so similar to the *che* genes of enteric bacteria we must ask the question, are they involved in chemotaxis of *M. xanthus*? Dworkin & Eide (1983) tried to detect chemotaxis in this organism but were unsuccessful. They hypothesized that chemotaxis to soluble nutrients might not be possible in *M. xanthus* because its rate of

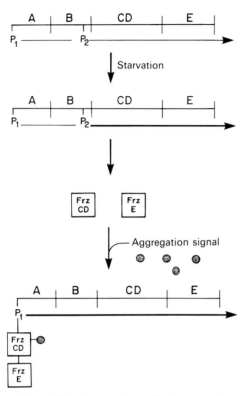

Fig. 10. Model of regulation of the *frz* gene cluster. During vegetative growth, there is weak transcription from both P_1, the promoter upstream of frzA, and P_2, the promoter upstream of frzCD. At the onset of starvation, transcription from P_2 was increased, leading to the synthesis of FrzCD and FrzE. These two proteins then cause increased transcription from P_1. In order for FrzCD to stimulate transcription, it may have to bind to some effector molecule. Transcription from P_1 may preclude transcription from P_2 (Weinberg & Zusman, 1989).

movement is slower than the rate of diffusion for many substances. However, if the chemoeffectors were not freely soluble, chemotaxis could occur in *M. xanthus*. The chemoeffectors may be bound to cell surfaces or to the material present in slime trails. We have observed biased cell movement in time lapse video micrographs that suggest chemotaxis. However, clear-cut chemoeffectors with proven taxis potential have yet to be identified. Their discovery will represent a major breakthrough in our understanding of cell–cell interactions in the myxobacteria.

ACKNOWLEDGEMENT

The research in my laboratory is supported by Public Health Service grant GM20509 from the National Institute of Health and by National Science Foundation grant DMB-8820799.

REFERENCES

Blackhart, B. D. & Zusman, D. R. (1985a). Cloning and complementation analyses of the 'frizzy' genes of *Myxococcus xanthus*. *Molecular and General Genetics*, **198**, 243–54.

Blackhart, B. D. & Zusman, D. R. (1985b). 'Frizzy' genes of *Myxococcus xanthus* are involved in the control of rate of reversal of gliding motility. *Proceedings of the National Academy of Sciences, USA*, **82**, 8767–70.

Burchard, R. P. (1974). Studies on gliding motility in *Myxococcus xanthus*. *Archives of Microbiology*, **99**, 271–80.

Burchard, R. P. (1981). Gliding motility of prokaryotes: ultrastructure, physiology, and genetics. *Annual Review of Microbiology*, **35**, 497–529.

Burchard, R. P. (1984). Gliding motility and taxes in *Myxobacteria: Development and Cell Interactions*, Rosenberg, E., ed. Springer-Verlag Inc, N.Y. pp. 139–61.

Dworkin, M. & Eide, D. (1983). *Myxococcus xanthus* does not respond chemotactically to moderate concentration gradients. *Journal of Bacteriology*, **154**, 437–42.

Dworkin, M., Keller, K. H. & Weisberg, D. (1983). Experimental observations consistent with a surface tension model of gliding motility of *Myxococcus xanthus*. *Journal of Bacteriology*, **155**, 1367–71.

Ely, B. & Shapiro, L. (1984). Regulation of cell differentiation in *Caulobacter crescentus* in *Microbial Development*, Losick, R. & Shapiro, L., eds, pp. 1–26. Cold Spring Harbor Laboratory, NY.

Fink, J. M. & Zissler, J. F. (1989). Defects in motility and development of *Myxococcus xanthus* lipopolysaccharide mutants. *Journal of Bacteriology*, **171**, 2042–8.

Gitt, M. A. Wang L. & Doi, R. H. (1985). A strong sequence homology exists between the major RNA polymerase σ factors of *Bacillus subtilis* and *Escherichia coli*. *Journal of Biological Chemistry*, **260**, 7178–85.

Glagoleva, T. N., Glagolev, A. N., Gusev, M. V. & Nikitiana, K. A. (1980). Proton motive force supports gliding in cyanobacteria. *FEBS Letters*, **117**, 49–53.

Hagen, D. C., Bretscher, A. P. & Kaiser, D. (1978). Synergisms between morphogenetic mutants of *Myxococcus xanthus*. *Developmental Biology*, **64**, 284–96.

Hodgkin, J. & Kaiser, D. (1977). Cell-to-cell stimulation of movement in nonmotile mutants of *Myxococcus*. *Proceedings of the National Academy of Sciences, USA*, **74**, 2938–42.

Hodgkin, J. & Kaiser, D. (1979a). Genetics of gliding motility in *Myxococcus xanthus* (Myxobacteriales): Genes controlling movement of single cells. *Molecular and General Genetics*, **171**, 167–76.

Hodgkin, J. & Kaiser, D. (1979b). Genetics of gliding motility in *Myxococcus xanthus* (Myxobacteriales): Two gene systems control movement. *Molecular and General Genetics*, **171**, 177–91.

Humphrey, B. A., Dickson, M. R. & Marshall, K. C. (1979). Physiochemical and *in situ* observations on the adhesion of gliding bacteria to surfaces. *Archives of Microbiology*, **120**, 231–8.

Kaiser, D. (1979). Social gliding is correlated with the presence of pili in *Myxococcus xanthus*. *Proceedings of the National Academy of Sciences, USA*, **76**, 5952–6.

Kaiser, D. (1986). Control of multicellular development: *Dictyostelium* and *Myxococcus*. *Annual Reviews of Genetics*, **20**, 539–66.

Kim, S. K. & Kaiser, D. (1990). C-Factor: A cell–cell signaling protein required for fruiting body morphogenesis of *M. xanthus*. *Cell*, **61**, 19–26.

Krikos, A., Conley, M. P., Boyd, A., Berg, H. C. & Simon, M. I. (1985). Chimeric chemosensory transducers of *Escherichia coli*. *Proceedings of the National Academy of Sciences, USA*, **82**, 1326–30.

Krikos, A., Mutoh, N., Boyd, A. & Simon, M. (1983). Sensory transduction of *E. coli* are composed of discrete structural and functional domains. *Cell*, **33**, 615–22.

Kroos, L., Hartzell, P., Stephens, K. & Kaiser, D. (1988). A link between cell movement and gene expression argues that motility is required for cell–cell signaling during fruiting body development. *Genes & Development*, **2**, 1677–85.

Kühlwein, H. & Reichenbach, H. (1968). 'Schwarmentwicklung und Morphogenese bei Myxobacterien: *Archangium-Myxococcus-Chondrococcus-Chondromyces*.' Wissenschaftlichen Film, Göttingen.

Lapidus, I. R. & Berg, H. C. (1982). Gliding motility of Cytophage sp. strain U67. *Journal of Bacteriology*, **151**, 384–98.

McBride, M. J., Weinberg, R. A. & Zusman, D. R. (1989). 'Frizzy' aggregation genes of the gliding bacterium *Myxococcus xanthus* show sequence similarities to the chemotaxis genes of enteric bacteria. *Proceedings of the National Academy of Sciences, USA*, **86**, 424–8.

McCleary, W. R., McBride, M. J. & Zusman, D. R. (1990). Developmental sensory transduction in *Myxococcus xanthus* involves methylation and demethylation of FrzCD. *Journal of Bacteriology*, **87**, 5898–902.

McCleary, W. R. & Zusman, D. R. (1990). FrzE of *Myxococcus xanthus* is homologous to both CheA and CheY of *Salmonella typhimurium*. *Proceedings of the National Academy of Sciences, USA*, in press.

Morrison, C. E. & Zusman, D. R. (1979). *Myxococcus xanthus* mutants with temperature-sensitive, stage-specific defects: Evidence for independent pathways in development. *Journal of Bacteriology*, **140**, 1036–42.

O'Connor, K. O. & Zusman, D. R. (1989). Patterns of cellular interactions during fruiting body formation in *Myxococcus xanthus*. *Journal of Bacteriology*, **171**, 6013–24.

Pate, J. (1988). Gliding motility in procaryotic cells. *Canadian Journal of Microbiology*, **34**, 459–65.

Pate, J. L. & Chang, L. E. (1979). Evidence that gliding motility in prokaryotic cells is driven by rotary assemblies in the cell envelopes. *Current Microbiology*, **2**, 59–64.

Radford, S. E., Lave, E. D., Perham, R. N., Martin, S. R. & Appella, E. (1989). Conformational flexibility and folding of synthetic peptides representing an interdomain segment of polypeptide chain in the pyruvate dehydrogenase multienzyme complex of *Escherichia coli*. *Journal of Biological Chemistry*, **264**, 767–75.

Ridgway, H. F., Wagner, R. M., Dawsey, W. T. & Lewin, R. A. (1975). Fine structure of the cell envelope layers of *Flexibacter polymorphus*. *Canadian Journal of Microbiology*, **21**, 1733–50.

Russo, A. F. & Koshland, D. E. Jr (1983). Separation of signal transduction and adaptation functions of the aspartate receptor in bacterial sensing. *Science*, **220**, 1016–20.

Shimkets, L. J. (1986). Role of cell cohesion in *Myxococcus xanthus* fruiting body formation. *Journal of Bacteriology*, **166**, 842–8.

Shimkets, L. J. (1987). Control of morphogenesis in myxobacteria. *CRC Critical Reviews in Microbiology*, **14**, 195–227.

Shimkets, L. J. & Kaiser, D. (1982). Induction of coordinated movements of *Myxococcus xanthus* cells. *Journal of Bacteriology*, **152**, 451–61.

Stanier, R. Y. (1942). A note on elasticotaxis in myxobacteria. *Journal of Bacteriology*, **44**, 405–12.
Stephens, K., Hartzell, P. & Kaiser, D. (1989). Gliding motility in *Myxococcus xanthus*: *mgl* locus, RNA, and predicted protein products. *Journal of Bacteriology*, **171**, 819–30.
Stephens, K. & Kaiser, D. (1987). Genetics of gliding motility in *Myxococcus xanthus*: molecular cloning of the *mgl* locus. *Molecular and General Genetics*, **207**, 256–66.
Stock, J. & Koshland, D. E. Jr (1984). Sensory adaptation mechanisms in swarm development. In *Microbial Development*, Losick, R. & Shapiro, L., eds., pp. 117–31. Cold Spring Harbor Laboratory, N.Y.
Stock, J. B., Ninfa, A. J. & Stock, A. M. (1989). Protein phosphorylation and regulation of adaptive responses in bacteria. *Microbiological Reviews*, **53**, 450–90.
Weinberg, R. A. & Zusman, D. R. (1989). Evidence that the *Myxococcus xanthus frz* genes are developmentally regulated. *Journal of Bacteriology*, **171**, 6174–86.
Zusman, D. R. (1982). Frizzy mutants: A new class of aggregation-defective developmental mutants of *Myxococcus xanthus*. *Journal of Bacteriology*, **150**, 1430–7.
Zusman, D. R. (1984). Cell–cell interactions and development in *Myxococcus xanthus*. *Quarterly Reviews of Biology*, **59**, 119–38.

SIGNAL TRANSDUCTION IN
HALOBACTERIUM HALOBIUM

D. OESTERHELT and W. MARWAN

Max-Planck-Institut für Biochemie, 8033 Martinsried, West Germany

INTRODUCTION

Considerable progress has been made in recent years in understanding the function of molecular components in eukaryotic and eubacterial signal transduction chains. Among well-known examples are G-proteins in eukaryotes, and methyl-accepting proteins in eubacteria (see the chapters by Segall and by Hazelbauer in this book). In contrast, almost nothing is known about signal transduction in archaebacteria, which constitute the third branch of the living world. The light-regulated swimming behaviour of *Halobacterium halobium* is the only example of an archaebacterial signal chain that is currently studied at the cellular and molecular level, and its biochemistry therefore deserves general interest for its relationship to eubacterial and eukaryotic signalling systems.

Despite evolutionary divergence, the halobacterial signal chain includes the same basic principles as are known for other sensory systems. These are: reception of the stimulus, amplification of the signal input, integration of signals caused by different types of sensory stimuli, and adaptation to the stimulus background. Since halobacteria integrate light and chemical stimuli for a common control of the flagellar motor switch, they provide a unique system for studying a branched signal chain and, if mutants defective in BR and HR are used, there will be no interference from the photosynthetic processes. A previous disadvantage was the lack of a gene transfer system but this has now been overcome by the development of a transformation system (Cline *et al.*, 1989). As we will show in this review, *Halobacterium* is currently the best understood system in prokaryotic photobiology with respect to identification of molecular components involved, and analysis of their action in the cell.

Halobacteria are facultatively anaerobic organisms. In the presence of oxygen, the cells grow using respiration but, under anaerobic conditions, they either ferment arginine which is metabolized via a reversed urea cycle (Hartmann *et al.*, 1980) or grow photosynthetically (Oesterhelt & Krippahl, 1983). The capacity for anaerobic growth is important since the oxygen tension in the natural habitat of brines and salt ponds can become very low under strong sunlight.

The entire photobiochemistry of halobacteria is based upon the action of retinal proteins. Two light energy converters, bacteriorhodopsin (BR) and halorhodopsin (HR) power the energy-driven processes of the cell and two light sensors, sensory rhodopsin-I and -II (SR-I, SR-II) mediate colour vision that aids the cell in finding the optimal photosynthetic microenvironment. Bacteriorhodopsin is a light-driven proton pump that supplies the ATP-synthase with a proton motive force and thus allows photophosphorylation. Halorhodopsin is a light-driven chloride pump which, in cooperation with a potassium uniport, mediates net salt uptake by the cells during growth.

The primary photochemical process in all four proteins is the isomerization of all-*trans* retinal to its 13-*cis* state which is followed by a conformational change of the protein. The light-independent re-isomerization to the all-*trans* state is catalysed by the protein and thermodynamically driven by part of the photon's energy which has been stored in the protein conformation. The major differences between the pumps and the sensors are a longer life-time of the intermediate 13-*cis* state of the sensors and the connection of ion release and binding to *cis-trans* isomerization in the pumps. In other words, retinal acts as a light-triggered switch in all four proteins and causes the translocation of an ion, in the case of bacteriorhodopsin and halorhodopsin, or the formation of a signalling state in the sensory rhodopsins. Several reviews in the physiology, biochemistry and biophysics of the four retinal proteins have been published (Oesterhelt & Tittor, 1989; Lanyi, 1984; Spudich & Bogomolni, 1988).

Halobacterium halobium cells are propelled by a monopolarly inserted flagellar bundle (Fig. 1). In the stationary growth phase, bipolarly flagellated cells also occur, but generally one of the flagellar bundles is much shorter than the other and thus contributes less to the overall movement of the cell (Alam & Oesterhelt, 1984). Each flagellar bundle consists of five to ten flagellar filaments (Houwink, 1956) that form a right-handed semi-rigid helix. Bacterial flagella are too thin to be visible in the light microscope under standard conditions. Fortunately, they can be visualized by high-intensity darkfield microscopy (Macnab, 1976; Reichert, 1909) so that their motion can be directly observed *in vivo*. Alam & Oesterhelt (1984) used this method to study the flagellation and swimming mechanism of *H. halobium*. Their experiments revealed that CW rotation of the right-handed flagellar bundles exerts a pushing force to the cell and propels it in the forward direction. In the CCW rotational mode, on the other hand, the bundle exerts a pulling force, and the cell swims with the flagellated pole at the front (Fig. 1(*c*)). Thus, in both rotational modes, a translational movement results. This is quite different to the motion of an *Escherichia coli* cell which only swims when the flagella are rotating CCW whereas CW rotation produces tumbling (Larsen *et al.*, 1974; Macnab, 1977).

In the absence of stimuli, the halobacterial cells spontaneously switch the rotational sense of their flagellar bundle. When the direction of rotation

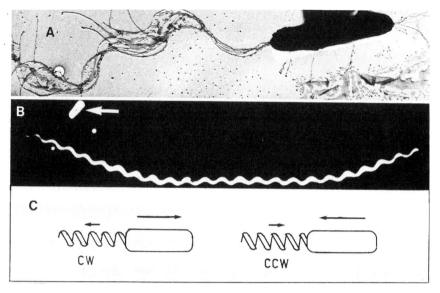

Fig. 1. Halobacteria swim by polarly inserted flagellar bundles. (*a*) Electron micrograph of a bipolarly flagellated cell. (*b*) Superflagella, formed by spontaneous aggregation of flagellar filaments lost from cells. The arrow points to a halobacterial cell. (*c*) Modes of rotation of the flagellar bundle in a monopolarly flagellated cell. The swimming direction resulting from either rotational sense is indicated by long arrows. (Adapted from Alam & Oesterhelt, 1984.)

changes, the bundle does not fly apart as is the case in *E. coli* but remains associated so that the cell continues swimming but in reversed direction (Alam & Oesterhelt, 1984). After a switching event, however, the cell does not necessarily take the original path. The resulting random walk keeps the cells evenly distributed in their suspending medium. The random walk may become biased because the flagellar motor switch is under the control of a signal chain that integrates the sensory input from the various receptors.

The experimental approaches have two aims. One is the identification of molecules involved in signalling, the other is to learn how these molecules interact within the cell. The first part of this contribution summarizes results on the molecular components of the signal chain that have so far been identified. The second part deals with kinetic studies on cellular behaviour that describe how the signalling system may function as a whole.

THE MOLECULAR COMPONENTS OF THE SIGNAL CHAIN

Several components of the halobacterial signal chain have been identified and partially characterized. These are the photoreceptors that receive the light stimuli, some molecules involved in signal transduction linking the receptors to the final target, the flagellar motor switch, and the flagellar proteins which mechanically transduce motor rotation into cell propulsion.

Flagellins

Halobacterial flagellar bundles consist of five to ten individual flagellar filaments (Houwink, 1956). Flagellar filaments lost from the cells spontaneously aggregate into thick superflagella which can grow to sizes of about 10 to 20 times that of an individual cell (Fig. 1(b)). These superflagella can be easily isolated from overproducing cultures by differential centrifugation (Alam & Oesterhelt, 1984). Polymorphic transitions from normal to a curly, a ring, and a straight form are induced by different pH values and heat treatments (Alam & Oesterhelt, 1987). They contain three protein species, Fla I (23.5 kDa), Fla II (26.5 kDa) and Fla III (31.5 kDa) (Alam & Oesterhelt, 1984) which were characterized as sulphated glycoproteins with N-glycosidically linked oligosaccharides (Wieland et al., 1985; Sumper, 1987). The glycoprotein nature of the flagellins is apparently instrumental in the flagellar function because a mutant strain producing superflagella, as a result of the loss of the filaments from the cell, has a changed glycosylation pattern. Halobacterial glycoproteins, including the flagellins, are glycosylated at the extracellular surface of the cell membrane (Sumper, 1987), therefore the flagellin polypeptides must by translocated across the cell membrane before glycosylation (Gerl & Sumper, 1988). This implies an export mechanism different from that of eubacteria.

Gerl & Sumper (1988) isolated a gene fragment of one of the flagellins in an expression vector using immunological probes. By use of this fragment as a probe, five highly homologous flagellin genes were identified and sequenced. The genes are arranged tandemly in a group of two and in a group of three at two different loci on the chromosome (Gerl & Sumper, 1988).

The same basic mechanism for motility, including superflagella formation, has also been described for peritrichously flagellated halophilic square bacteria (Alam et al., 1984).

Sensory rhodopsins

Two sensory rhodopsins, SR-I and SR-II, were detected spectroscopically in the cell membrane of strains defective in BR and HR (Bogomolni & Spudich, 1982; Spudich & Spudich, 1982; Wolff et al., 1986; Scherrer et al., 1987). SR-II is constitutively present in a copy number of about 400 per cell. SR-I is induced during growth of the culture, presumably by a drop in oxygen tension, and then reaches a level of expression at about 4000 copies per cell (Otomo et al., 1989).

Upon photoactivation, both sensory rhodopsins undergo a photocycle

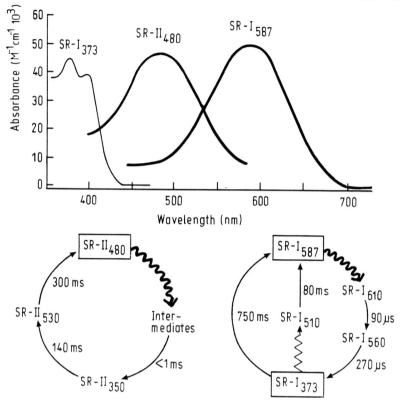

Fig. 2. The halobacterial photoreceptors. *Top:* The absorption spectra of the physiologically active photoreceptor species are shown. SR-I_{373} and SR-II_{480} act as repellent receptors, SR-I_{587} acts as attractant receptor. Note that the relative amount of SR-I increases during growth of a culture (Otomo et al., 1989). *Bottom:* The photocycles of SR-I and SR-II are shown. Intermediates that exhibit photoreceptor function are boxed. The subscripts of the photocycle intermediates indicate their absorption maximum. The wavy arrows indicate photochemical reactions, the other arrows indicate thermal reactions with the half-life times. (The figure was redrawn from Spudich & Bogomolni, 1988.)

and thermally return to the ground state in the subsecond range (Fig. 2). SR-I is a photochromic pigment: If orange light is absorbed by the ground state, a metastable intermediate (SR-I_{373}) is formed which returns to the ground state either thermally or photochemically by absorption of a near UV photon through an alternative path (Spudich & Spudich, 1982; Bogomolni & Spudich, 1987).

SR-I mediates the attractant response of the cells to orange light though its ground state SR-I_{587} and the repellent response to near UV light through its metastable SR-I_{373} intermediate (Spudich & Bogomolni, 1984; Takahashi et al., 1985a; Tomioka et al., 1986; Marwan & Oesterhelt, 1990). SR-II, formerly called P_{480} or phoborhodopsin is a blue light receptor that induces a photophobic response of the cell similar to that caused by SR-I_{373}

(Takahashi et al., 1985b; Wolff et al., 1986; Marwan & Oesterhelt, 1987). More details about the physiological functions of these receptors are presented below.

SR-I has been isolated from halobacterial membranes after solubilization in lauryl-maltoside. In the presence of retinal, detergent and salt, the native protein was obtained in pure form by sucrose density gradient centrifugation, hydroxyapatite chromatography and gel filtration (Schegk & Oesterhelt, 1988). The apparent molecular weight of the molecule is 24 kDa as analysed by SDS gel electrophoresis and 49 kDa by sedimentation and size-exclusion chromatography, suggesting that the native protein occurs as a dimer. The purified SR-I has an absorption maximum of 580 nm which is 7 nm blue-shifted compared to the membrane-bound state. The action spectrum for the formation of $SR-I_{373}$ coincided with the absorption spectrum.

The primary structure of SR-I has been determined by protein sequencing and by cloning and sequencing of the SR-I gene (Blanck et al., 1989). The protein consists of 239 amino acids that are arranged in seven transmembrane helices as suggested by hydropathy plots. A sequence identity of 14% with BR and HR was found but no significant homology with bovine rhodopsin. A striking feature of the secondary structure in SR-I is the lack of the large cytoplasmic domains found in animal rhodopsins, suggesting that the signalling state might be transduced within the membrane (see below). The retinal is bound to the lysine residue 205 in helix G. In contrast to BR and HR, the photocycle in sensory rhodopsins does not involve electrogenic events (Bogomolni & Spudich, 1982; Spudich & Spudich, 1982; Ehrlich et al., 1984; Spudich & Bogomolni, 1988).

Methyl-accepting taxis proteins

It has been shown by various groups that methyl accepting proteins occur in *H. halobium* (Schimz, 1981, 1982; Bibikov et al., 1982; Spudich et al., 1988; Alam et al., 1989). Proteins that receive their methyl groups from S-adenosylmethionine can be radiolabelled specifically *in vivo* by supplying methyl-^{3}H methionine to cells in which protein synthesis is blocked by, e.g. puromycin. Analysis by electrophoresis and fluorography revealed methyl-^{3}H labelled protein species between 90 and 135 kDa (Alam et al., 1989) and 65 to 150 kDa (Spudich et al., 1988). The label that was released from the protein by mild alkali treatment was shown to be a volatile compound as expected for a carboxymethylester. Chemostimuli caused distinct changes on specific bands whereas no changes could be detected upon photostimulation. The release of labelled volatile methyl groups in intact labelled cells could, however, be measured upon photostimulation in a flow apparatus where cells on a filter were washed continuously with buffer containing an excess of non-radioactive methionine (Kehry et al., 1984).

Chemo- and photostimulation of halobacteria both caused a transient increase in the net rate of demethylation (Alam et al., 1989). This increased rate is always observed independently of whether the stimulus is an attractant or a repellent. In *Escherichia coli*, responses to positive and negative stimuli consist of transient decreases and increases in methylesterase activity respectively (Kehry et al., 1984). The same demethylation phenomenon as in *Halobacterium* has also been seen in *Bacillus subtilis* (Thoelke et al., 1987) and is explainable by simple kinetic assumptions.

Several arguments support the view that methyl-accepting proteins are involved in the sensory transduction machinery of *Halobacterium* (Spudich et al., 1988; Alam et al., 1989).

1. Demethylation is transiently increased by photo- and chemostimuli that are known to control behaviour.
2. Mutant strains (Pho 71, Pho 72 and M 402; Sundberg et al., 1985; Spudich et al., 1988; Alam et al., 1989) which show neither photo- nor chemoresponses lack almost all the methyl-labelled bands in the 70 to 135 kDa region.
3. In cells of strain Pho 81 which is deleted in SR-I and SR-II (Sundberg et al., 1985) but performs normal chemotaxis, no light-induced release of volatile methyl groups can be observed whereas the demethylation response to chemostimuli is unaltered (Alam et al., 1989).

One of the methyl-accepting proteins seems to be specifically involved in phototransduction by SR-I. This 94 kDa protein was shown to be present in increased amounts in an overproducer of SR-I but absent in a strain that contained only SR-II and lacked SR-I. It is also absent in a mutant lacking both photoreceptors (Spudich et al., 1988).

Currently, it is unclear whether the methylation system is involved in adaptation as is the case in enteric bacteria (Stewart & Dahlquist, 1987) or alternatively in the excitation system. Treatment of cells with ethionine which should lower intercellular pools of the methyl donor S-adenosyl methionine causes increased response times to both 360 nm step-up (repellent) and 360 nm step-down (attractant) stimuli. The repellent response, therefore, is weakened and the attractant response is enhanced by ethionine although the frequency of spontaneous motor switching remains constant (Schimz & Hildebrand, 1987). These results suggest that ethionine suppresses stimulus-induced motor switching, in general, but it does not necessarily argue for impaired adaptation as the repellent response should have been enhanced. The observation that stimulus-induced demethylation persists over periods roughly equivalent to adaptation times is currently the only argument for assigning methylation to adaptation.

Two antagonistic stimuli, i.e. attractant and repellent, delivered at the same time, cancel demethylation, therefore a feedback loop from the integrated signal to the demethylation system can be postulated (Spudich et al., 1989). Further experimental work will be necessary to clarify the role of methylation in halobacterial signalling.

Fumaric acid

The smooth swimming mutant strain M415 is defective in spontaneous and stimulus-induced flagellar motor switching. Cells of this strain can be cured after permeabilization by mild ultrasonication in the presence of a cell extract from wild-type cells. This somatic complementation was used to identify the active compound, called the switch factor, and this was purified from extracts of wild-type cells (Marwan et al., 1990). Chemical group analysis suggested that the switch factor was fumarate. Indeed, fumaric acid, when used in the complementation assay, replaces the switch factor and enables the cell to respond to blue light by motor switching. Fumaric acid is active at the level of only a few molecules per cell. Quantitative mass spectrometric analysis with deuterated fumaric acid as internal standard showed that the amount of fumarate present in 'switch factor' preparations correlated well with the minimal content of active compound as determined by the complementation assay. It therefore seems probable that a major part of the activity is due to fumaric acid. It cannot, however, be excluded that a derivative of fumaric acid is also involved. The switch factor is bound to the membrane fraction of the cell, and it is released upon heat treatment, presumably by denaturation of a membrane-associated protein which binds the factor *in vivo*.

When the switch factor or fumarate is applied to cells in high concentrations, not only stimulated responses but also spontaneous stopping responses of the flagellar motor are observed. This suggests that the factor acts in a part of the signal chain located beyond the point of stimulus integration.

When a suspension of wild-type cells is irradiated with blue light (480 nm) and the cells rapidly lysed with an appropriate amount of distilled water, the concentration of 'free' switch factor in the lysate increases as assayed with the complementation test when compared to untreated cells. These experiments revealed that the switch factor is released from the membrane by light and suggests that sensory rhodopsin-II controls its cytoplasmic concentration (Marwan & Oesterhelt, unpublished results).

Motor switching is also induced by a decrease in orange light intensity (step-down), a response mediated by SR-I (see below). Interestingly, a step-down in orange light intensity also causes a release of the switch factor (Marwan & Oesterhelt, unpublished results). This demonstrates that its concentration is controlled by SR-I as well as SR-II, and it further supports the idea that the switch factor carries the integrated information from SR-I, SR-II and other receptors. A plausible working hypothesis is that the cytoplasmic concentration of the switch factor controls the flagellar motor switch either directly or mediated by other components of the signal chain. Since its concentration is controlled by various external signals, the switch factor can be regarded as a second messenger.

G-proteins and cyclic GMP

In crude preparations of *H. halobium* membranes, addition of cGMP causes a slow acidification of the suspension indicating that it is hydrolysed to GMP (Schimz et al., 1989). Orange light (>550nm) partially inhibits hydrolysis in membranes of cells that contain only SR-I and SR-II but are deleted in BR and HR. Since the hydrolysis of cGMP is accelerated by GTP, GTP-γ-S and fluoroaluminate, components that are activators of G-proteins, it was suggested that a phosphodiesterase is controlled by a G-protein. Indeed, a protein with a molecular mass of 59 kDa has been detected on immunoblots with antisera produced against a conserved peptide sequence of the GTP binding site in most G-proteins (α-subunit). However, no detectable cross-reaction with an antiserum against purified transducin or conserved peptides of the β subunit was observed (Schimz et al., 1989).

The hypothesis that a G-protein-controlled phosphodiesterase might be involved in sensory transduction in *Halobacterium* emerged from behavioural studies with inhibitors which are known to alter cGMP concentration in eukaryotes. Dibutyryl-cGMP and IBMX (3'-isobutyl-1-methylxanthine), both inhibitors of the phosphodiesterase, lengthened spontaneous and attractant intervals in cells by a factor of two. AlF_4^-, on the other hand, which activates phosphodiesterase via a G-protein and hence reduces the cGMP level in eukaryotic cells, shortened the spontaneous and attractant interval in *Halobacterium* (Schimz & Hildebrand, 1987). From these studies, it seems that low cGMP levels might activate, and high levels might inhibit the flagellar motor switch. This is consistent with the observation mentioned above that attractant light inhibits the hydrolysis of cGMP in crude cell extracts.

BEHAVIOURAL KINETICS

Measuring bacterial behavior

To explore the intracellular function of the signal chain, it is necessary to carry out behavioural measurements on intact cells. For this purpose, the use of a computerized system for recording and evaluating swimming tracks of halobacteria (Fig. 3) is a tremendous advantage. Computerized cell tracking makes spontaneous and light-induced behaviour an objectively measurable physical entity (Sundberg et al., 1986; Berg et al., 1987; Sager et al., 1988; Marwan & Oesterhelt, 1990). Since many cells can be tracked simultaneously, this method allows for a fast and easy acquisition of a

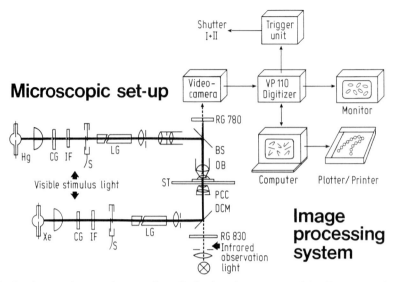

Fig. 3. Microscopic set-up for recording of behavioural responses to light. Hg, mercury lamp; CG, coloured glass; IF, interference filter; S, shutter; LG, light guide; BS, beam splitter; OB, objective; ST, stage; PCC, phase contrast condenser; DCM, dichroitic mirror; RG830, RG780, far-red cut-off filters; Xe, xenon lamp.

huge body of behavioural data. As pointed out in the following paragraphs, these data can be quantitatively analysed in terms of chemical kinetics. This information not only gives access to an analysis of the signal chain *in toto* but also allows the assignment of biochemical results on isolated molecules to physiological events. Because computerized cell tracking is a relatively new and fast-evolving technique, we will briefly describe our set-up and the algorithm used (Fig. 3).

The bacteria are suspended in normal growth medium with arginine added as a fermentative energy source. A specimen is prepared by putting a drop of the cell suspension onto a slide which is then covered by a cover slip. To avoid any streaming of the cells caused by evaporation of the suspension, the slip is sealed with a molten mixture of paraffin and vaseline. The specimen is placed on a microscope stage and the cells are observed by infrared observation light which is detected by a video camera. This allows for the observation of unstimulated cells. The light stimulus is produced by an arc lamp, filtered through coloured glass and an interference filter, and projected via a light guide through the objective or, alternatively, through the phase contrast condenser, onto the cells. The stimulating light is controlled by electronic shutters that are connected to the image processing system and allow for delivery of defined light stimuli. The core of the image processing system that records behavioural responses, is a commercially available digitizer which sends the digitized frames to a monitor and to a computer for further analysis. In addition, the digitizer controls a trigger unit which opens and closes the shutters. Taking the video data, the computer calculates the centre of each cell in each frame and then reconstructs the swimming paths of the cells by an appropriate combination of the centroids in successive frames. This yields the two-dimensional spatial coordinates as a function of time for each cell. Finally, these sets of coordinates are analysed in order to check whether or not a cell has switched its flagellar motor. This is managed by a PASCAL program which was specifically designed for the analysis of halobacterial motility (Marwan & Oesterhelt, 1990). The program compares the position of the cell at any given time with the position of the cell a given number of frames before and afterwards. A reversal (motorswitching event) is detected by the system when the angle which is included by the three positions is smaller than a certain threshold. The algorithm allows for easy discrimination between curved swimming and motor switching and yields a high signal to noise ratio. To calibrate the system for systematic errors introduced by the threshold and, e.g. path crossings or other artefacts, a behavioural mutant that is defective in flagellar motor switching was used as a control (see under fumaric acid).

Spontaneous motor switching

Halobacteria periodically spontaneously switch the direction of rotation of their flagellar motor, even in the absence of stimuli. The frequency distribution of intervals between spontaneous motor switching events is similar for switching from CW to CCW and from CCW to CW (Hildebrand & Schimz, 1985; Marwan et al., 1987). The frequency of spontaneous reversals depends highly on temperature (Marwan & Oesterhelt, 1987). The time elapsing between two successive switching events in a given cell at e.g. 20°C ranges from few seconds to several minutes. The length of a given interval is not correlated to the length of the preceding one (McCain et al., 1987). The interval distribution is fitted by a symmetrical four-state model or an asymmetrical three-state model of the flagellar motor switch (Marwan & Oesterhelt, 1987; McCain et al., 1987). Both models describe motor switching as the consequence of stochastic transitions within a sequence of kinetic states of the switch, which behaves as a single molecule. This is quite analogous to the models established to describe the opening and closing of ion channels. These receptors also proceed through a sequence of defined states, thermodynamically driven by ligand-binding. A four-state model has also recently been described for the flagellar motor switch in *Salmonella typhimurium* (Kuo & Koshland, 1989).

Schimz & Hildebrand (1985) have proposed an alternative model where spontaneous switching is induced by level-crossings of a regulatory substance which oscillates in its concentration. Light stimuli are supposed

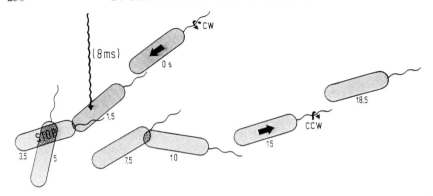

Fig. 4. Photophobic response of *Halobacterium* to a blue light flash. The cell swims forward by CW rotation of its flagellar bundle. After being hit by a 8 ms blue light flash, the cell continues to swim for about two seconds, then switches rotation of its flagellar motor and, as consequence, reverses its swimming direction. The sequence was redrawn from video recordings. Numbers indicate time in seconds, the thick arrows indicate the swimming direction of the cell.

to cause behavioural responses by shifting the phase of the oscillation (Schimz & Hildebrand, 1985). The concept of this oscillator model implies that the cellular concentration of the regulatory substance is sufficiently high so that its changes in concentration are due to the oscillating system and not due to a stochastic fluctuation in its copy number. The transition from periodic to chaotic behaviour in response to exponential light stimulus ramps is regarded as a support for the hypothesis that the behaviour of the cells is controlled by an oscillator (Schimz & Hildebrand, 1989). However, the same result would be obtained if, for example, none of the components of the signal chain themselves but a metabolic intermediate, which is somehow linked to a component of the signal chain, would oscillate in its concentration.

Photosensory and chemosensory signals are processed by integration

In the absence of stimuli, the spontaneously switching in direction of rotation of the flagellar motors results in a random walk. Attractant stimuli suppress spontaneous motor switching and thus lengthen swimming intervals. Repellent stimuli, on the other hand, activate the motor switch leading to a stimulus-induced reversal which is also called a phobic response (Fig. 4). Repellents and attractants (light and chemicals) can act either as an attractant stimulus or as a repellent stimulus, depending on whether they are applied or removed. Application of a repellent to a cell acts as repellent stimulus. If the repellent stimulus is continuously present, the cell adapts, i.e. returns to its pre-stimulus behaviour. Removal of the repellent, after the cell has adapted, is then an attractant stimulus. The symmetrical situation is true for attractant stimuli.

Blue, near and far UV light act as repellents (with maxima in the action spectrum at 280, 373 and 480 nm; Hildebrand & Dencher, 1975; Wolff et al., 1986; Marwan & Oesterhelt, 1987; Wagner et al., 1989). Green to orange light is an attractant (maximal effectiveness at 565 and 590 nm; Hildebrand & Dencher, 1975; Sperling & Schimz, 1980; Traulich et al., 1983). The response to orange and near UV light is mediated by SR-I, the response to blue light by SR-II (Takahashi et al., 1985b; Wolff et al., 1986; Marwan & Oesterhelt, 1987; Otomo et al., 1989).

Attractant and repellent stimuli are integrated by the cell if applied simultaneously, independently of their nature (Spudich & Stoeckenius, 1979; Hildebrand & Schimz, 1986). If, for example, blue step-up (repellent stimulus) and green step-up (attractant stimulus) are simultaneously applied, the repellent action of the blue light is cancelled, provided the appropriate intensity ratio of both light colours was chosen. The same is true for a symmetrically inversed stimulus program, i.e. blue step-down (attractant) and green step-down (repellent). If, on the other hand, two attractant stimuli are applied, e.g. blue step-down and green step-up, the attractant effect is enhanced compared to the effect produced if only one of the stimuli is applied. Integration also occurs between chemical stimuli and light (Spudich & Stoeckenius, 1979).

The chemical basis for integration is completely unknown. It is also unknown whether the transduction paths from the different receptors merge at one or more points into a common signal chain that finally tunes the flagellar motor switch.

Motor switching is induced by a diffusible signal

The possibility that light-induced changes in the membrane potential might act as a trigger for motor switching has been discussed in the literature. Although it has been shown that SR-I action is non-electrogenic in both vesicles and intact cells (Ehrlich et al., 1984), a membrane potential change might still be involved in signal transduction. Direct experimental evidence from impaling a halobacterial cell with microelectrodes is unlikely to be successful. A feasible experimental strategy, however, is the application of local light stimuli to individual halobacterial cells while direct observing the behaviour of the flagellar movement. Standard halobacterial cells are only about $3\,\mu m$ in length but the addition of aphidicolin, an inhibitor of the eukaryotic DNA polymerase to the nutrient broth, allows non-septate cells of up to $40\,\mu m$ to be obtained (Forterre et al., 1984). These cells are either monopolarly or bipolarly flagellated and have been used for microbeam irradiation with a light spot of about $2.3\,\mu m$ in diameter (Oesterhelt & Marwan, 1987). Mono- and bipolarly flagellated cells responded only when the flagella-carrying pole was irradiated. None of the flagellar motors switched when the middle part of the cell was stimulated but both ends of a bipolarly flagellated cell were shown to be equally sensitive to blue light. The response time, i.e. the time elapsing after stimulus application until the flagellar motor switches is independent of the direction of rotation during stimulation. These experiments on cells from 8 to 22 μm, have shown that stimulation of one flagellated pole does not lead to a response of the second flagellar bundle beyond a distance of 15 μm. Furthermore, a flagellar motor did not switch when the stimulus was applied at a distance of more than 5 μm from the motor. This behaviour was independent of the rotational mode of the flagellar bundle at the stimulated pole and independent of the movement of the cell. This behaviour is quite different from that of the bipolarly flagellated *Rhodospirillum rubrum* where only the flagellar bundle ahead of the moving bacterial cell responds to a stimulus, followed almost immediately by the response of the second flagellar bundle, which does not itself respond when stimulated (Buder, 1917).

The short-range signalling in halobacteria excludes membrane potential changes from being directly involved in phototransduction by both SR-I and SR-II, since a potential electric signal would, according to the cable equation, decay only over a distance of 420 μm to 1/e of its original value (Oesterhelt & Marwan, 1987). These experiments suggested that a diffusible chemical messenger exists which causes the flagellar motor to switch. This switch factor has been shown to be fumaric acid (see above).

From the microbeam irradiation experiments, it cannot be concluded that the photoreceptors are localized at the poles of the cell. Experiments to map the SR-I distribution over the membrane by fluorescent antibodies are now in progress.

The signal is formed in a photocatalytic way

The time span between successive spontaneous reversals is on average about 10–30 s depending on strain and temperature. Saturating flashes of blue light induce motor switching events within 1–2 s. Since the motor responds after the light has been switched off, the signal must be formed in the dark (Marwan & Oesterhelt, 1987). The use of light flashes instead of continuous (step-up) stimuli thus allows the separation of light perception from the processes of signal formation.

The time between the flash and the stop response of the cell is called the response time t_R. It is a function of photon exposure (irradiance integrated over the flash time) and increases from a minimal value under light saturation to about 2 to 5 seconds (t_{max}) as the response time to very weak flashes. On lowering the flash intensity further, not every flash induces a response and the probability for a flash to cause a response depends on flash intensity (see below).

The response of halobacterial cells to blue light stimuli consisting of single and double flashes with varied time duration (τ_1, τ_2), intensity (I) and dark interval (D), can be quantitatively described by the equation

$$t_R = t_{min} + \frac{b}{I(\tau_1 + \tau_2)} + \frac{\tau_2}{(\tau_1 + \tau_2)} D.$$

This equation also describes the response to single flashes when either D or τ_2 are set to zero:

$$t_R = t_{min} + \frac{b}{I\tau}$$

The empirical constant b depends on the photoreceptor concentration and t_{min} is the response time at light saturation. According to these equations, the following observations were made:

1. If a flash of a given duration, intensity and constant wavelength is split into two flashes of half the original duration, the response time, t_R measured depends on the dark period (D) between the two flashes.
2. If two flashes of unequal time of duration, but with a constant dark period between them, were applied, a longer t_R is observed when the short flash is applied first, and a shorter t_R value when the long flash is applied first.

This behaviour demonstrated that the signal production depends on time and that the time t_R at which a critical signal concentration is reached depends on the length of time between the excitation of the receptors and the response.

Attraction by SR-I is governed by the life-time of the SR-I_{373} intermediate rather than by photocycling

McCain et al. (1987) reconstituted sensory opsin-I in retinal-deficient cells with pentaenal, a retinal analogue that lacks a part of the β-ionone ring. This substitution slows down the photocycle of SR-I by decreasing the thermal decay rate of SR-I_{373}. The attractant response of these cells to orange light was drastically increased. Therefore, attractant signalling is mediated by a signalling state of SR-I and not by the frequency of photocycling.

Poisson stimulus-response curves

An effective tool for measuring the functional photoreceptor number in a cell is the evaluation of Poisson stimulus response curves (Hecht et al., 1942; Hegemann & Marwan, 1988; Marwan et al., 1988). In Poisson stimulus-response curves, the probability of an organism responding to a flash stimulus of given intensity is plotted against the logarithm of photon exposure (i.e. intensity integrated over the flash time). In such plots, a number n determines the shape (steepness) of the sigmoidal curve obtained. The value of n is the minimal number of photons which have to be absorbed by the photoreceptors of a cell to cause its response. The steeper the curve, the more photons (increasing n) are required to produce the response. On the other hand, the number of active receptor molecules present in the cell determines the position of the curve with respect to the logarithmically scaled abscissa. The equations relevant to this type of analysis can be found in Marwan *et al.* (1988). If the gain of the signal chain is decreased, at constant photoreceptor concentration, the response curve becomes steeper (increase in n; Otomo *et al.*, 1989). If, on the other hand, the photoreceptor concentration in the cell varies and the gain remains constant, a parallel shift of the curve would be found whereas the slope would remain constant. The validity of this method was confirmed by a combined application of spectroscopic and behavioural measurements to growing halobacterial cell populations. It was shown that the time course of receptor concentration and gain of the signal chain during growth of the culture developed as predicted by Poisson stimulus-response plots (Otomo *et al.*, 1989). Using these plots, it was shown that, at minimum, one activated SR-II molecule is able to cause a response by the cell (Marwan *et al.*, 1988).

Kinetics of near UV reception by sensory rhodopsin-I

It has been proposed that $SR-I_{373}$ acts as a near UV receptor (Bogomolni & Spudich, 1982). The assignment of this intermediate to its physiological role as UV receptor was essentially based on the observation that the response to near UV is increased by orange background light and that, under blue background light, irradiation with orange light is a repellent rather than an attractant stimulus (Spudich & Bogomolni, 1984; Takahashi *et al.*, 1985*a*). Both of these observations could be due to two principally different mechanisms: 1. to a photochromic receptor where the near UV receptor species is produced by orange light as is the case in SR-I, or 2. to a second orange light absorbing photoreceptor which tunes the gain of the signal chain. This alternative possibility deserves attention since two different photoreceptor systems are postulated to mediate near UV and orange light reception respectively (Schimz & Hildebrand, 1988). Furthermore, the 94 kDa protein was shown to bind retinal (Spudich *et al.*, 1988) and hence could be regarded as a candidate for an additional photoreceptor species.

There are three arguments in addition to those provided by Spudich & Bogomolni (1984) supporting the identity of the near UV receptor and $SR-I_{373}$ (Marwan & Oesterhelt, 1990):

1. Orange light acts by increasing the number of functional near UV receptors rather than by tuning the gain of the signal chain. This was concluded from the finding that constant orange background light produces a parallel shift of Poisson stimulus–response curves whereas the slope remains unchanged.
2. The steady state concentration of the near UV receptor at different orange background light irradiance fits the calculated intensity dependence of the SR-I_{373} intermediate quantitatively. It can be calculated from the simplified photocycle of SR-I:

$$SR_{587} \underset{k}{\overset{\varphi}{\rightleftarrows}} SR_{373}$$

where only the rate-limiting steps are considered:

$$[SR_{373}] = \frac{I\sigma\varphi[SR]}{k + I\sigma\varphi}$$

In this equation, I is the intensity of the orange background light, φ is the quantum yield, σ is the absorption cross-section of the ground state SR_{587} and [SR] is the total amount of SR-I.
3. The half-life time for the thermal decay of the near UV receptor was determined from the rate constant k as 4.4 ± 0.8 s at 25 °C which is compatible to that found spectroscopically for the SR_{373} intermediate in de-energized intact cells (Manor et al., 1988).

CONCLUSIONS

Signal transduction in halobacteria is the first example of an archaebacterial sensory system to be studied in detail. Its analysis at a physiological, biophysical and molecular level has produced knowledge which is partly unique, e.g. the first characterization of a prokaryotic photoreceptor with photochromic properties, and partly confirms elements in signalling which resemble those of eukaryotes and eubacteria. Chemical and photostimuli (and possibly other different stimuli) are sensed by the cell, amplified and integrated. The common denominator, so far, for all the signalling states is a diffusible molecule, fumarate, which mediates the target response, i.e. the switch of the flagellar motor. The quantitation of the entire phototactic signal chain and the stepwise elucidation of the molecules participating in the different branches are continuously improving our understanding of the system and have started to contribute to general knowledge on biological signalling.

REFERENCES

Alam, M., Claviez, M., Oesterhelt, D. & Kessel, M. (1984). Flagella and motility behaviour of square bacteria. *EMBO Journal*, **3**, 2899–903.

Alam, M. & Oesterhelt, D. (1984). Morphology, function and isolation of halobacterial flagella. *Journal of Molecular Biology*, **176**, 459–75.

Alam, M. & Oesterhelt, D. (1987). Purification, reconstitution and polymorphic transition of halobacterial flagella. *Journal of Molecular Biology*, **194**, 495–9.

Alam, M., Lebert, M., Oesterhelt, D. & Hazelbauer, G. L. (1989). Methyl-accepting taxis proteins in *Halobacterium halobium*. *EMBO Journal*, **8**, 631–9.

Berg, H. C., Block, S. M., Conley, M. P., Nathan, A. R., Power, J. N. & Wolfe, A. J. (1987). Computerized video analysis of tethered bacteria. *Review of Scientific Instruments*, **58**, 418–23.

Bibikov, S. I., Baryshev, V. A. & Glagolev, A. N. (1982). The role of methylation in the taxis of *Halobacterium halobium* to light and chemo-effectors. *FEBS Letters*, **146**, 255–8.

Blanck, A., Oesterhelt, D., Ferrando, E., Schegk, E. S. & Lottspeich, F. (1989). Primary structure of sensory rhodopsin I a prokaryotic photoreceptor. *EMBO Journal*, **8**, 3963–71.

Bogomolni, R. A. & Spudich, J. L. (1982). Identification of a third rhodopsin-like pigment in phototactic *Halobacterium halobium*. *Proceedings of the National Academy of Sciences, USA*, **79**, 6250–4.

Bogomolni, R. A. & Spudich, J. L. (1987). The photochemical reactions of bacterial sensory rhodopsin-I. *Biophysics Journal*, **52**, 1071–5.

Buder, J. (1917). Zur Kenntnis phototaktischer Richtungsbewegungen. *Jahrbuch Wissenschaftliche Botanik*, **58**, 105–220.

Cline, S. W., Schalkwyk, L. C. & Doolittle, W. F. (1989). Transformation of the archaebacterium *Halobacterium volcanii* with genomic DNA. *Journal of Bacteriology*, **171**, 4987–91.

Ehrlich, B. E., Schen, C. R. & Spudich, J. L. (1984). Bacterial rhodopsins monitored with fluorescent dyes in vesicles and *in vivo*. *Journal of Membrane Biology*, **82**, 89–94.

Forterre, P., Elie, C. & Kohiyama, M. (1984). Aphidicolin inhibits growth and DNA synthesis in halophilic archaebacteria. *Journal of Bacteriology*, **159**, 800–2.

Gerl, L. & Sumper, M. (1988). Halobacterial flagellins are encoded by a multigene family. Characterization of five flagellin genes. *Journal of Biological Chemistry*, **263**, 13246–51.

Hartmann, R., Sickinger, H.-D. & Oesterhelt, D. (1980). Anaerobic growth of Halobacteria. *Proceedings of the National Academy of Sciences, USA*, **77**, 3821–5.

Hecht, S., Shlaer, S. & Pirenne, M. H. (1942). Energy, quanta and vision. *Journal of General Physiology*, **25**, 819–40.

Hegemann, P. & Marwan, W. (1988). Single photons are sufficient to trigger movement responses in *Chlamydomonas reinhardtii*. *Photochemistry & Photobiology*, **48**, 99–106.

Hildebrand, E. & Dencher, N. (1975). Two photosystems controlling behavioural responses of *Halobacterium halobium*. *Nature, London*, **257**, 46–8.

Hildebrand, E. & Schimz, A. (1985). Behavioral pattern and its sensory control in *Halobacterium halobium*. In M. Eisenbach & M. Balaban, eds., *Sensing and Response in Microorganisms*, pp. 129–42. Elsevier: Amsterdam.

Hildebrand, E. & Schimz, A. (1986). Integration of photosensory signals in *Halobacterium halobium*. *Journal of Bacteriology*, **167**, 305–11.
Houwink, A. L. (1956). Flagella, gas vacuoles and cell-wall structure in *Halobacterium halobium*; an electron microscope study. *Journal of General Microbiology*, **15**, 146–50.
Kehry, M. R., Doak, T. G. & Dahlquist, F. W. (1984). Stimulus-induced changes in methylesterase activity during chemotaxis in *Escherichia coli*. *Journal of Biological Chemistry*, **259**, 11828–35.
Kuo, S. C. & Koshland, D. E. (1989). Multiple kinetic states of the flagellar motor switch. *Journal of Bacteriology*, **171**, 6279–87.
Lanyi, J. (1984). Bacteriorhodopsin and related light-energy converters. In Ernster, L., ed. *Comparative Biochemistry Bioenergetics*, pp. 315–35, Elsevier, Amsterdam.
Larsen, S. H., Reader, R. W., Kort, E. N., Tso, W.-W. & Adler, J. (1974). Change in direction of flagellar rotation is the basis of the chemotactic response in *Escherichia coli*. *Nature, London*, **249**, 74–7.
Macnab, R. M. (1976). Examination of bacterial flagellation by dark-field microscopy. *Journal of Clinical Microbiology*, **4**, 258–65.
Macnab, R. M. (1977). Bacterial flagella rotating in bundles: A study in helical geometry. *Proceedings of the National Academy of Sciences, USA*, **74**, 221–5.
Manor, D., Hasselbacher, C. A. & Spudich, J. L. (1988). Membrane potential modulates photocycling rates of bacterial rhodopsins. *Biochemistry*, **27**, 5843–8.
McCain, D. A., Amici, L. A. & Spudich, J. L. (1987). Kinetically resolved states of the *Halobacterium halobium* flagellar motor switch and modulation of the switch by sensory rhodopsin I. *Journal of Bacteriology*, **169**, 4750–8.
Marwan, W. & Oesterhelt, D. (1987). Signal formation in the halobacterial photophobic response mediated by a fourth retinal protein (P 480). *Journal of Molecular Biology*, **195**, 333–42.
Marwan, W., Alam, M. & Oesterhelt, D. (1987). Die Geißelbewegung halophiler Bakterien. *Naturwissenschaften*, **74**, 585–91.
Marwan, W., Hegemann, P. & Oesterhelt, D. (1988). Single photon detection in an archaebacterium. *Journal of Molecular Biology*, **199**, 663–4.
Marwan, W. & Oesterhelt, D. (1990). Quantitation of photochromism of sensory rhodopsin-I by computerized tracking of *Halobacterium halobium* cells. *Journal of Molecular Biology*, in press.
Marwan, W., Schäfer, W. & Oesterhelt, D. (1990). Signal transduction in *Halobacterium* depends on fumarate. *EMBO Journal*, **9**, 355–62.
Oesterhelt, D. & Krippahl, G. (1983). Phototrophic growth of Halobacteria and its use for isolation of photosynthetically deficient mutants. *Annals in Microbiology* (Inst. Pasteur), **134B**, 137–50.
Oesterhelt, D. & Marwan, W. (1987). Change of membrane potential is not a component of the photophobic transduction chain. *Journal of Bacteriology*, **169**, 3515–20.
Oesterhelt, D. & Tittor, J. (1989). Two pumps, one principle: light-driven ion transport in halobacteria. *Trends in Biochemical Sciences*, **14**, 57–61.
Otomo, J., Marwan, W., Oesterhelt, D., Desel, H. & Uhl, R. (1989). Biosynthesis of the two halobacterial light sensors P480 and SR and variation in gain of their signal transduction chains. *Journal of Bacteriology*, **171**, 2155–9.
Reichert, K. (1909). Über die Sichtbarmachung der Geißeln und die Geißelbewegung der Bakterien. *Centralblatt für Bacteriologie Abteilung*, **I**, 51.
Sager, B. M., Sekelsky, J. J., Matsumura, P. & Adler, J. (1988). Use of a computer to assay motility in bacteria. *Analytical Biochemistry*, **173**, 271–7.

Schegk, E. S. & Oesterhelt, D. (1988). Isolation of a prokaryotic photoreceptor: sensory rhodopsin from halobacteria. *EMBO Journal*, **7**, 2925–33.

Scherrer, P., McGinnis, K. & Bogomolni, R. A. (1987). Biochemical and spectroscopic characterization of the blue-green photoreceptor in *Halobacterium halobium*. *Proceedings of the National Academy of Sciences, USA*, **84**, 402–6.

Schimz, A. (1981). Methylation of membrane proteins is involved in chemosensory and photosensory behaviour of *Halobacterium halobium*. *FEBS Letters*, **125**, 205–7.

Schimz, A. (1982). Localization of the methylation system involved in sensory behaviour of *Halobacterium halobium* and its dependence on calcium. *FEBS Letters*, **139**, 283–6.

Schimz, A. & Hildebrand, E. (1985). Response regulation and sensory control in *Halobacterium halobium* based on an oscillator. *Nature, London*, **317**, 641–3.

Schimz, A. & Hildebrand, E. (1987). Effects of cGMP, calcium and reversible methylation on sensory signal processing in halobacteria. *Biochimica & Biophysica Acta*, **923**, 222–32.

Schimz, A. & Hildebrand, E. (1988). Photosensing and processing of sensory signals in *Halobacterium halobium*. *Botanica Acta*, **101**, 111–17.

Schimz, A. & Hildebrand, E. (1989). Periodicity and chaos in the response of *Halobacterium* to temporal light gradients. *European Biophysics Journal*, **17**, 237–43.

Schimz, A., Hinsch, K.-D. & Hildebrand, E. (1989). Enzymatic and immunological detection of a G-protein in *Halobacterium halobium*. *FEBS Letters*, **249**, 59–61.

Sperling, W. & Schimz, A. (1980). Photosensory Retinal Pigments in *Halobacterium halobium*. *Biophysics of Structure and Mechanism*, **6**, 165–9.

Spudich, J. L. & Bogomolni, R. A. (1984). Mechanism of colour discrimination by a bacterial sensory rhodopsin. *Nature, London*, **312**, 509–13.

Spudich, E. N., Hasselbacher, C. A. & Spudich, J. L. (1988). Methyl-accepting protein associated with bacterial sensory rhodopsin-I. *Journal of Bacteriology*, **170**, 4280–5.

Spudich, J. L. & Bogomolni, R. A. (1988). Sensory rhodopsins of halobacteria. *Annual Reviews in Biophysics and Biophysical Chemistry*, **17**, 193–215.

Spudich, E. N. & Spudich, J. L. (1982). Control of transmembrane ion fluxes to select halorhodopsin-deficient and other energy-transduction mutants of *Halobacterium halobium*. *Proceedings of the National Academy of Sciences, USA*, **79**, 4308–12.

Spudich, J. L. & Stoeckenius, W. (1979). Photosensory and Chemosensory Behavior of *Halobacterium halobium*. *Photobiochemistry & Photobiophysics*, **1**, 43–53.

Spudich, E. N., Takahashi, T. & Spudich, J. L. (1989). Sensory rhodopsins I and II modulate a methylation/demethylation system in *Halobacterium halobium* phototaxis. *Proceedings of the National Academy of Sciences, USA*, **86**, 7746–50.

Stewart, R. C. & Dahlquist, F. W. (1987). Molecular components of bacterial chemotaxis. *Chemical Reviews*, **87**, 997–1025.

Sumper, M. (1987). Halobacterial glycoprotein synthesis. *Biochimica & Biophysica Acta*, **906**, 69–79.

Sundberg, S. A., Bogomolni, R. A. & Spudich, J. L. (1985). Selection and properties of phototaxis-deficient mutants of *Halobacterium halobium*. *Journal of Bacteriology*, **164**, 282–7.

Sundberg, S. A., Alam, M. & Spudich, J. (1986). Excitation signal processing times in *Halobacterium halobium* phototaxis. *Biophysical Journal*, **50**, 895–900.

Takahashi, T., Mochizuki, Y., Kamo, N. & Kobatake, Y. (1985*a*). Evidence that the long-lifetime photointermediate of S-Rhodopsin is a receptor for negative phototaxis in *Halobacterium halobium*. *Biochemical & Biophysical Research Communications*, **127**, 99–105.

Takahashi, T., Tomioka, H. Kamo, N. & Kobatake, Y. (1985*b*). A photosystem other than PS 370 also mediates the negative phototaxis of *Halobacterium halobium*. *FEMS Microbiology Letters*, **28**, 161–4.

Thoelke, M. S., Bedale, W. A., Nettleton, D. O. & Ordal, G. W. (1987). Evidence for an intermediate methyl-acceptor for chemotaxis in *Bacillus subtilis*. *Journal of Biological Chemistry*, **262**, 2811–16.

Tomioka, H., Takahashi, T., Kamo, N. & Kobatake, Y. (1986). Action spectrum of the photoattractant response of *Halobacterium halobium* in early logarithmic growth phase and the role of sensory rhodopsin. *Biochimica & Biophysica Acta*, **884**, 578–84.

Traulich, B., Hildebrand, E., Schimz, A., Wagner, G. & Lanyi, J. (1983). Halorhodopsin and photosensory behavior in *Halobacterium halobium* mutant strain L-33. *Photochemistry & Photobiology*, **37**, 577–9.

Wagner, G., Rothärmel, T. & Traulich, B. (1989). Retinal-opsin-dependent detection of short-wavelength ultraviolet radiation (UV-B) and endogenous bias on direction of flagellar rotation in tethered *Halobacterium halobium* cells. *Plenum Publications*, in press.

Wieland, F., Paul, G. & Sumper, M. (1985). Halobacterial flagellins are sulfated glycoproteins. *Journal of Biological Chemistry*, **260**, 15180–5.

Wolff, E. K., Bogomolni, R. A., Scherrer, P., Hess, B. & Stoeckenius, W. (1986). Color discrimination in halobacteria: Spectroscopic characterization of a second sensory receptor covering the blue-green region of the spectrum. *Proceedings of the National Academy of Sciences, USA*, **83**, 7272–6.

MUTATIONAL STUDIES OF AMOEBOID CHEMOTAXIS USING *DICTYOSTELIUM DISCOIDEUM*

JEFFREY E. SEGALL

Department of Anatomy and Structural Biology, Albert Einstein College of Medicine, 1300 Morris Park Avenue, Bronx, NY 10461, USA

INTRODUCTION

Dictyostelium discoideum provides a potent system for dissecting the biochemical mechanisms that embody amoeboid behaviour (see Spudich, 1987). Cells are easily cloned and grown, enabling both behavioural studies of homogeneous cell populations as well as purification of important proteins. Strains with haploid genotypes are stable and undergo all stages of the life cycle, allowing the experimenter to generate mutations easily. Complementation groups can be established using a parasexual genetic system. Molecular genetic manipulations can be performed using both integrating and non-integrating transformation vectors. These technical advantages are combined with a developmental life cycle in which chemotaxis plays an important role.

This review will first introduce the stages of the life cycle in which chemotaxis could be important, and summarize the actual properties of amoeboid movement and chemotaxis displayed by *Dictyostelium* amoebae. A review of mutational studies forms the main body of the review, followed by a discussion of the strategies used in amoeboid chemotaxis.

Life cycle of Dictyostelium
(*for more details see* Bonner, 1967; Loomis, 1975, 1982; Raper, 1984)

The natural habitat for *Dictyostelium* amoebae is the soil, and the natural food source is bacteria (Fig. 1). Growth phase amoebae respond chemotactically to many bacterial species, enabling them to seek out their food directly (Konijn, 1969). The chemoattractant used appears to be folate secreted by the bacteria (Pan *et al.*, 1972). When the food source disappears, the amoebae begin to starve. Starvation induces a developmental cycle that involves several stages of transcription and translation of gene products, co-ordinated with morphogenetic movement and aggregation of the cells.

The first stage of development constitutes aggregation of the amoebae into a mass of up to 10^5 cells. Starvation induces synthesis of a cyclic adenosine 3'-5'-monophosphate (cAMP) signalling system (Klein et al., 1985; Gerisch, 1987), consisting of adenylate cyclase and receptors that stimulate both the cyclase and the chemotaxis system. The basal cyclase activity leads to levels of extracellular cAMP that bind to receptors, which in turn stimulate the cyclase activity. This generates a positive feedback loop leading to a burst in cyclase activity and extracellular cAMP, termed cAMP relay. At high or constant concentrations of cAMP, the system appears to desensitize, and cyclase activity drops. Extracellular phosphodiesterases secreted by the cells degrade the cAMP (Barra et al., 1980; Faure et al., 1989), and then the system resensitizes, allowing a new burst of cAMP to be produced. The net result for cells shaken in suspension is a periodic oscillation in the concentration of extracellular cAMP. In a sheet of cells on a surface, each burst becomes a wave of increases in cAMP concentration moving out from the point of initiation. The waves of cAMP stimulate chemotactic movement towards the initiation point, or aggregation centre. The net result is formation of streams of amoebae moving towards the centre and, eventually, construction of an aggregate of up to 10^5 cells.

This aggregate can then form a migrating structure termed a pseudoplasmodium, or, as its appearance suggests, a slug. The 1 mm long slug moves as a single multicellular organism, and is oriented by light and temperature gradients (Fisher et al., 1984; Poff et al., 1986). However, the amoebae comprising the slug remain individual cells. Communication between the cells must then be responsible for the co-ordination necessary to produce control of slug movement. Both cAMP and NH_3 are implicated in this behaviour, which could rely upon responses of the individual amoebae (Bonner et al., 1989, Schaap, 1986).

The thermotactic and phototactic capabilities of the slug are probably important in finding cracks or openings in the soil through which a fruiting body may emerge. The final stage of the life cycle is formation of a 1–4 mm high fruiting body. This structure consists of two cell types: stalk and spore cells. The basal disc and stalk are formed from cells that deposit cellulose cell walls, vacuolize, and die. At the tip of the stalk, the rest of the cells form a globular mass of spores that can resist environmental extremes. When conditions improve, the amoebae shed their spore coats, and emerge to repeat the life cycle. The function of the fruiting body is most likely to aid in dispersing the spores. The forming fruiting body secretes NH_3, and build-up of NH_3 on one side due to a nearby surface appears to cause an increase in movement and stalk formation on that side. This results in the fruiting body avoiding surfaces as it is formed (Bonner et al., 1986;

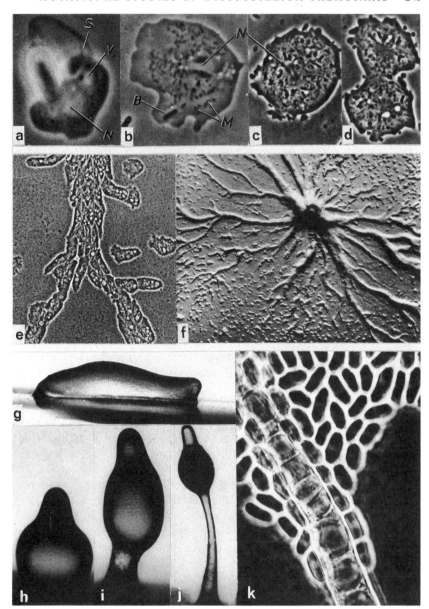

Fig. 1. Stages of *Dictyostelium* development. (*a*)–(*d*) Single cell stage. An amoeba germinates from a spore (*a*), leaving behind a spore case (S). The cell grows by phagocytosis of bacteria (*b*) and divides (*c*, *d*) with a generation time on bacteria of 3 hours. In the cell can be seen an interphase nucleus (N), mitochondria (M), and a freshly ingested *E. coli* bacterium (B), as well as the contractile vesicle (V) which is responsible for osmoregulation. During the aggregation stage, (*e*) individual cells form streams which can be seen under lower magnification (*f*) to coalesce into aggregates. (*g*) Slug stage. (*h*)–(*k*) Culmination stage – formation of a fruiting body. From Gerisch (1986).

Feit & Sollito, 1987). Both the co-ordination of fruiting body formation and the avoidance reaction could involve chemotactic responses.

Motility and chemotaxis of Dictyostelium *amoebae*
(for recent reviews, see Janssens & van Haastert, 1987; McRobbie, 1986; van Haastert, in press; Zigmond & Devreotes, 1988)

Dictyostelium amoebae are about 10 μm in diameter and move at speeds varying between 5 and 20 μm/minute (Eckert *et al.*, 1977; Potel & MacKay, 1979; Varnum *et al.*, 1986). The two major types of extensions of the cell surface that are seen are slender (0.1–0.2 μm wide), long filopods and broader, more oval-shaped pseudopods or lobopods (Taylor & Condeelis, 1979). The formation of filopods can be stimulated by chemoattractants but the role of filopods in motility is unclear. Cell translocation typically begins with the production of a pseudopod which, in phase contrast or in differential interference contrast, appears clear of intracellular organelles. After a delay, cytoplasmic vesicles and organelles will begin to flow into the pseudopod. The rear of the cell usually reduces in size as it moves up. There are often multiple pseudopods on a single cell which appear to complete for dominance with one eventually winning and the others dying out. The cell movement then follows the remaining pseudopod. The presence of multiple pseudopods is in contrast to leukocytes; see Wilkinson (this volume).

The outer edges of the pseudopods seem homogeneous in the light microscope (have a hyaline zone) because they are filled with actin filaments in a dense meshwork. Electron (Bennett & Condeelis, 1984; Spudich & Spudich, 1982) and light microscopic studies (Eckert & Lazarides, 1978; Fukui *et al.*, 1987; Hall *et al.*, 1988; Rubino *et al.*, 1984) indicate that filamentous actin is concentrated in the clear zones of pseudopods, and, to a lesser extent, in a cortex underlying the membrane in other parts of the cell. Immunofluorescence localization studies reveal that myosin II is present mainly at the rear and sides of the cell, and myosin I (Fukui *et al.*, 1989) is at the advancing edges of pseudopods. Microtubules radiate out from a microtubule organizing centre near the nucleus, and probably stop short of the actin meshwork of the pseudopod (Fukui *et al.*, 1990; Ross *et al.*, 1986; Rubino *et al.*, 1984).

Stimulation of starved cells with a micropipette filled with cAMP leads to the rapid extension of a pseudopod (within seconds) towards the micropipette (Fig. 2; Futrelle *et al.*, 1982; Gerisch *et al.*, 1975; Swanson & Taylor, 1982). Although there is a bias towards generation of pseudopods from the front of the cell, pseudopods can be induced from all regions of the cell with varying delays and even while the flow of cytoplasm is opposite to the direction of formation of the pseudopod. As in unstimulated motility, the flow of cytoplasmic vesicles into the pseudopods occurs at varying times

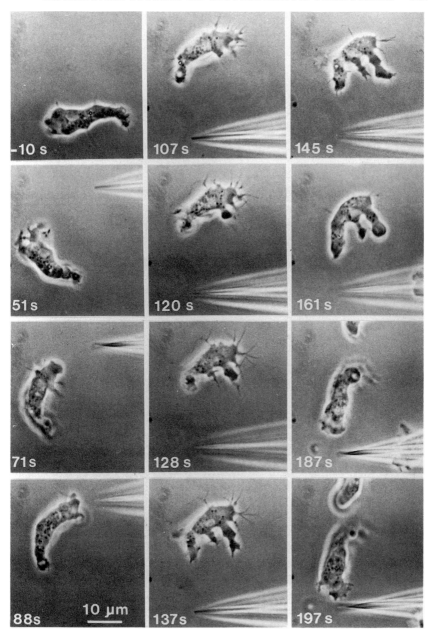

Fig. 2. Typical chemotactic responses of a *Dictyostelium* cell to a micropipette filled with 10^{-4} M cAMP. At the beginning, the cell was moving towards the left. The micropipette was introduced at 0 seconds, and the cell turned towards the tip maintaining its initial front. At 100 seconds, the pipette was moved to a new position, and the cell now responded by producing multiple pseudopods. After a phase of competition, the original rear end became established as the new front. In this particular case an aggregation competent cell of mutant HG1132, defective in the F-actin cutting protein severin was used. The response of this cell was indistinguishable from wild-type chemotaxis. (From Segall & Gerisch, 1989.)

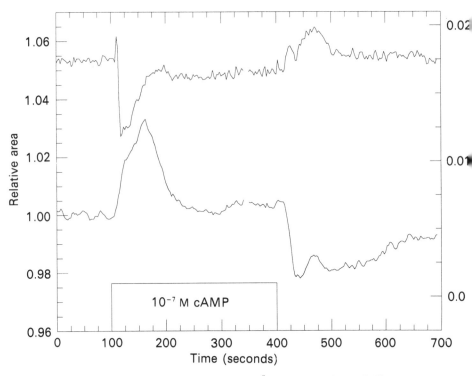

Fig. 3. Responses of starved wild-type cells to 10^{-7} M cAMP in a flow cell. The upper curve is the motility and the lower curve is the area. The solution flowing through the chamber was switched from buffer to cAMP at 100 seconds and back to buffer at 400 seconds. Data from different experiments were aligned such that the times of switching were the same. The break in the record is because the length of time exposed to cAMP varied. Area is the total number of pixels in the image that are covered by cells, divided by the average value during the pre-stimulus time. Motility is the number of pixels between two images that switch from being covered by cells to not covered, and vice versa (with area changes subtracted off), divided by the total area. (From Segall, 1988.)

(up to 10 seconds) after the initial formation of the pseudopod. In less steep gradients, quantification of pseudopod activity indicates that as many pseudopods are produced in the direction of the gradient as away (Varnum-Finney et al., 1987). However, pseudopods in the direction of the gradient are more likely to generate cell translocation.

Temporal changes in chemoattractant concentration also have effects on cell movement (Fig. 3; Segall, 1988; Varnum-Finney et al., 1988). Rapid, uniform increases in concentration result in a brief increase in motility peaking at 10 seconds after stimulation, followed by a longer-lasting decrease seen 20–30 seconds after stimulation, which slowly adapts. The area covered by the cell is first reduced under some circumstances, but, most consistently seen, is an increase in area that is maximal around 60 seconds after stimulation. Removal of chemoattractant produces an increase in motility and decrease in area. This complex series of responses may reflect multiple signalling systems in the cell. With slow temporal waves in cAMP concentration (Varnum et al.,1985), the third and fourth waves of a series of four waves produce increases in speed while the concentration is rising, combined with reduced turning rates. Intracellular vesicle movements are also inhibited by shifts to high cAMP concentrations (Wessels et al., 1989).

Biochemical changes induced by chemoattractants are described in detail by Newell et al. in this volume. Briefly, addition of cAMP induces rapid increases in inositol trisphosphates (IP_3) (Europe-Finner et al., 1989; van Haastert et al., 1989) and cytoskeletal actin (Hall et al., 1988; McRobbie & Newell, 1983) peaking at 5–10 seconds. Extracellular calcium flows into the cells (Bumann et al., 1984) and intracellular calcium increases (Abe et al., 1988). This is followed by an increase in cGMP which is maximum at 10–15 seconds (Mato et al., 1977; Wurster et al., 1977). There is a reduction in myosin associated with the cytoskeleton at 10–15 seconds (Berlot et al., 1985; Yumura & Fukui; 1985), followed by an increase at around 30 seconds (Dharmawardhane et al., 1989). Phosphorylation of myosin heavy chains first decreases and then increases along with that of the light chains, reaching a crest 30–60 seconds after stimulation (Berlot et al., 1985). A cAMP receptor becomes phosphorylated with a half-time of 1–2 minutes (Klein et al., 1987). Changes in extracellular potassium (Aekerle et al., 1985) and pH (Malchow et al., 1978) occur, and may reflect intracellular changes. The phosphorylation of the receptor and changes in extracellular calcium have been shown to remain in the continued presence of attractant. Most of the other changes appear to decline, or adapt.

MUTANT STUDIES

Mutants have been identified through varied methods. Screening chemically mutagenized populations for altered colony morphology has yielded the *fgdA, stmF* and *mot*1 mutants. Direct selection for chemotaxis mutants generated the *folA* and HB5 mutants. Screening of phagocytosis mutants for motility defects yielded the *motA* mutant. A transformation vector has been used to produce anti-sense mRNA for investigating the role of the *ras* protein, myosin II heavy chain, and cAMP receptors, while gene disruption or chemical mutagenesis was used to generate mutants lacking myosin I or II heavy chain, α-actinin, severin, and the 120 kD gelation factor. The mutants will be described in the order that their respective proteins are thought to be activated in the signal transduction pathway – starting with the receptors and ending with the motility apparatus.

Receptors

Cell-surface binding studies have suggested that there are multiple receptors for cAMP as well as for folate (for recent review, see Janssens & van Haastert, 1987). For each compound, the high affinity receptors are termed the B receptors, and lower affinity receptors the A receptors. Pharmacological studies indicate that the B receptors mediate chemotactic responses, while the A receptors are involved with stimulation of adenylate cyclase (de Wit *et al.*, 1985). It is not yet clear whether the differences between the A and B binding characteristics reflect different gene products, post-translational modification of a protein coded for by a single gene, or interactions of a single receptor with different G proteins.

A cAMP receptor (CAR1) has been isolated and the gene encoding it cloned (Klein *et al.*, 1988). The sequence indicates that the protein has seven transmembrane sequences and has regions of homology with rhodopsin, suggesting that it is a G protein-linked receptor. Anti-sense expression suppressed production of the receptor, and also blocked aggregation. It is still unresolved whether CAR1 is responsible for both the A and B cAMP binding sites. Southern blotting reveals that there are several homologous genes which may also express cAMP receptor proteins (C. Saxe, P. Devreotes & A. Kimmel, personal communication). It is uncertain whether the anti-sense experiment suppressed these as well. Gene disruption experiments are presently under way to resolve this question (P. Devreotes, personal communication).

The folate receptors have not yet been isolated. A mutant (termed the *fol*A mutant) was isolated that was defective in chemotaxis to folate and had dramatically altered folate B receptor binding (Segall *et al.*, 1987, 1988). In the wild-type, the B-receptors appear to undergo a cycle of changes in affinity upon binding folate (de Wit & Bulgakov, 1986*a, b*). In the absence of folate, the B receptors are in a state with low affinity for folate (about 50 nM), from which folate rapidly dissociates (de Wit & van Haastert, 1985). Upon binding folate, the sites redistribute between two higher affinity states, termed B^S and B^{SS}, which have slow and extremely slow rates, respectively, for dissociation of folate. In the wild-type, at steady state in the presence of folate, roughly 70% of the receptors binding folate are in the B^S state, and 30% in the B^{SS} state. In the mutant, over 90% of the receptors are in a high affinity state (Fig. 4(*a*)) from which folate dissociates 10 × more slowly than from the wild-type B^{SS} state (Fig. 4(*b*)), and has correspondingly higher affinity. The folate B receptor appears to be in the high affinity state even in the absence of folate. Surprisingly, in the mutant, changes in motility and area as well as changes in intracellular cGMP in response to step increases in folate required 10–100-fold *higher* concentrations of folate. The responses generated at high folate concentrations could be due to a small number of normal receptors, or binding

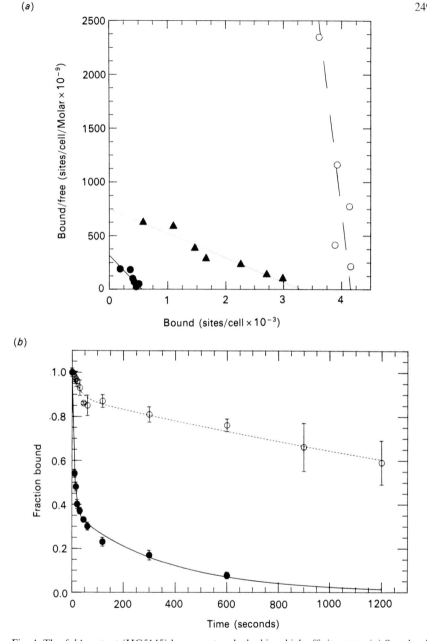

Fig. 4. The *folA* mutant (HG5145) has receptors locked in a high affinity state. (a) Scatchard plot of binding of N_{10}-methylfolate to total B sites (filled triangles) or B^{SS} sites (filled circles) on AX2 (parent strain), or to B^{SS} sites on HG5145 (open circles, total B^S sites on HG5145 was similar). The solid line is for a K_d of 1.8 nM with 555 sites, the dotted line is for a K_d of 4.4 nM, 3300 sites, and the broken line is for a K_d of 0.22 nM, 4150 sites. (b) Dissociation kinetics of N_{10}-methylfolate from total B sites on AX2 (solid circles) or HG5145 (open circles). Data were fitted to the sum of two exponentials. For the slower dissociating site (B^{SS}) the initial value and rate constant are 0.37, $2.8 \times 10^{-3} s^{-1}$ (AX2) and 0.89, $3.2 \times 10^{-4} s^{-1}$ (HG5145). (From Segall *et al.*, 1988 with the permission of the Company of Biologists.)

of folate at high concentrations to a pterin chemotaxis receptor (van Haastert et al., 1982a). The high affinity receptor in the mutant does not transduce folate binding into chemotactic responses.

The folA mutant indicates that the folate B receptors are important for folate chemotaxis, consistent with previous studies correlating binding of folate analogues to the B receptors with their chemotactic potency. In addition, it separates binding of folate to the receptor from signal transduction. The receptor appears to be locked in a high affinity state, suggesting that cycling of the receptor states is part of the signal transduction process. The actual site of the folA mutation is still unknown. cAMP chemotaxis of the mutant is normal, indicating the defect is limited to the folate chemotaxis pathway, consistent with it being in the receptor itself or in a protein that interacts with the receptor.

Proteins interacting with receptors

Mutants have been identified that are defective in a protein that probably binds directly to the cAMP receptor – the α subunit of a G protein (Firtel et al., 1989). The first mutants of this type were produced by chemical mutagenesis and were isolated as aggregation-defective strains that were not responsive to cAMP pulses (Coukell, 1975). These 'frigid' (fgd) mutants were grouped into five complementation groups (Coukell et al., 1983). The fgdA complementation group was distinguished by appearing to initiate the developmental program but having very poor chemotaxis to cAMP. cAMP binding to mutant cells is 30% of wild-type, indicating that the mutants should be able to respond to cAMP. However, cells do not respond chemotactically, and IP_3, cGMP and cAMP production stimulated by cAMP is less than 10% of wild-type (Kesbeke et al., 1988; P. van Haastert, personal communication). In addition, consistent with the frigid phenotype, genes that are induced by cAMP in the wild-type are not induced in the fgdA mutants (Mann et al., 1988). Thus the fgdA defect identifies a common pathway for cAMP chemotaxis, cAMP relay, and control of gene expression. Several experiments suggested that the fgdA mutants might be defective in a G protein (Kesbeke et al., 1988). Compared to wild-type membranes, in membranes from the fgdA mutants, guanine nucleotides did not have as strong effects on cAMP binding, nor did a cAMP analogue stimulate binding of GTPγS as strongly. Mutant membranes also had reduced GTPase activity.

These suggestive results were confirmed when *Dictyostelium* genes homologous to mammalian G_α genes were cloned (Pupillo et al., 1989). Using oligonucleotide probes to conserved sequences, originally two G_α subunits were cloned. On the basis of hybridization studies, currently, five α subunit genes have been identified (R. Firtel, personal communication). G_β subunit genes have also been identified (P. Lilly & P. N. Devreotes, personal com-

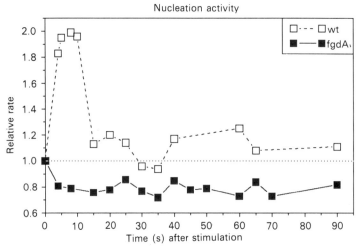

Fig. 5. Kinetics of actin nucleation responses of starved *fgdA* cells (solid symbols and lines) and parental wild-type cells (open symbols, dashed lines) after stimulation with 2'deoxycAMP at 0 seconds. (From Hall et al., 1989.)

munication). $G_\alpha 1$ is expressed in growth phase cells and decreases after 12 hours of starvation, $G_\alpha 2$ is expressed during aggregation, $G_\alpha 3$ a little later, and then $G_\alpha 4$ and $G_\alpha 5$ at the end of development. *FgdA* mutants carry mutations in $G_\alpha 2$. One mutant, HC85, carries a 2.2 kb deletion in the gene, resulting in a null mutation (Kumagai et al., 1989). Transformation of HC85 with the wild-type gene restores the wild-type phenotype, proving that the *fgdA* locus is the $G_\alpha 2$ gene (R. Firtel, personal communication).

This result establishes that chemotactic responses to cAMP, cAMP relay, and cAMP control of gene expression are all funneled through the $G_\alpha 2$ protein. This protein probably interacts directly with the cAMP B receptor. Binding studies with cAMP indicate that the B^{SS} state is lacking in the *fgdA* mutants, suggesting that binding of the G protein containing $G_\alpha 2$ to the receptor generates this state (Kesbeke et al., 1988). Thus it appears that the receptor directly interacts with the $G_\alpha 2$ protein, which then triggers a signalling pathway involving IP_3 and cGMP and leading to chemotaxis, stimulation of adenylate cyclase, and gene regulation. The point(s) at which cAMP relay and gene regulation diverge from the chemotaxis pathway is not known. However, not all receptor-mediated processes require the $G_\alpha 2$ protein. In *fgdA* mutants, cAMP addition still stimulates phosphorylation of the cloned cAMP receptor (Kesbeke et al., 1988), downregulation of the receptor, and activation of an inhibitor of nucleation activity in cell extracts (Fig. 5; Hall et al., 1989a,b).

In addition, folate chemotaxis still occurs in *fgdA* mutants (Coukell et al., 1983; B. E. Snaar-Jagalski & P. van Haastert, personal communication). This is consistent with chemotaxis to folate being maximal in growth phase cells, which express very little of the $G_\alpha 2$ protein. On the other hand, $G_\alpha 1$ is expressed during growth phase, and is the logical candidate for a G subunit which could mediate responses to folate. This turns out not to be the case. Both $G_\alpha 1$ anti-sense transformants and a gene disruption mutant are chemotactic to folate and show increases in cGMP and cAMP when stimulated by folate (R. Firtel, personal communication).

Another protein thought to be closely coupled to receptors is the *ras* protein. *Dictyostelium* contains as least two *ras* genes that are differentially expressed during development (Reymond *et al.*, 1984; Robbins *et al.*, 1989). *DdrasG* is expressed during growth and early development, while *Ddras* is first expressed during the slug stage (Weeks & Pawson, 1987). It was not possible to express high levels of anti-sense mRNA in cells, suggesting that expression of *Ddras* may be necessary for cell viability (Reymond *et al.*, 1985). Therefore transformation with a mutated *Ddras* sequence carrying a Gly12–Thr12 substitution (which activates *ras* in other cell types) was performed (Reymond *et al.*, 1986). Compared to controls transformed with the normal Gly12 *Ddras*, transformants with Thr12 *Ddras* show subtle changes in chemotactic responses (van Haastert *et al.*, 1987). Chemotaxis is delayed and requires higher concentrations of cAMP compared to the control. cGMP production is reduced, apparently because the guanylate cyclase is inactivated more rapidly. *In vitro* studies of cAMP binding to membranes suggest that the Thr12 *Ddras* may be constitutively active (Luderus *et al.*, 1988). There are also changes in the rate of incorporation of tritiated inositol into IP_3 and IP_6, suggesting a more rapid turnover of inositol lipids (Europe-Finner *et al.*, 1988; P.van Haastert personal communication). Though the direct effect of the *ras* mutation is still unclear, the result is a partial reduction in chemotaxis.

Intracellular messengers

Addition of cAMP stimulates production of IP_3 (Europe-Finner *et al.*, 1989; van Haastert *et al.*, 1989), and, presumably, diacylglycerol as well (Berridge & Irvine, 1989). As is discussed by Newell *et al.* in this volume (see also Newell *et al.*, 1988), the IP_3 may induce Ca release from intracellular stores (Europe-Finner & Newell, 1986), which might then stimulate a guanylate cyclase (Small *et al.*, 1986). Both folate and cAMP have been shown to stimulate synthesis of intracellular cGMP (Mato *et al.*, 1977; Wurster *et al.*, 1977). The increase in cGMP concentration is maximal 10–15 seconds after stimulation, and declines back to prestimulus levels within 45 seconds. A specific cGMP phosphodiesterase is responsible for the reduction in cGMP concentration (Dicou & Brachet, 1980; van Haastert *et al.*, 1982*b*).

Mutants with reduced activities of this phosphodiesterase have been isolated. They were identified as one of six complementation groups that have the phenotype of producing abnormally long streams of cells during aggregation, and are termed streamer (*stm*) mutants (Ross & Newell, 1979; 1981). Members of the *stmF* complementation group were found to have abnormally large and long-lasting increases in intracellular cGMP in response to chemoattractant. Compared to the parental strain, *stmF* mutants had peak concentrations of cGMP that were twice as high, and the concentration remained above the peak value seen in the parental strain for at least 60 seconds. Measurements of cGMP-specific phosphodiesterase activity revealed that the remaining activity was 0.4%–2% of wild-type for various *stmF* alleles (van Haastert *et al.*, 1982c). The defect is most likely in the structural gene for the phosphodiesterase (Coukell & Cameron, 1986).

The behavioural effect of this extended presence of cGMP within the cell is an lengthened period of cell elongation. In stable gradients of cAMP, this appears to have minimal effects (Fisher *et al.*, 1989). However, during aggregation, the concentration of cAMP is constantly changing. As each wave of cAMP travels outward, wild-type cells elongate and move towards the aggregation center for about 100 seconds. In the mutant, this stage is lengthened to as long as 500 seconds. Although the mutant cells are elongated, cell translocation appears to be suppressed (G. McNamara, personal communication). The time required to aggregate is then increased, resulting in long streams of cells moving towards the aggregate. Correlated with the increase in size and duration of the cGMP pulse in *stmF* cells is an increased association of myosin with the Triton-insoluble cytoskeleton (Liu & Newell, 1988). cGMP may regulate the state of myosin association with the cytoskeleton (by activating a kinase or phosphatase, for example), and in that way control cell movement (see also Newell *et al.* in this volume).

Motility and cytoskeletal mutants

One method for isolating motility mutants involved selecting for temperature-sensitive phagocytosis mutants, assuming that processes required for phagocytosis and motility would overlap (Clarke, 1978). These mutants were then screened for strains that were defective for movement at the restrictive temperature (27°C). Six strains were examined in more detail (Kayman *et al.*, 1982). All strains required several hours exposure to 27°C before the defect in motility occurred. After becoming non-motile, the rate of oxygen consumption of the mutants remained 50–100% of wild-type, indicating that a defect in energy metabolism was not the cause of the motility change. However, the motility phenotype may be due to a mutation that has pleiotropic effects. Several mutants no longer grew on axenic medium. One mutant that was studied in more detail expressed discoidin, a species-specific lectin, at the restrictive temperature (Kayman *et al.*, 1988).

In different genetic backgrounds, the discoidin synthesis began up to 3 hours after the cells became non-motile, indicating that the discoidin synthesis could not be the direct cause of the reduced motility. Thus phagocytosis defects can affect cell movement, but these may be due to pleiotropic effects that complicate subsequent analysis.

Motility mutants can also be identified as strains that produce smaller plaques on plates. A dynamic morphology study of one mutant revealed a lack of polarity and more rounded cell shape (Soll et al., 1988). Pseudopods were smaller, and tended to occur randomly over the cell surface, leading to little net translocation. Staining for actin filaments revealed a shell of filaments lying under the cell membrane with no net polarization of the cell, whereas in the wild-type the majority of the filamentous actin is in the pseudopods.

Defects in adhesion to a surface could also affect cell motility. Starving *Dictyostelium* cells secrete lectins called discoidins. Discoidin I contains a peptide sequence that has been implicated in cell-substrate adhesion, gly–arg–gly–asp. Mutants defective in the production of discoidin I fail to form streams (Springer et al., 1984; Crowley et al., 1985). However, the mutants can still move. The result is that only small aggregates are formed, apparently by random collision. It is unclear whether the reduction in discoidin expression affects cell motility or another aspect of the aggregation process. Adhesion defects are also present in phagocytosis mutants which adhere poorly to plastic surfaces (Vogel, 1987; Waddell et al., 1987). The motility of these mutants has not been studied in detail.

Myosin mutants

Because of the prominence of the actin/myosin network in amoeboid cells, myosin II (the classic double-headed myosin), has been thought to play a critical role in generation of cell movement. The role of myosin II has been subjected to the ultimate test by selectively removing it and looking at the effect on cell behaviour (for review, see Spudich, 1989). Three types of mutants have been generated: ones expressing anti-sense mRNA which express <1% of the normal amount of myosin II (Knecht & Loomis, 1987), gene replacement mutants which express a truncated myosin II which cannot form filaments but still has ATPase activity (De Lozanne & Spudich, 1987), and null mutants in which the entire gene has been removed (Manstein et al., 1989).

The data from all three mutants are consistent. Myosin II is not absolutely required for movement (Peters et al., 1988; Wessels et al., 1988). Mutants move at speeds ranging from 30–60% of the wild-type. They tend to be rounder than the wild-type, and their pseudopods are smaller. (Fig. 6). On average, their chemotactic orientation (the fraction of cell translocation

MUTATIONAL STUDIES ON *DICTYOSTELIUM* CHEMOTAXIS

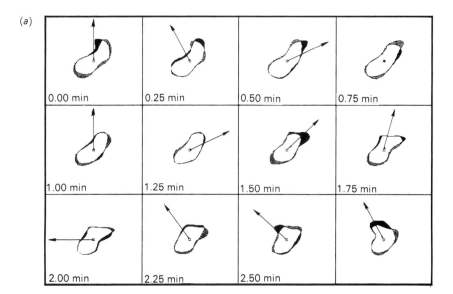

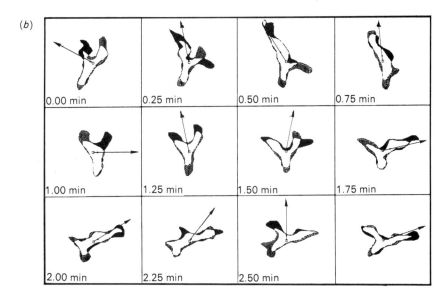

Fig. 6. Difference pictures for an average HMM amoeba (*a*) and an average parental cell (*b*). Cell outlines were plotted for a 0.25 minute interval with the later image overlapping the earlier image. The filled region represents expansion zones, hatched or grey regions contraction zones, and clear regions common zones. The small square and small circle represent the centroids of the first and second image, and the directional vector drawn through the consecutive centroids is represented by an arrow. (From Wessels *et al.*, 1988.)

which is towards the source of chemoattractant) was 20% of that of the parental strain. Pseudopods are clearly produced in the myosin II mutants, and a micropipette filled with cAMP induces oriented production of pseudopods in the direction of the micropipette. Overall cell polarity in the absence of a strong stimulus is reduced, however, and multiple pseudopods are more prevalent. Addition of chemoattractant as a temporal stimulus does not produce the reduction in motility seen in the parental strain, but the area increase afterwards still occurs (Fukui et al., 1990; D. Wessels & D. R. Soll, personal communication; Segall, unpublished). As in the wild-type, there is a broad distribution of cell characteristics. About 15% of the cells move at roughly the mean speed of the parental strain, and are 70% as efficient in chemotaxis. Nevertheless, for the cell population as a whole, there is a clear shift to lower speeds and poorer chemotaxis.

Intracellular particle movement in the null mutant was altered (Wessels & Soll, personal communication). In the parent strain, particles moved at 0.71 μm/s and had a strong directionality, with the majority moving parallel to the direction of movement of the cell. Addition of cAMP caused a 70% reduction in speed and a loss of directionality. In the null mutant, particle speed was 35% of the parental and showed no directionality. Addition of cAMP had no effect.

There are general changes in the cell cortex of the mutants (Pasternak et al., 1989). Cortical tension measured with a 'cell poker' was 70% of the parental strain. Cortical tension doubled during capping of ConA receptors in the parental strain, while, in the null mutant, ConA receptors did not cap but rather appeared to internalize in small patches. Cortical tension associated with this process increased only 25%.

A number of other changes in cell properties are observed in the myosin II mutants (De Lozanne & Spudich, 1987; Fukui et al., 1990; Knecht & Loomis, 1987). The most dramatic is a defect in cytokinesis. During mitosis, although the nucleus can still divide, cell division no longer can occur in suspension, and cells become a syncytium which eventually lyses. If the cells are on a surface, cytokinesis can now occur by an unusual mechanism. The multinucleated cells produce multiple fronts, and these independent fronts can literally pull the cell apart, relying on 'traction-mediated cytofission'. This could reflect a primitive mechanism for cytokinesis which was used before myosin II was present in a contractile ring.

Finally, the myosin II mutants also show developmental defects (Knecht & Loomis, 1988). Aggregation is delayed and altered, with discontinuous, broken streams being formed. Aggregates form, but no further morphological transitions occur. Slug and fruiting body formation appear to be blocked. Developmentally regulated genes are expressed normally, suggesting that the signals for starting formation of these structures are present. When mixed with parental cells, myosin II mutants can form spores in the fruiting body, indicating that the mutants can undergo the entire process. It is possible that the physical forces that must be generated to produce a slug or fruiting body require the generation of cortical tension by myosin II.

The conclusion drawn from the studies of the myosin II mutants is that myosin is not absolutely needed for cell motility or pseudopod extension. On average, the mutants move more slowly and orient less well, but both processes still occur. Models of cell motility which exclusively rely upon myosin II-induced contractions for generating cell movement are inadequate. Myosin II does increase cortical tension and could, in this way, aid in suppressing or inhibiting pseudopod extension as well as generating cell contraction and vesicle flow.

The surprising observation that myosin II is not required for motility or chemotaxis has stimulated the search for other myosin-like molecules. Such force generators could be important for the cell behaviours that are present in the myosin II mutants. Single-headed myosins, termed myosin I molecules, have been isolated from *Acanthamoeba*, intestinal brush border, and *Dictyostelium* (for review, see Korn & Hammer, 1988). Clones from *Dictyostelium* encoding at least three different proteins have been partially sequenced (Jung *et al.*, 1989; Titus *et al.*, 1989). Currently, a mutant lacking one of the proteins has been generated (Jung & Hammer, 1989). The mutant can undergo the entire developmental cycle, producing fruiting bodies. The cells are elongated, but they cannot maintain a polarized morphology as well as the parental strain (Jakota, Wessels, Murray & Soll, personal communication). Preliminary results indicate that phagocytosis is reduced, and orientation of particle movement is random, but is still slowed down upon addition of cAMP. The myosin I molecule missing in this mutant may be responsible for maintaining pseudopod integrity, but not the actual generation of pseudopods. The search for a myosin-related molecule necessary for pseudopod formation (if one exists) continues.

Mutants in actin binding proteins

The results with the myosin mutants indicate that myosin molecules may not be responsible or necessary for the generation of pseudopods. Actin is the major component in pseudopods, can polymerize into filaments, and could generate pseudopods by polymerizing and forming an expanding gel (Condeelis *et al.*, in press; Oster, 1988; Tilney & Inoue, 1982). Actin polymerization can be controlled by ionic conditions and by various actin binding proteins. Actin binding proteins that have been isolated from *Dictyostelium* include monomer binding proteins, nucleation proteins, filament severing proteins, crosslinkers, and bundling proteins (for review, see Schleicher *et al.*, 1988). Pseudopod formation may be due to the control of the activities of these proteins in a spatially defined manner.

To determine the roles of the actin binding proteins, mutants lacking specific proteins are being isolated. Mutants generated from chemical muta-

Table 1. *Chemotaxis of mutants lacking actin binding proteins.*

Strain	Defect	Speed (μm/min)	Orientation	Turning (rad^2/min)	Number of experiments
AX2	parent	11.8 ± 2.0	0.29 ± 0.06	0.84 ± 0.16	9
HG1132	severin	12.1 ± 1.8	0.39 ± 0.08	0.83 ± 0.04	2
HG1264	120 kD	9.9 ± 2.5	0.34 ± 0.04	0.71 ± 0.13	6
HG1130	α-actinin	11.1 ± 1.4	0.22 ± 0.06	0.97 ± 0.22	3

Axenically grown cells were starved for 6 hours and then observed in a linear stable gradient of 25 nM cAMP per mm as described by Fisher *et al.*, 1989. Speed is the rate of centroid translocation, orientation is the fraction of cell movement in the direction of the gradient, and turning rate measures cell direction changes in terms of a rotational diffusion coefficient.

genesis have been isolated by screening single colonies with antibodies to the proteins, or if the gene for the proteins is available, by anti-sense mRNA production or gene disruption (Gerisch *et al.*, 1989; Segall & Gerisch, 1989). Currently, mutants have been isolated in three actin binding proteins: α-actinin (Noegel & Witke, 1988; Wallraff *et al.*, 1986), the 120 kD gelation factor (Brinke, Witke, *et al.*, unpublished results), and severin (Andre *et al.*, 1989). α-actinin and the 120 kD gelation factor are the two most active actin filament cross-linkers in cell extracts. α-actinin is calcium regulated. Severin cleaves and caps actin filaments. Both chemically mutagenized and gene disruption mutants have been isolated.

None of the mutants shows any major defects in motility or chemotaxis (Table 1). Starved cells elongate, and respond chemotactically to micropipettes filled with cAMP. Quantitative measurement of cell orientation in linear spatial gradients also reveals no large difference between the mutants and the parental strain. The growth rates of the mutants are within 5% of the parental strain. Development appears normal as well. There are a number of possible explanations for these results (see also Bray & Vasiliev, 1989). First, the proteins may be completely redundant. There are sequence homologies between actin binding proteins that could reflect similar functions. A 27 amino acid sequence in the 120 kD gelation factor has been identified as an actin binding site and similar sequences are found in filamin, β-spectrin, and α-actinin (Bresnick *et al.*, 1990; Noegel *et al.*, 1989*b*). It may be necessary to mutate all these proteins (or their *Dictyostelium* homologues) to get a strong effect. Secondly, the assays used so far may not be sensitive enough to detect the defects in the mutants. In terms of natural selection, a 5% difference in fitness could be a strong selective force but would be difficult to quantitate. Thirdly, it may be that the wrong properties are being tested in the mutants; the functions of the proteins may be completely different from those which have been assumed.

Other mutants

An intriguing mutant, HB5, has been isolated that is temperature-sensitive for chemotaxis (Barclay & Henderson, 1977; 1982). Cells of this mutant can be grown and starved at the permissive temperature, but then chemotaxis to cAMP is blocked when they are shifted to the restrictive temperature. cAMP receptors are normal, indicating that the cells develop to the correct stage. Mixtures of mutant with wild-type cells do not result in fruiting bodies with mutant spores, indicating that the defect is not in an extracellular protein. The temperature sensitivity lasts until slug formation, then slug movement and fruiting body formation are normal. If mutant slugs are bisected at the restrictive temperature, the front half will form a fruiting body. Unlike the wild-type, the rear half of a mutant slug cannot regulate to reform a tip and form a slug or fruiting body – it remains as an aggregate. This result suggests that the chemotaxis system used for aggregation is important for the initial organization necessary for forming a slug, but is not necessary for slug movement or fruiting body formation. The precise biochemical defect has not yet been localized.

Mutant strains defective in slug phototactic behaviour have also been tested for cAMP chemotactic responses (Fisher et al., 1989). No defect was found, supporting the inference from the HB5 study that control of slug movement is distinct from the cAMP chemotaxis system used during aggregation. More detailed studies with defined mutations will be necessary to confirm this.

A POSSIBLE MECHANISM FOR AMOEBOID CHEMOTAXIS

A conclusion that has been drawn from the behaviour of wild-type *Dictyostelium* cells is that there are probably multiple effector or response systems present (see also Gerisch et al., 1975; Mato et al., 1978; Segall, 1988; Swanson & Taylor, 1982; Varnum-Finney et al., 1987; Devreotes & Zigmond, 1989). The separation of the responses of cytoplasmic vesicles from the extension of pseudopods indicates more than one control system. The distinct kinetics of motility and area changes in response to temporal upshifts also suggests multiple responses. The results from the mutants identified thus far support this, indicating that amoeboid chemotaxis involves a complex signal transduction system. The cAMP receptor antisense, *folA* and *fgdA* mutants show dramatically reduced chemotactic responses, showing that the initial steps – binding of chemoattractant to its receptor and coupling the receptor to the cell interior – are absolutely necessary. All the other mutations appear to only partially affect chemotaxis, if at all. Especially surprising are the results with the actin binding protein mutants. The flow of information appears to be fan-shaped: narrow at the receptors and spreading out until a number of cytoskeletal elements are reached

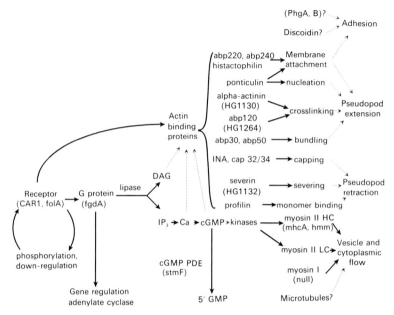

Fig. 7. Information flow in amoeboid chemotaxis. A tentative scheme encompassing signalling pathways significant for chemotaxis in *Dictyostelium*. Solid lines indicate steps for which there is experimental evidence, dashed lines indicate potential interactions. Mutants lacking particular proteins are indicated in parentheses. See Luna & Condeelis (in press) and Noegel *et al.* (1989a), for references to specific actin binding proteins.

(Fig. 7). This suggests that there could be several parallel pathways involved in chemotaxis.

Even if a single receptor is sufficient for cAMP or folate chemotaxis, studies with the *fgdA* mutant indicate that there is already a divergence of signals directly after receptor binding. As mentioned above, *fgdA* mutants still show receptor down-regulation, receptor modification, and cAMP stimulated production of an inhibitor of actin nucleation activity. The receptor down-regulation and modification could be due to adaptation processes that depend upon receptor conformations produced by cAMP binding. However, the inhibition of nucleation activity reveals another pathway coupling cAMP binding to the cytoskeleton. This inhibitor is cytosolic, and thus could spread throughout the cell, if its rate of inactivation is not too fast.

There is also the possibility of multiple signal pathways occurring after the $G_\alpha 2$ protein. *FgdA* mutants are blocked in chemotaxis, activation of adenylate cyclase, and gene regulation. It is not known whether all these responses derive from a single pathway. As described by Newell *et al.* in this volume, cAMP stimulates IP_3 production, and they discuss the evidence for a pathway involving IP_3-stimulated Ca release leading to synthesis of cGMP, and resulting in control of the association of myosin II with the cytoskeleton. However, production of IP$_3$ should also result in production of diacylglycerol. Diacylglcerol can also be a messenger molecule, and is known to activate protein kinases (Nishizuka, 1988). It could initiate a parallel pathway involved in regulating actin polymerization.

Multiple outputs which these signals could control include proteins that affect different aspects of actin or myosin organization in the cell. The state of actin *in vitro* can be altered by nucleation factors, nucleation inhibitors, cross-linkers, severing, or bundling proteins. Stimulation of actin polymerization, followed by cross-linking into a gel, could generate the forces for producing pseudopods. Myosin II activity is regulated by the presence of actin, by phosphorylation of the heavy chain (which reduces thick filament formation), and by phosphorylation of the regulatory light chain (which stimulates ATPase activity). Although the major focus is on actin and myosin, it is still possible that microtubules also are involved in producing the optimal chemotactic response (de Priester *et al.*, 1988; Cappuccinelli & Ashworth, 1976).

A model for the control of amoeboid behaviour must explain on the biochemical level how multiple outputs are controlled to generate spatially directed motion. The hypothesis detailed below assumes there are (at least) two intracellular signals generated when a receptor binds ligand (Fig. 8; see also Condeelis *et al.*, in press; Fisher in press). One signal is a localized messenger that does not spread far from its source (is strongly buffered or enzymatically degraded as it diffuses). The concentration of this signal will reflect the occupancy of nearby receptors. The second signal can diffuse throughout the cell. The concentration of the second signal at any point in the cell will then tend to reflect the occupancy of receptors over the entire cell. These two signals contain in principle the information necessary for generating spatially directed cell motion. The first signal represents the local receptor occupancy, the second signal the average global occupancy. A comparison of the two signals at a particular point in the cytoskeleton can indicate whether that point is on the upward side of the chemoattractant gradient. If it is, pseudopod production should be stimulated. If not, retraction should be induced.

For the scheme described above, potential short-range messengers include Ca, IP$_3$, or diacylglycerol. Short range messengers could stimulate pseudopod formation. Stimulation of actin polymerization is a likely function of the short range messenger. A chemoattractant-stimulated actin nucleation activity is present in the cytoskeleton, and would have the restricted spatial range necessary for a localized extension of the cytoskeleton. Potential long-range messengers include the inhibitor of actin nucleation and cGMP. Such messengers should suppress pseudopod extension. The estimated enzyme activities of the cGMP phosphodiesterase (Van Haastert *et al.*, 1982*b*) are too low to localize cGMP to less than 18 microns (about

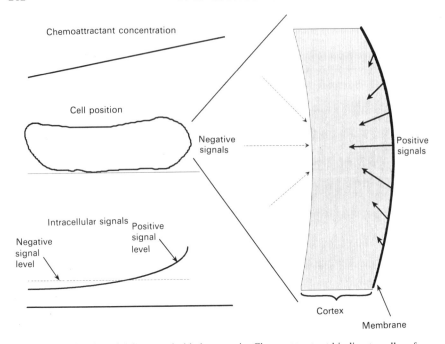

Fig. 8. Two-signal model for amoeboid chemotaxis. Chemoattractant binding to cell surface receptors generates two signals: a long-range signal that inhibits pseudopod formation (negative signal, dashed lines) and a local signal that stimulates pseudopod formation (positive signal, solid lines). *Left side*: spatial distribution of these intracellular signals in a cell migrating in a spatial gradient of chemoattractant. *Right side*: expanded view of a region of the cell cortex, with arrows indicating the relative significance of different sources of the signals.

1 cell length), thus making it a good candidate for a long range messenger. Myosin II is a possible target. Activation of myosin II activity by phosphorylation of the regulatory light chain could stiffen the cortex or cause contraction leading to cytoplasmic flow towards the front of the cell. The short range messengers could block this activity at the front of the cell by locally stimulating phosphorylation of the heavy chains causing dissociation of thick filaments. The heavy chain kinase is cytoskeletally associated (Berlot *et al.*, 1987), and therefore restricted in its range of action.

Thus there is the potential for a complex series of regulatory activities, where activated receptors generate local stimulation of pseudopod extension via localized activation of actin polymerization and depolymerization of myosin. The receptors could also produce long-range messengers that activate myosin by phosphorylating the light chains. The result would be a balance between locally stimulated expansion and globally regulated contraction. Where receptor occupancy is higher than the average over the cell, expansion would be favoured. Elsewhere, contraction would occur.

The mechanism of adaptation in this process is complex. There are multiple time scales of adaptation, as reflected by the time courses of receptor phosphorylation, IP_3, cGMP, myosin heavy and light chain phosphorylation, and motility and area changes. Receptor phosphorylation could explain adaptation by shutting off the signals produced by the receptor upon binding ligand. However, receptor phosphorylation is relatively slow (half-time of 1–2 minutes), and therefore cannot explain the rapid rises and falls in IP_3 and cGMP concentrations, and actin nucleation activity.

The role of adaptation is also complex. The basic function of adaptation is typically to remove the response to a basal level of stimulus, allowing the system to respond to small changes in stimulus. Most of the responses (but not receptor phosphorylation) return to less than about 10% of the peak values in the continued presence of the stimulus. This suggests that maximal cell responses may occur when cells are presented with spatial gradients that are increasing with time. This is precisely the sort of stimulus that occurs during aggregation – waves of cAMP travel outward, and cells move accurately towards aggregation centers. Cells do not follow the wave as it moves outward, probably because high concentrations of cAMP cause an inhibition of cell motility that lasts for several minutes (Segall, 1988; Varnum & Soll, 1984). In a stable gradient, cell movement may provide a temporal stimulus as well.

CONCLUSIONS

The mutational studies summarized in this review suggest that amoeboid chemotaxis relies on multiple signals generated when a chemoattractant binds to its receptor. This divergence of signals after the receptor contrasts with the eubacterial chemotaxis system, in which there appears to be a single signal transduction chain leading from a receptor type to the flagellum (Bourret et al., 1989; P. Matsumura, this volume; Stock et al., 1990). The motility of bacteria is simpler in that the swimming behaviour is regulated by the direction of rotation of the flagellum. Although the flagellum with its flagellar motor is an extremely complex organelle, the relevant output is in effect binary – either clockwise or counterclockwise rotation of the flagellum. The regulation of this output is performed by an elegantly designed system that measures temporal changes in chemoattractant concentration (Schnitzer et al., this volume; Segall et al., 1986). The result is a single pathway under negative feedback control that relies on the proper functioning of all of its components.

These two characteristics – a *single* output coupled to a system measuring *temporal* changes, distinguish bacterial behaviour from amoeboid chemotaxis. Control of amoeboid motility appears to involve *multiple* outputs in a system that measures both *temporal* and *spatial* changes. The multiple outputs involve at least actin and myosin II – removal of myosin II blocks some cAMP-induced responses but not all. Actin could play a critical role in pseudopod extension, although most of the evidence is circumstantial. In addition, the spatial localization of these two components inside the cell differs – actin in pseudopods, myosin II more in the rear. The chemotaxis system can control and direct this distribution, resulting in spatially organized cytoskeletal rearrangement in response to a spatial gradient. But temporal changes in chemoattractant concentration lead to multiple cell responses and adaptation, as well. Thus multiple outputs are part of the responses of the cells to both temporal and spatial gradients.

The conclusion from this analysis is that further insights into the mechanisms of amoeboid chemotaxis and motility may require studying partial defects rather than complete loss of the phenotype. A system which has branched signalling pathways with multiple outputs may show only partial defects on the overall behaviour when only a single output is removed. The challenge for future studies will be to synthesize such results in order to produce a clear understanding of how amoeboid chemotaxis works.

ACKNOWLEDGEMENTS

I wish to thank the authors and journals who kindly allowed me to reproduce figures from their articles, and Dr John Condeelis for his comments on an earlier version of this review.

REFERENCES

Abe, T., Maeda, Y. & Iijima, T. (1988). Transient increase of the intracellular calcium concentration during chemotactic signal transduction in *Dictyostelium discoideum* cells. *Differentiation*, **39**, 90–6.

Aekerle, S., Wurster, B. & Malchow, D. (1985). Oscillations and cyclicAMP-induced changes of the K^+ concentration in *Dictyostelium discoideum*. *EMBO Journal*. **4**, 39–43.

Andre, E., Brink, M., Gerisch, G., Isenberg, G., Noegel, A., Schleicher, M., Segall, J. E., & Wallraff, E. A. (1989). A *Dictyostelium* mutant deficient in severin, an F-actin fragmenting protein, shows normal motility and chemotaxis. *Journal of Cell Biology*. **108**, 985–95.

Barclay, S. L. & Henderson, E. J. (1977). A method for selecting aggregation-defective mutants of *Dictyostelium discoideium*. In *Development and Differentiation in the Cellular Slime Moulds*, P. Cappuccinelli and J. M. Ashworth, eds. pp. 291–6, Elsevier: North-Holland, Amsterdam.

Barclay, S. L. & Henderson, E. J. (1982). Thermosensitive development and tip regulation in a mutant of *Dictyostelium discoideum*. *Proceedings of the National Academy of Sciences, USA*, **79**, 505–9.

Barra, J., Barrand, P, Blondelet, M. H. & Brachet, P. (1980). *pds*A, a gene involved in the production of active phosphodiesterase during starvation of *Dictyostelium discoideum* amoebae. *Molecular & General Genetics* **177**, 607–13.

Bennett, H. & Condeelis, J. (1984). Decoration with myosin subfragment-1 disrupts contacts between microfilaments and the cell membrane in isolated *Dictyostelium* cortices. *Journal of Cell Biology* **99**, 1434–40.

Berlot, C. H., Spudich, J. A. & Devrotes, P. N. (1985). Chemoattractant-elicited increases in myosin phosphorylation in *Dictyostelium*. *Cell*, **43**, 307–14.

Berlot, C. H., Devreotes, P. N. & Spudich, J. A. (1987). Chemoattractant-elicited increases in *Dictyostelium* myosin phosphorylation are due to changes in myosin localization and increases in kinase activity. *Journal of Biological Chemistry* **262**, 3918–26.

Berridge, M. J. & Irvine, R.F. (1989). Inositol phosphates and cell signalling. *Nature, London*, **341**, 197–205.

Bresnick, A. R., Warren, V. & Condeelis, J. (1990). Identification of a short sequence essential for actin binding by *Dictyostelium* ABP-120. *Journal of Biological Chemistry*, **265**, 9236–40.

Bonner, J. T. (1967). *The Cellular Slime Molds*. Princeton University Press: Princeton, New Jersey.

Bonner, J. T., Suthers, H. B. & Odell, G. M. (1986). Ammonia orients cell masses and speeds up aggregating cells of slime moulds. *Nature, London*, **323**, 630–2.

Bonner, J. T., Har, D. & Suthers, H. B. (1989). Ammonia and thermotaxis: further evidence for a central role of ammonia in the directed cell mass movements of *Dictyostelium discoideum*. *Proceeding of the National Academy of Sciences, USA*, **86**, 2733–6.

Bourret, R. B., Hess, J. F., Borkovidh, K. A., Pakula, A. A. & Simon, M. I, (1989). Protein phophorylation, chemotaxis and the two-component regulatory system of bacteria. *Journal of Biological Chemistry*, **264**, 7085–8.

Bray, D. & Vasiliev, J. (1989). Networks from mutants. *Nature, London*, **338**, 203–4.

Bresnick, A. R., Warren, V. & Condeelis, J. (1990). Identification of a short sequence essential for actin binding by *Dictyostelium* ABP-120. *Journal of Biological Chemistry*, **16**, 9236–40.

Bumann, J., Wurster, B. & Malchow, D. (1984). Attractant-induced changes and oscillations of the extracellular calcium concentration in suspensions of differentiating *Dictyostelium* cells. *Journal of Cell Biology*, **98**, 173–8.

Cappuccinelli, P. & Ashworth, J. M. (1976). The effect of inhibitors of microtubule and microfilament function on the cellular slime mould *Dictyostelium discoideum*. *Experimental Cell Research*, **103**, 387–93.

Clarke, M. (1978). A selection method for isolating motility mutants of *Dictyostelium discoideum*. In *Cell Reproduction*, E. R. Dirksen D.M. Prescott & C. F. Fox, ed., pp. 621–9, Academic Press: New York.

Condeelis, J., Bresnick, A., Demma, M., Dharmawardhane, S., Eddy, R., Hall, A. L., Sauterer, R. & Warren, V. (in press). Mechanisms of amoeboid chemotaxis: an evaluation of the cortical expansion model.

Coukell, M. B. (1975). Parasexual genetic analysis of aggregation-deficient mutants of *Dictyostelium discoideum*. *Molecular and General Genetics*, **142**, 119–35.

Coukell, M. B., Lappano, S. & Cameron, A. M. (1983). Isolation and characterization of cAMP unresponsive (frigid) aggregation-deficient mutants of *Dictyostelium discoideum*. *Developmental Genetics*, **3**, 283–97.

Coukell, M. B. & Cameron A. M. (1986). Characterization of revertants of *stmF* mutants of *Dictyostelium discoideum*: evidence that *stm*F is the structural gene of the cGMP-specific phosphodiesterase. *Developmental Genetics*, **6**, 163–77.

Crowley, T. E., Nellen, W., Gomer, R. H. & Firtel, R. A. (1985). Phenocopy of discoidin I-minus mutants by antisense transformation in *Dictyostelium*. *Cell*, **43**, 633–41.

De Lozanne, A. & Spudich, J. (1987). Disruption of the *Dictyostelium* myosin heavy chain by homologous recombination. *Science*, **236**, 1087–92.

De Priester, W., Riegman, P. & Vonk, A. (1988). Effects of microtubule-disrupting agents on chemotactic events in *Dictyostelium discoideum*. *European Journal of Cell Biology*, **46**, 94–7.

De Wit, R. J. W. & Bulgakov, R. (1986*a*) Folate chemotactic receptors in *Dictyostelium discoideum*. I. Ligand-induced conversion between four receptor states. *Biochimica & Biophysica Acta*, **886**, 76–87.

De Wit, R. J. W. & Bulgakov, R. (1986*b*). Folate chemotactic receptors in *Dictyostelium discoideum*. II. Guanine nucleotides alter the rates of interconversion and the proportioning of four receptor states. *Biochimica & Biophysica Acta* **886**, 88–95.

De Wit, R. J. W. & van Haastert, P. J. M. (1985). Binding of folates to *Dictyostelium discoideum* cells. Demonstration of five classes of binding sites and their interconversion. *Biochimica & Biophysica Acta*, **814**, 199–213.

De Wit, R. J. W., Bulgakov, R., Pinas, J. E. & Konijn, T. M. (1985). Relationships between the ligand specificity of cell surface folate binding sites, folate degrading enzymes and cellular responses in *Dictyostelium discoideum*. *Biochimica & Biophysica Acta* **814**, 214–26.

Devreotes, P. N. & Zigmond, S. H. (1989). Chemotaxis in eukaryotic cells: a focus on leukocytes and *Dictyostelium*. *Annual Reviews of Cell Biology*, **4**, 649–86.

Dharmawardhane, S. Warren, V., Hall, A. L. & Condeelis, J. (1989). Changes in the association of actin-binding proteins with the actin cytoskeleton during chemotactic stimulation of *Dictyostelium discoideum*. *Cell Motility & the Cytoskeleton*, **13**, 57–63.

Dicou, E. & Brachet, P. (1980). A separate phosphodiesterase for the hydrolysis of cyclic guanosine 3,5-monophosphate in growing *Dictyostelium discoideum* amoebae. *European Journal of Biochemistry*, **109**, 507–14.

Eckert, B. S. & Lazarides, E. (1978). Localisation of actin in *Dictyostelium discoideum* amebas by immunofluorescence. *Journal of Cell Biology*, **77**, 714–21.

Eckert, B. S., Warren, R. H. & Rubin, R. W. (1977). Structural and biochemical aspects of cell motility in amebas of *Dictyostelium discoideum*. *Journal of Cell Biology*, **72**, 339–50.

Europe-Finner, G. N., Gammon, B., Wood, C. A. & Newell, P. C. (1989). Inositol tris- and polyphosphate formation during chemotaxis of *Dictyostelium*. *Journal of Cell Science*, **93**, 585–92.

Europe-Finner, G. N., Luderus, M. E. E., Small, N. V., van Driel, R., Reymond, C. D., Firtel, R. A., & Newell, P. C. (1986). Mutant *ras* gene induces elevated levels of inositol tris- and hexakisphosphates in *Dictyostelium*. *Journal of Cell Science*, **89**, 13–20.

Europe-Finner, G. N. & Newell, P. C. (1986). Inositol, 1,4,5-trisphosphate induces calcium release from a non-mitochondrial pool in amoebae of *Dictyostelium*. *Biochimica & Biophysica Acta*, **88**, 335–40.

Faure, M., Podgorski, G. J., Franke, J. & Kessin, R. H. (1989). Rescue of a *Dictyostelium discoideum* mutant defective in phosphodiesterase. *Developmental Biology*, **131**, 366–72.

Feit, A. & Sollito, A. (1987). Ammonia is the gas used for the spacing of fruiting bodies in the cellular slime mold, *Dictyostelium discoideum*. *Differentiation*, **33**, 193–6.

Firtel, R. A., van Haastert, P. J. M., Kimmel, A. R. & Devreotes, P. N. (1989). G protein linked signal transduction pathways in development: *Dictyostelium* as an experimental system. *Cell*, **58**, 235–9.

Fisher, P. R. (in press). Pseudopodium activation and inhibition signals in chemotaxis by *Dictyostelium discoideum* amoebae. *Seminars in Cell Biology*.
Fisher, P. R., Dohrmann, E. & Williams, K. L. (1984). Signal processing *Dictyostelium discoideum* slugs. *Modern Cell Biology*, **3**, 197–248.
Fisher, P. R., Merkl, R. & Gerisch, G. (1989). Quantitative analysis of cell motility and chemotaxis in *Dictyostelium discoideum* by using an image processing system and a novel chemotaxis chamber providing stationary chemical gradients. *Journal of Cell Biology*, **108**, 973–84.
Fukui, Y., De Lozanne, A. & Spudich, J. A. (1990). Structure and function of the cytoskeleton of a *Dictyostelium* myosin-minus mutant. *Journal of Cell Biology*, **110**, 367–78.
Fukui, Y., Lynch, T. J., Brzeska, H. & Korn, E. D. (1989). Myosin I is located at the leading edges of locomoting *Dictyostelium* amoebae. *Nature, London*, **341**, 328–31.
Fukui, Y., Yumura, S. & Yumura, T. K. (1987). Agar-overlay immunofluorescence: high-resolution studies of cytoskeletal components and their changes during chemotaxis. *Methods in Cell Biology*, **28**, 347–56.
Futrelle, R. P., Traut, J. & McKee, W. G. (1982). Cell behavior in *Dictyostelium discoideum*: preaggregation response to localized cyclic AMP responses. *Journal of Cell Biology*, **92**, 807–21.
Gerisch, G. (1986). *Dictyostelium discoideum*. In *Cellular and Molecular Aspects of Developmental Biology*. (M. Fougereau and R. Stora, eds.) Elsevier Science Publishers.
Gerisch, G. (1987). Cyclic AMP and other signals controlling cell development and differentiation in *Dictyostelium*. *Annual Reviews in Biochemistry*, **56**, 853–79.
Gerisch, G., Malchow, D., Huesgen, A., Nanjundiah, V., Roos, W., Wick, U. & Huelser, D. (1975). Cyclic-AMP reception and cell recognition in *Dictyostelium discoideum*. In *Developmental Biology: ICN-UCLA Symposia on Molecular and Cellular Biology* (D. McMahon and C. F. Fox, eds.) **2**, 76–88.
Gerisch, G. Segall, J. E. & Wallraff, E. (1989). Isolation and behavioral analysis of mutants defective in cytoskeletal proteins. *Cell Motility and the Cytoskeleton*, **14**, 75–9.
Hall, A. L., Schlein, A. & Condeelis, J. (1988). Relationship of pseudopod extension to chemotactic hormone-induced actin polymerization in amoeboid cells. *J.Cellular Biochemistry*, **37**, 285–99.
Hall, A. L., Warren, V. Dharmawardhane, S. & Condeelis, J. (1989a). Identification of actin nucleation activity and polymerization inhibitor in ameboid cells: their regulation by chemotactic stimulation. *Journal of Cell Biology*, **109**, 2207–13.
Hall, A. L., Warren, V. & Condeelis, J. (1989b). Transduction of the chemotactic signal to the actin cytoskeleton of *Dictyostelium discoideum*. *Developmental Biology*, **136**, 517–25.
Janssens, P. M. W. & van Haastert, P. J. M. (1987). Molecular basis of transmembrane signal transduction in *Dictyostelium discoideum*. *Microbiological Reviews*, **51**, 396–418.
Jung, G. & Hammer, J. A. III (1989). Creation of *Dictyostelium* myosin I mutants. *Journal of Cell Biology*, **109**, 281a.
Jung, G., Saxe, C. L., III, Kimmel, A. R. & Hammer, J. A. III (1989). *Dictyostelium discoideum* contains a gene encoding a myosin I heavy chain. *Proceedings of the National Academy of Sciences, USA*, **86**, 6186–90.
Kayman, S. D., Birchman, R. & Clarke, M. (1988). MotA1552, a mutation of *Dictyostelium discoideum* having pleotropic effects on motility and discoidin I regulation. *Genetics*, **118**, 425–36.

Kayman, S. C., Reichel, M. & Clarke, M. (1982). Motility mutants of *Dictyostelium discoideum*. *Journal of Cell Biology*, **92**, 705–11.
Kesbeke, F. Snaar-Jagalska, B. E. & van Haastert, P. J. M. (1988). Signal transduction in *Dictyostelium fgd*A mutants with a defective interaction between surface cAMP receptors and a GTP-binding regulatory protein. *Journal of Cell Biology*, **107**, 521–8.
Klein, P., Fontana, D., Knox, B, Theibert, A. & P. Devreotes (1985) cAMP receptors controlling cell-cell interactions in the development of *Dictyostelium*. *Cold Spring Harbor Symposia in Quantitative Biology*, **50**, 787–99.
Klein, P. S., Sun, T. J., Saxe, C. L., III, Kimmel, A. R., Johnson, R. L. & Devreotes, P. N. (1988). A chemoattractant receptor controls development in *Dictyostelium discoideum*. *Science*, **241**, 1467–72.
Klein, P., Vaughan, R., Borleis, J. & Devreotes, P. (1987). The surface cyclic AMP receptor in *Dictyostelium*: levels of ligand-induced phosphorylation, solubilization, identification of primary transcript, and developmental regulation of expression. *Journal of Biological Chemistry*, **262**, 358–64.
Knecht, D. A. & Loomis, W. F. (1987). Antisense RNA inactivation of myosin heavy chain gene expression in *Dictyostelium discoideum*. *Science*, **236**, 1081–6.
Knecht, D. A. & Loomis, W. F. (1988). Developmental consequences of the lack of myosin heavy chain in *Dictyostelium discoideum*. *Developmental Biology*, **128**, 178–84.
Konijn, T. M. (1969). Effect of bacteria on chemotaxis in the cellular slime molds. *Journal of Bacteriology*, **99**, 503–9.
Korn, E. D. & Hammer, J. A. III (1988). Myosins of nonmuscle cells. *Annual Reviews in Biophysics & Biophysical Chemistry*, **17**, 23–45.
Kumagai, A., Pupillo, M., Gundersen, R. Miake-Lye, R., Devreotes, P. N. & Firtel, R. A. (1989). Regulation and function of G-α protein subunits in *Dictyostelium Cell*, **57**, 265–75.
Liu, G. & Newell, P. C. (1988). Evidence that cyclic GMP regulates myosin interaction with the cytoskeleton during chemotaxis of *Dictyostelium*. *Journal of Cell Science*, **90**, 123–9.
Loomis, W. F. (1975). Dictyostelium discoideum, *A Developmental System*. Academic Press, Inc.: New York.
Loomis, W. F. (ed.) (1982). *The development of* Dictyostelium discoideum. Academic Press, Inc: New York.
Luderus, M. E., Reymond, C. D., van Haastert, P. J. M. & van Driel, R. (1988). Expression of a mutated *ras* gene in *Dictyostelium discoideum* alters the binding of cyclic AMP to its chemotatic receptor. *Journal of Cell Science*, **90**, 701–6.
Luna, E. J. & Condeelis, J. S. (in press). Actin-associated proteins in *Dictyostelium discoideum*. *Developmental Genetics*.
Malchow, D., Nanjundiah, V. & Gerisch, G. (1978). pH oscillations in cell suspensions of *Dictyostelium discoideum*: their relation to cyclic AMP signals. *Journal of Cell Science*, **30**, 319–30.
Mann, S. K. O., Pinko, C. & Firtel, R. A. (1988). cAMP regulation of early gene expression in signal transduction mutants of *Dictyostelium*. *Developmental Biology*, **130**, 294–303.
Manstein, D. J., Titus, M. A., De Lozanne, A. & Spudich, J. A. (1989). Gene replacement in *Dictyostelium*: generation of myosin null mutants. *EMBO Journal*, **8**, 923–32.
Mato, J. M. van Haastert, P. J. M., Krens, F. A. & Konijn, T. M. (1978). Chemotaxis in *Dictyostelium discoideum*: effect of concanavalin A on chemoattractant

mediated cyclic GMP and light scattering decrease. *Cell Biology International Reports.* **2**, 163–70.
Mato, J. M. van Haastert, P. J. M., Krens, F. A. Rhijnsburger, E. H., Dobbe, F. C., & Konijn, T. M. & P. M. (1977). Cyclic AMP and folic acid stimulated cyclic GMP accumulation in *Dictyostelium discoideum. FEBS Letters.* **79**, 331–6.
McRobbie, S. J. (1986). Chemotaxis and cell motility in the cellular slime molds. *CRC Critical Reviews in Microbiology*, **13**, 335–75.
McRobbie, S. J. & Newell, P. C. (1983). Changes in actin associated with the cytoskeleton following chemotactic stimulation of *Dictyostelium discoideum. Journal of Cell Science*, **68**, 139–51.
Newell, P., Europe-Finner, G. N., Small, N. V. & Liu, G. (1988). Inositol phosphates, G-proteins and *ras* genes involved in chemotactic signal transduction of *Dictyostelium. Journal of Cell Science*, **89**, 123–7.
Nishizuka, Y. (1988). The molecular heterogeneity of protein kinase C and its implications for cellular regulation. *Nature, London*, **344**, 661–5.
Noegel, A. A., Leiting, B, Witke, W., Gurniak, C., Harloff, C., Hartmann, H, Wiesmueller, E. & Schleicher, M. (1989*a*). Biological roles of actin-binding proteins in *Dictyostelium discoideum* examined using genetic techniques. *Cell Motility & Cytoskeleton*, **14**, 69–74.
Noegel, A. A., Rapp, S., Lottspeich, F., Schleicher, M., & Stewart, M. (1989*b*) The Dictyostelium gelation factor shares a putative actin binding site with alpha-actinins and dystrophin and also has a rod domain containing six 100-residue motifs that appear to have a cross-β conformation. *Journal of Cell Biology*, **109**, 607–18.
Noegel, A. A. & Witke, W. (1988). Inactivation of the alpha-actinin gene in *Dictyostelium. Developmental Genetics*, **9**, 531–8.
Oster, G. (1988). Biophysics of the leading lamella. *Cell Motility Cytoskeleton* **10**, 164–71.
Pasternak, C., Spudich, J. A. & Elson, E. L. (1989). Capping of surface receptors and concomitant cortical tension are generated by conventional myosin. *Nature, London*, **341**, 549–51.
Pan, P. Hall, E. M. & Bonner, J. T. (1972). Folic acid as second chemotactic substance in the cellular slime moulds. *Nature New Biology*, **237**, 181–2.
Peters, D. J. M., Knecht, D. A., Loomis, W. F., DeLozanne A., Spudich, J. & van Haastert, P. J. M. (1988). Signal transduction, chemotaxis and cell aggregation in *Dictyostelium discoideum* cells without myosin heavy chain. *Developmental Biology*, **128**, 158–63.
Poff, K. L., Fontana, D. R., Hader, D. P. & Schneider, M. J. (1986). An optical model for phototactic orientation in *Dictyostelium discoideum* slugs. *Plant Cell Physiology* **27**, 533–39.
Potel, M. H. & Mackay, S. A. (1979). Preaggregative cell motion in *Dictyostelium. Journal of Cell Science*, **36**, 281–309.
Pupillo, M., Kumagai, A., Pitt, G. S., Firtel, R. A. & Devreotes, P. N. (1989). Multiple alpha subunits of guanine nucleotide-binding proteins in *Dictyostelium. Proceedings of the National Academy of Sciences*, USA **86**, 4892–6.
Raper, K. B. (1984). *The Dictyostelids*. Princeton University Press: Princeton, New Jersey.
Reymond, C. D., Gomer, R. H., Mehdy, M. C. & Firtel, R. A. (1984). Developmental regulation of a *Dictyostelium* gene encoding a protein homologous to mammalian *ras* protein. *Cell*, **39**, 141–8.
Reymond, C. D., Gomer, Nellen, W., Theibert, A., Devreotes, P. & Firtel, R. A. (1986). Phenotypic changes induced by a mutated *ras* gene during the develop-

ment of *Dictyostelium* transformants. *Nature, London*, **323**, 340–3.
Reymond, C. D., Nellen, W. & Firtel, R. A. (1985). Regulated expression of *ras* gene constructs in *Dictyostelium* transformants. *Proceedings of the National Academy of Sciences, USA*, **82**, 7005–9.
Robbins, S. M., Williams, J. G., Jermyn, K. A., Spiegelman, G. B. & Weeks, G. (1989). Growing and developing *Dictyostelium* cells express different *ras* genes. *Proceedings of the National Academy of Sciences, USA*, **86**, 938–42.
Roos, U. P., De Brabander, M. & Nuydens, R. (1986). Cell shape and organization of F-actin and microtubules in randomly moving and stationary amebae of *Dictyostelium discoideum*. *Cell Motility & Cytoskeleton*, **6**, 176–85.
Ross, F. M. & Newell, P. C. (1979). Genetics of aggregation pattern mutations in the cellular slime mould *Dictyostelium discoideum*. *Journal of General Microbiology*, **115**, 289–300.
Ross, F. M. & Newell, P. C. (1981). Streamers: chemotactic mutants of *Dictyostelium discoideum* with altered cGMP metabolism. *Journal of General Microbiology*, **127**, 339–50.
Rubino, S., Fighetti, M, Unger, E. & Cappuccinelli, P. (1984). Location of actin, myosin and microtubular structures during directed locomotion of *Dictyostelium* amoebae. *Journal of Cell Biology*, **98**, 382–90.
Schaap, P. (1986). Regulation of size and pattern in the cellular slime moulds. *Differentiation*, **33**, 1–16.
Schleicher, M., Andre, E., Hartmann, H. & Noegel, A. A. (1988). Actin-binding proteins are conserved from slime molds to man. *Developmental Genetics*, **95**, 521–30.
Segall, J. E. (1988). Quantification of motility and area changes of *Dictyostelium discoideium* amoebae in response to chemoattractants. *Journal of Muscle Research and Cell Motility* **9**, 481–90.
Segall, J. E., Block, S. M. & Berg, H. C. (1986). Temporal comparisons in bacterial chemotaxis. *Proceedings of the National Academy of Sciences, USA*, **83**, 8987–91.
Segall, J. E., Bominaar, A. A., Wallraff, E. & de Wit, R. J. W. (1988). Analysis of a *Dictyostelium* chemotaxis mutant with altered chemoattractant binding. *Journal of Cell Science*, **91**, 479–89.
Segall, J. E., Fisher, P. R. & Gerisch, G. (1987). Selection of chemotaxis mutants of *Dictyostelium discoideum*. *Journal of Cell Biology*, **104**, 151–61.
Segall, J. E. & Gerisch, G. (1989). Genetic approaches to cytoskeleton function and the control of cell motility. *Current Opinion in Cell Biology: Cytoplasm and Cell Motility*. **1**: 44–50.
Small, N. V., Europe-Finner, G. N. & Newell, P. C. (1986). Calcium induces cyclic GMP formation in *Dictyostelium*. *FEBS Letters*, **203**, 11–14.
Soll, D. R., Voss, E., Varnum-Finney, B & Wessels, D. (1988). 'Dynamic morphology system': a method for quantitating changes in shape, pseudopod formation and motion in normal and mutant amoebae of *Dictyostelium discoideum*. *Journal of Cell Biochemistry*, **37**, 177–92.
Springer, W. R., Cooper, D. N. W. & Barondes, S. H. (1984). Discoidin I is implicated in cell–substratum attachment and ordered cell migration of *Dictyostelium discoideum* and resembles fibronectin. *Cell*, **39**, 557–664.
Spudich, J. A. (1987). *Dictyostelium discoideum*: molecular approaches to cell biology. *Methods in Cell Biology*, Volume 28.
Spudich, J. A. (1989). In pursuit of myosin function. *Cell Regulation*, **1**, 1–11.
Spudich, J. A. & Spudich, A. (1982). Cell motility. In *The development of* Dictyostelium discoideum. W. F. Loomis, ed., pp. 169–94 Academic Press: New York.

Stock, J. B., Stock, A. M. & Mottonen, J. M. (1990). Signal transduction in bacteria. *Nature, London*, **344**, 395–400.

Swanson, J. A. & Taylor, D. L. (1982). Local and spatially coordinated movements in *Dictyostelium discoideum* amoebae during chemotaxis. *Cell*, **28**, 225–32.

Taylor, D. L. & Condeelis, J. S. (1979). Cytoplasmic structure and contractility in amoeboid cells. *International Reviews of Cytology*, **56**, 57–144.

Tilney, L. & Inoue, S. (1982). Acrosomal reaction of thyone sperm II. the kinetics and mechanism of acrosomal process elongation. *Journal of Cell Biology* **93**, 820–7.

Titus, M. A., Warrick, H. M. & Spudich, J. A. (1989). Multiple actin-based motors in *Dictyostelium*. *Cell Regulation*, **1**, 55–63.

Van Haastert, P. (in press) Signal transduction and the control of development in *Dictyostelium discoideum*. *Seminars in Developmental Biology*.

Van Haastert, P. J. M., de Wit, R. J. W. & Konijn, T. M. (1982a) Antagonists of chemoattractants reveal separate receptors for cAMP, folic acid and pterin in *Dictyostelium*. *Experimental Cell Research*, **140**, 453–6.

Van Haastert, P. J. M., de Vries, M. J., Penning, L. C., Roovers, E., van der Kaay, J., Erneux, C. & van Lookeren Campagne, M. M. (1989). Chemoattractant and guanosine 5'-thiotriphosphate induce the accumulation of inositol 1, 4, 5-trisphosphate in *Dictyostelium* cells that are labelled with ^{3}H-inositol by electroporation. *Biochemical Journal*, **258**, 577–86.

Van Haastert, P. J. M. Vanwalsum, H, Vandermeer, R, Bulgakov, R. & Konijn, T. M. (1982b) Specificity of the cyclic GMP-binding activity and of a cyclic GMP-dependent cyclic GMP phosphodiesterase in *Dictyostelium discoideum*. *Journal of Molecular and Cellular Endocrinology*, **25**, 171–82.

van Haastert, P. J. M., Kesbeke, F., Reymond, C. D., Firtel, R. A., Luderus, E. & van Driel, R. (1987). Aberrant transmembrane signal transduction in *Dictyostelium* cells expressing a mutated *ras* gene. *Proceedings of the National Academy of Sciences*, USA **84**, 4905–9.

van Haastert, P. J. M., van Lookeren Campagne, M. M. & Ross, F. M. (1982c). Altered cGMP-phosphodiesterase activity in chemotactic mutants of *Dictyostelium discoideum*. *FEBS Letters*, **147**, 149–52.

Varnum, B., Edwards, K. B. & Soll D. R. (1985). *Dictyostelium* amoebae alter motility differently in response to increasing versus decreasing temporal gradients of cAMP. *Journal of Cell Biology*, **101**, 1–5.

Varnum, B., Edwards, K. B. & Soll, D. R. (1986). The developmental regulation of single-cell motility in *Dictyostelium discoideum*. *Developmental Biology*, **113**, 218–27.

Varnum, B. & Soll, D. R. (1984). Effects of cAMP on single cell motility in *Dictyostelium*. *Journal of Cell Biology*, **99**, 1151–5.

Varnum-Finney, B., Schroeder, N. A. & Soll, D. R. (1988). Adaptation in the motility response to cAMP in *Dictyostelium discoideum*. *Cell Motility and the Cytoskeleton*, **9**: 9–16.

Varnum-Finney, B., Voss, E. & Soll, D. R. (1987). Frequency and orientation of pseudopod formation of *Dictyostelium discoideum* amoebae chemotaxing in a spatial gradient: further evidence for a temporal mechanism. *Cell Motility and the Cytoskeleton*, **8**, 18–26.

Vogel, G. (1987). Endocytosis and recognition mechanisms in *Dictyostelium discoideum*. *Methods in Cell Biology*, **28**, 129–37.

Waddell, D. R., Duffy, K. & Vogel, G. (1987). Cytokinesis is defective in *Dictyostelium* mutants with altered phagocytic recognition, adhesion and vegetative cell cohesion properties, *Journal of Cell Biology*, **105**, 2293–300.

Wallraff, E., Schleicher, M., Modersitzki, M, Rieger, D., Isenberg, G. & Gerisch, G. (1986). Selection of *Dictyostelium* mutants defective in cytoskeletal proteins: use of an antibody that binds to the ends of α-actinin rods. *EMBO Journal*, **5**, 61–7.

Weeks, G. & Pawson, T. (1987). The synthesis and degradation of *ras*-related gene products during growth and differentiation in *Dictyostelium discoideum*. *Differentiation*, **33**, 207–13.

Wessels, D., Soll, D. R., Knecht, D. Loomis, W. F. DeLozanne, A., & J. Spudich (1988) Cell motility and chemotaxis in *Dictyostelium* amebae lacking myosin heavy chain. *Developmental Biology*, **128**, 164–77.

Wessels, D., Schroeder, N. A., Voss, E., Hall, A. L., Condeelis, J. & Soll, D. R. (1989). cAMP-mediated inhibition of intracellular particle movement and actin reorganization in *Dictyostelium*. *Journal of Cell Biology*, **109**, 2841–51.

Wurster, B., Schubiger, K., Wick, U. & Gerisch, G. (1977). Cyclic GMP in *Dictyostelium discoideum*. Oscillations and pulses in response to folic acid and cyclic AMP signals. *FEBS Letters*, **76**, 141–4.

Yumura, S. & Fukui, Y. (1985). Reversible cyclic AMP-dependent change in distribution of myosin thick filaments in *Dictyostelium*. *Nature, London*, **314**, 194–6.

CHEMOTAXIS OF *DICTYOSTELIUM*: THE SIGNAL TRANSDUCTION PATHWAY TO ACTIN AND MYOSIN

P. C. NEWELL, G. N. EUROPE-FINNER, G. LIU, B. GAMMON AND C. A. WOOD

Department of Biochemistry, University of Oxford, South Parks Road, Oxford OX1 3QU, UK

INTRODUCTION

The cellular slime mould *Dictyostelium* exists in its vegetative state as a population of amoebae which feed on bacteria and lead a unicellular independent existence. When the food source is exhausted, however, the amoebae move together to form aggregates of between 100 and 100 000 cells which develop into organisms (about 3 mm long) that are able to move over the substratum at 1 mm h^{-1} in response to shallow heat gradients or weak sources of light. The amoebae are attracted into the aggregate by chemotaxis, the chemoattractant being, in the case of *D. discoideum*, cyclic AMP. This nucleotide is emitted by cells at the centre of the aggregate and attracts those further out to move towards the centre. Rather than forming a large radial gradient, the cyclic AMP is produced as a series of pulses which are relayed outwards by the amoebae at the same time as they move inwards towards the aggregate centre (Bonner, 1971; Newell, 1981; Gerisch, 1982).

Most species of the genus *Dictyostelium* employ cyclic AMP for chemotaxis during aggregation (Konijn *et al.*, 1967) and produce highly specific receptors on their cell surfaces that bind this nucleotide in the *anti* configuration (Van Haastert & Kien, 1983). A few other species are known, however, to employ different small molecules (e.g. pteridines by *D. lacteum*) (Van Haastert *et al.*, 1982a) with appropriate cell surface receptors to match. Members of the genus *Polyspondylium* (for example, *P. violaceum* and *P. pallidum*) use a dipeptide called 'glorin' (a blocked derivative of glutamic acid and ornithine) as their chemoattractant (Shimomura, Suthers & Bonner, 1982), showing a close parallel, in this respect, to the leucocyte chemotactic peptide fMet–Leu–Phe (Schiffmann, Corcoran & Wahl, 1975).

The current model of the two signal transduction pathways that connect the cyclic AMP receptors on the cell surface of *D. discoideum* to the systems

Abbreviations: Ins(1,4,5)P$_3$, Inositol (1,4,5)trisphosphate; InsP$_6$, Inositol hexakisphosphate; PtdIns, Phosphatidylinositol; PtdIns4P, phosphatidylinositol 4-phosphate; PtdIns (4,5)P$_2$, phosphatidylinositol(4,5)bisphosphate

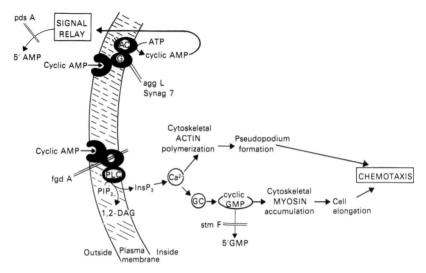

Fig. 1. Signal transduction pathways leading from the cell surface cyclic AMP receptors to regeneration of the cyclic AMP signal (signal relay), and to events of chemotaxis such as pseudopodium formation and cell elongation. *Abbreviations*: G, G-protein; AC, adenylate cyclase; PLC, phospholipase C; 1,2-DAG, 1,2-diacyl glycerol; InsP$_3$, inositol(1,4,5)trisphosphate; PIP$_2$, phosphatidylinositol(4,5) bisphosphate; GC, guanylate cyclase. *Mutant phenotypes: pdsA*, lacking functional cyclic AMP phosphodiesterase (Barra *et al.*, 1980); *fgdA*, non-functional G(alpha)2 subunit of G-protein (Coukell, Lappano & Cameron, 1983; Kesbeke, Snaar-Jagalska & Van Haastert, 1988; Kumagai *et al.*, 1989); *aggL* and *synag 7*, inability to relay due to deficiency in the G-protein/adenylate cyclase coupling (Glazer & Newell, 1981; Snaar-Jagalska & Van Haastert, 1988); *stmF*, deficient in cyclic GMP-specific phosphodiesterase (streamer F mutants) (Ross & Newell, 1979, 1981; Van Haastert, Van Lookeren Campagne & Ross, 1982*b*; Coukell & Cameron, 1986). (From Newell *et al.* (1988, 1990).)

for cyclic AMP relay and chemotactic movement inside the cells is shown in Figure 1. The signal relay system uses adenylate cyclase activated through a G-protein that is coupled to a cell surface cyclic AMP receptor. The cyclic AMP that is produced (with a peak at about 60 s) is released from the cells and activates the cyclic AMP receptors on cells nearby, thereby causing relay of the cyclic AMP signal outwards from the signalling centre throughout the monolayer population of starving amoebae as they aggregate. (For review, see Janssens & Van Haastert, 1987).

This review is concerned with the signal transduction pathway that links the cyclic AMP receptor with chemotactic movement. In brief, it involves the production of inositol phosphates, release of Ca^{2+} from non-mitochondrial stores, formation of cyclic GMP, and association of F-actin and myosin II with the cytoskeleton. These events occur very rapidly compared to those involved in signal relay. Peaks of Ins(1,4,5)P$_3$ formation and actin polymerization are observed within 5 s of a cyclic AMP stimulus and peaks of cyclic GMP and cytoskeletal myosin II follow at 10 s and 25 s respectively (Wurster *et al.*, 1977; McRobbie & Newell, 1984; Small, Europe-Finner

& Newell, 1987; Europe-Finner & Newell, 1987a; Liu & Newell, 1988; Europe-Finner et al., 1989).

CELL SURFACE CYCLIC AMP RECEPTORS

The cell surface cyclic AMP receptors have been studied from many different angles for a number of years and a great deal is known about the binding kinetics and molecular transitions that occur when cyclic AMP binding occurs (reviewed in Newell, 1986b). More recently, the cDNA for a cell-surface cyclic AMP receptor has been cloned and sequenced (Klein et al., 1988). Libraries of lambda-gt-11 were screened with cyclic AMP receptor-specific antiserum and clones were obtained, some of which included the entire coding sequence. The nucleotide sequence of cDNA encompassing the coding region revealed an open-reading frame encoding a polypeptide of 392 amino acids with a calculated molecular size of 44 243 D. The NH_2-terminal two-thirds of the coding region was found to be enriched in hydrophobic residues in clusters of 21 to 25 amino acids, while the last 130 amino acids were predominantly hydrophilic and enriched in serines and threonines. The pattern was found to be remarkably similar to the profile of bovine rhodopsin, the beta-adrenergic receptor and several other receptors, the cyclic AMP receptor and bovine rhodopsin sharing 22% amino acid identity over 270 residues in the NH_2-terminal regions of the two receptors. This structural relationship of the cyclic AMP receptor to bovine rhodopsin is supported by immunological data. Antiserum to bovine rhodopsin recognizes the cyclic AMP receptor in immunoblots of crude membranes from aggregation stage amoebae in which the receptor is 0.1% of the total protein. Phosphorylation of the cyclic AMP receptor has been correlated with desensitization, in the same way that phosphorylation has been correlated with deactivation of rhodopsin. On the basis of hydropathy analysis, and the structural and functional similarity to bovine rhodopsin, Klein et al. (1988) proposed a model for the structure of the cyclic AMP receptor in which there were seven transmembrane domains with the NH_2-terminus extracellular and the serine-rich COOH-terminal tail intracellular and the site of ligand-induced phosphorylation.

CHANGES IN INOSITOL POLYPHOSPHATES

The first intimation that inositol phosphates were involved in the chemotactic signal transduction pathway came from work with saponin-permeabilized amoebae. When 5 μM $Ins(1,4,5)P_3$ was added to permeabilized cells, it induced a peak of cyclic GMP formation which was closely similar in its size and duration to that produced by cyclic AMP acting via the cell surface receptors (Europe-Finner & Newell, 1985). Inositol 1,4-bisphosphate (the breakdown product of $Ins(1,4,5)P_3$) had no effect, while Ca^{2+} (which is liberated from internal stores in many cell types by $Ins(1,4,5)P_3$ (see below)

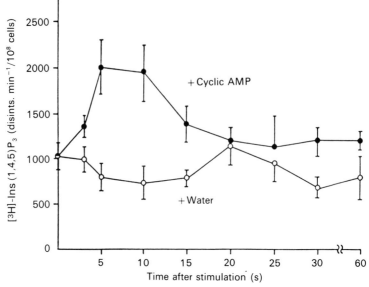

Fig. 2. Time course of Ins(1,4,5)P$_3$ accumulation in amoebae of *D. discoideum* after stimulation with cyclic AMP at 22°C. Amoebae were incubated on filters with ^{3}H-labelled inositol (sp. act. 41.0 Ci mmol^{-1}) for 4h at 22°C then harvested and incubated shaken in suspension for 2h. Amoebae were stimulated with 1 μM cyclic AMP (●--●) or water (○--○) and extracts made at the times shown. The Ins(1,4,5)P$_3$ that accumulated was determined for each time point after HPLC separation of the extracts. Results are the means of six experiments (cyclic AMP) and four experiments (water control). Error bars represent S.E.M. (From Europe-Finner *et al.* (1989).)

mimicked the action of Ins(1,4,5)$_3$, with a half-maximal response at 60 μM (Small, Europe-Finner & Newell, 1986).

More direct evidence came from studying the changes in tritium-labelled inositol phosphates in the cell, initially using the separation technique of Dowex column anion exchange chromatography (Europe-Finner & Newell, 1987a,b), and more recently using HPLC, which gives better resolution of the inositol phosphate isomers (Europe-Finner *et al.*, 1989; Van Haastert *et al.*, 1989).

Using Partisil SAX HPLC columns it was possible to assay specifically for tritium-labelled Ins(1,4,5)P$_3$ in cell extracts from amoebae that had been labelled for several hours with tritiated inositol. When labelled amoebae were stimulated with cyclic AMP, it was found that there was an approximately twofold increase in their content of tritium-labelled Ins(1,4,5)P$_3$ within 5s of the stimulus (Fig. 2) (Europe-Finner *et al.*, 1989; Van Haastert *et al.* 1989). Other peaks from the HPLC column also changed after cyclic AMP stimulation of the amoebae, such as an unidentified peak called P3 which increased more slowly than Ins(1,4,5)P$_3$ to a maximum at about 20s. (P3 runs just after Ins(1,4,5)P$_3$ on the HPLC column and may be an $InsP_3$ isomer.) Other inositol phosphates such as $InsP_6$ did not change appreciably over a time scale of 60 seconds following cyclic AMP stimulation.

BREAKDOWN OF $INS(1,4,5)P_3$: THE INOSITOL PHOSPHATE CYCLE

The enzymes involved in catalysing the breakdown of $Ins (1,4,5)P_3$ and regeneration of the starting components of the inositol phosphate cycle have been studied in detail by Van Lookeren Campagne et al. (1988). In mammalian systems, the $Ins(1,4,5)P_3$ is rapidly broken down by a specific phosphatase that removes the phosphate from the 5-position to produce $Ins(1,4)P_2$ which is then further dephosphorylated to Ins4P and finally to Inositol. In Dictyostelium the predominant route of dephosphorylation was found to be different, with the (lithium-sensitive) 1-phosphatase being the predominant enzyme activity. This produced $Ins(4,5)P_2$ as the main product initially formed and this was then degraded (together with any $Ins(1,4)P_2$ formed) to Ins4P and then inositol.

EVIDENCE FOR G-PROTEINS

Evidence for the involvement of a G-protein linking the cyclic AMP receptor to the inositol phosphate signal transduction pathway is derived from several different types of experiment. Using isolated membrane preparations, Van Haastert and colleagues have reported that GTP and GTP analogues (such as GTPγS) decrease the binding affinity of cyclic AMP to the cell surface receptors by increasing the dissociation rate of the cyclic AMP-receptor complex: effects of this type are characteristic of systems controlled by G-proteins (Van Haastert et al., 1986; Janssens et al., 1986). Other experiments employing saponin-permeabilized amoebae have demonstrated that addition of GTPγS is able to mimic the effects of direct stimulation of the receptors with cyclic AMP, the level of labelled $Ins(1,4,5)P_3$ rising to a maximum within 5s (Fig. 3) (Europe-Finner & Newell, 1987a; Van Haastert et al., 1989; Europe-Finner et al., 1989). Using a very different approach, other workers have identified the gene that appears to code for a G-protein involved with signal transduction. Use was made of 'FgdA'; mutants which are defective in chemotaxis (Coukell, Lappano & Cameron, 1983). These mutants are unable to activate adenylate cyclase or guanylate cyclase in response to cyclic AMP stimulation, and isolated membranes prepared from them do not show the decrease in affinity for cyclic AMP in the presence of GTP or GTP analogues as was found for wild type cells (Kesbeke, Snaar-Jagalska & Van Haastert, 1988). Using DNA transformation experiments and G-alpha-specific antibodies, Kumagai et al. (1989) have demonstrated that the defective gene in these 'FgdA' mutants codes for a G-protein alpha subunit designated G-alpha2. This gene was found to be developmentally regulated, being

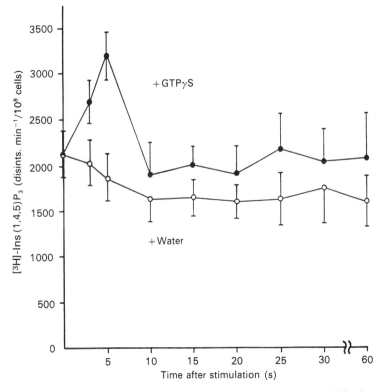

Fig. 3. Stimulation of Ins(1,4,5)P_3 formation by GTPγS in saponin-permeabilized amoebae. Amoebae labelled for 4 h at 22 °C with ^{3}H-inositol (sp. act. 99.5 Ci mmol^{-1}) were rendered permeable by treatment with saponin in the presence of 5 mM ATP and were then stimulated with 68 μM GTPγS (final concentration) (●--●), or water (○--○) and cell extracts made at the times indicated for determination of Ins(1,4,5)P_3 by HPLC. Results are the means of four experiments (GTPγS) and three experiments (water control). Error bars represent S.E.M. (From Europe-Finner et al., 1989.)

expressed at very low levels during the growth phase of the wild type and maximally during aggregation.

INOSITOL LIPIDS

In mammalian systems it is well established that the Ins(1,4,5)P_3 which is formed in response to hormonal activation is produced from the plasma membrane lipid PtdIns(4,5)P_2 (also called PIP$_2$). The following evidence suggests that this lipid is also the precursor for Ins(1,4,5)P_3 in *Dictyostelium*. The PtdIns(4,5)P_2 in the amoebae was labelled with ^{32}P-phosphate, and the extent of accumulation of this label in the various lipids was then determined by extraction with acidified chloroform/methanol followed by deacylation to liberate and then identify the glycerophospho derivatives (Europe-Finner et al., 1989). By this technique, the changes in PtdIns(4,5)P_2 were

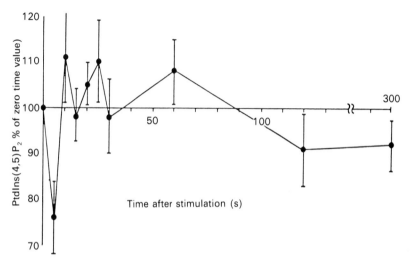

Fig. 4. Changes in PtdIns(4,5)P_2 in amoebae in response to stimulation with cyclic AMP at 22°C. Amoebae were incubated in liquid suspension culture for 5 h in the presence of ^{32}P (10 µCi ml^{-1}) then washed and resuspended without label and samples stimulated with 1 µM cyclic AMP. After intervals of 5 to 300 s, reactions were terminated with ice-cold chloroform/methanol/12 M HCl (20:40:1 (v/v)) and the lipid-soluble fractions extracted. The inositol lipids were deacylated and the water-soluble glycerophospho derivatives separated by anion exchange chromatography. The data are the mean of five experiments with error bars representing S.E.M. (From Europe-Finner et al., 1989.)

assayed in amoebae before and after stimulation of the amoebae with cyclic AMP (Fig. 4). It was found that, within 5 s of stimulation, the level of radioactivity of the PtdIns(4,5)P_2 had fallen to 76 ± 8% of the zero time value. By 10 s the level of PtdIns(4,5)P_2 had returned to a value close to the starting value and remained at this level for over 300 s. The changes in PtdIns(4,5)P_2 were similar to (but in the opposite direction to) those for Ins(1,4,5)P_3. These results are similar to those found for higher cells and support the hypothesis that in *Dictyostelium* activation of phosphoinositidase C (inositol lipid-specific phospholipase C) cleaves PtdIns(4,5)P_2 to produce Ins(1,4,5)P_3.

THE INVOLVEMENT OF THE *D. DISCOIDEUM* RAS GENE

There is now strong evidence that *D. discoideum* possesses two *ras* genes that show close homology to each other and to the human *ras* gene (Pawson et al., 1985; Weeks & Pawson, 1987; Weeks, Lima & Pawson, 1987; Reymond et al., 1984, 1985, 1986; Robbins et al., 1989).

The *ras* gene most studied codes for a protein (p23) that reacts with monoclonal antibodies to the p21$^{v\text{-}ras}$ protein of Harvey murine sarcoma virus. Reymond et al. (1986) used site-directed mutagenesis to mutate the gene to produce a threonine substitution at position 12 (*ras*-Thr$_{12}$) in place of the normal glycine at this position (ras-Gly_{12}) and transformed *D. discoideum* amoebae with plasmids bearing the mutant or normal strains. The phenotype for chemotaxis was found to be normal for the amoebae containing multiple copies of the ras-Gly_{12} gene but was impaired in the amoebae with multiple copies of the ras-Thr_{12} gene (Reymond *et al.*, 1986; Van Haastert *et al.*, 1987). The adenylate cyclase relay system was unaffected but the ability to produce cyclic GMP in response to chemotactic stimuli was reduced in the ras-Thr_{12} strain.

When the levels of labelled inositol phosphates were determined by HPLC, it was found that amoebae transformed with ras-Gly_{12} were normal in their formation of inositol phosphates. Those transformed with the ras-Thr_{12} gene, however, showed a three- to five-fold raised level of labelled $Ins(1,4,5)P_3$ and $InsP_6$ in unstimulated cells, and stimulation with cyclic AMP did not greatly raise the level further (Europe-Finner *et al.*, 1988). It appears likely, therefore, that the ras gene has some function in the control of inositol pathway intermediates. Recent investigations of Wood & Newell (unpublished) into the incorporation of $^{32}PO_4$ into PtdIns4P and $PtdIns(4,5)P_2$ (the precursors of $Ins(1,4,5)P_3$) in strains transformed with either ras-Gly_{12} or ras-Thr_{12} indicated that the ^{32}P in PtdIns4P and $PtdIns(4,5)P_2$ reaches a plateau at approximately 1.5 hours in strains transformed with the normal ras-Gly_{12} in a similar manner to untransformed controls. However, in ras-Thr_{12}-transformed cells the incorporation did not plateau but continued for the 8 hours of the experiment. During this time, the amount of ^{32}P incorporated was two- three fold higher in both PtdIns4P and $PtdIns(4,5)P_2$. One interpretation of these and other related data from the ras-Thr_{12}-transformed cells is that in normal cells the ras protein has a regulatory effect on the step that converts PI to PtdIns4P, although other explanations cannot, as yet, be ruled out.

INTRACELLULAR Ca^{2+} RELEASE BY $INS(1,4,5)P_3$

There are at least two distinct Ca^{2+} pools in *D. discoideum*. One of these is thought to be mitochondrial because of its sensitivity to mitochondrial inhibitors and shows an affinity for Ca^{2+} in the μM concentration range. In contrast, the other Ca^{2+} pool was found to be insensitive to the mitochondrial inhibitors and had a lower capacity but higher affinity for Ca^{2+} (in the nM range). The non-mitochondrial (possibly microsomal) pool was found to be sensitive to $Ins(1,4,5)P_3$. When 5 μM $Ins(1,4,5)P_3$ was added to saponin-permeabilized amoebae it induced a rapid release of Ca^{2+} from this pool (Fig. 5) but had no effect on the mitochondrial pool (Europe-Finner & Newell, 1986a). The effect was not shown by addition of inositol 1,4-bisphosphate. This effect of $Ins(1,4,5)P_3$ is precisely analogous to its role in higher cells and strongly suggests that $Ins(1,4,5)P_3$ has a similar role in regulating cytosolic Ca^{2+} as for mammalian signal transduction.

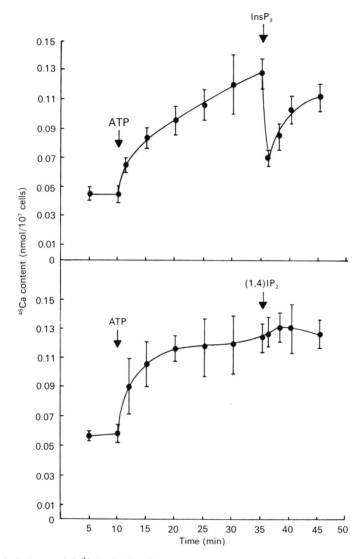

Fig. 5. Release of Ca^{2+} by $Ins(1,4,5)P_3$ in saponin-permeabilized amoebae. At time zero 1 μCi ml^{-1} (final concentration) of $^{45}Ca^{2+}$ was added to a 2.5 ml suspension of saponin-permeabilized amoebae (2×10^7 ml^{-1}) incubated in the presence of antimycin A (10 μM), an ATP regenerating system, and with the free Ca^{2+} set at 180 nM using 1 mM EGTA buffer. After 10 min, 1.5 mM ATP was added and, at 35 min, 100 μl of $Ins(1,4,5)P_3$ (upper curve) or $Ins(4,5)P_2$ as control (lower curve) to a final concentration of 5 μM. Bars indicate S.E.M. of ten ($Ins(1,4,5)P_3$) or six ($Ins(4,5)P_2$) independent experiments. (Data from Europe-Finner & Newell, 1986a.)

More recently, Abe, Maeda & Iijima (1989) using the Ca^{2+} indicators quin 2 and fura 2 (loaded into *D. discoideum* amoebae by electroporation)

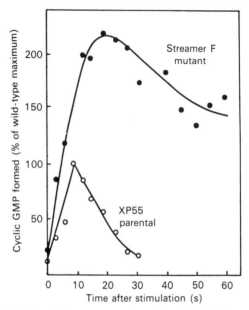

Fig. 6. Changes in the cellular concentration of cyclic GMP in response to a 50 nM pulse of cyclic AMP given at time zero to amoebae of strain XP55 (○--○), and streamer F mutant NP368 (●--●), after 8 h of starvation. Streamer F mutant NP377 gave similar results. The data represent the means of four experiments. (Data from Ross & Newell, 1981.)

have demonstrated that cyclic AMP stimulation produces a four-fold rise in cytosolic Ca^{2+} within 1–2 s, strongly supporting a role for this ion in this signal transduction pathway.

STREAMER F MUTANTS AND THE FORMATION OF CYCLIC GMP

Evidence for cyclic GMP being a part of the signal transduction comes from its transient formation in response to various chemoattractants in several species (Mato et al., 1977; Wurster et al., 1977; Van Haastert & Konijn, 1982) and its formation in large amounts in the 'streamer F' mutants which show aberrant chemotaxis (Ross & Newell, 1979, 1981). In the wild-type strain, cyclic GMP is transiently formed in response to a cyclic AMP signal with a peak at 10 s, but, in the streamer F mutants, the cyclic GMP is not destroyed so rapidly but persists for approximately five-fold longer (Fig. 6). The persistence of the cyclic GMP results from a defect in the gene for the cyclic GMP-specific phosphodiesterase. (Ross & Newell, 1981; Van Haastert, Van Lookeren Campagne & Ross, 1982b; Coukell & Cameron, 1986). The most obvious effect of this defective gene on the visible phenotype is that the amoebae remain in the elongated state during chemotaxis for approximately 5-fold longer than the parental wild-type. The molecular link between cyclic GMP formation and the observed changes in cell shape is thought to involve myosin II attachment to the cytoskeleton, as described in a later section below.

The connection between the cell surface cyclic AMP receptors and guanylate cyclase is not fully understood. When amoebae were made permeable with saponin, and treated with 5 μM Ins(1,4,5)P$_3$ or 60 μM Ca^{2+}, they responded by forming a peak of cyclic GMP which closely resembled that produced by cyclic AMP in both its timing and duration (Europe-Finner & Newell, 1985; Small et al., 1986). Guanylate cyclase has been reported, however, not to be activated by Ca^{2+} in vitro (Padh & Brenner, 1984; Janssens & Jong, 1988). This may be due to loss of such activation during the enzyme extraction procedure or it possibly indicates that another unknown step lies between Ca^{2+} and guanylate cyclase.

ASSOCIATION OF F-ACTIN WITH THE CYTOSKELETON

In unstimulated amoebae, about one-third of the actin in the cell is associated with the cytoskeleton and is in the form of filamentous F-actin. The other two-thirds is free monomeric G-actin. Within 3–5 seconds of addition of 100 μM cyclic AMP to responsive populations of amoebae, there is a very rapid increase in the amount of polymeric F-actin associated with the cytoskeleton and the F/G actin ratio changes to roughly two-thirds cytoskeletal F-actin and only one-third free G-actin (McRobbie & Newell, 1983, 1984b; Newell, 1986a; Condeelis et al., 1988; Liu & Newell, 1988). (Fig. 7). This effect has been observed using three different actin assays: (i) by using the Triton-insoluble F-actin pellet assay (McRobbie & Newell, 1984b, 1985), (ii) by using the DNase-1 inhibition assay for G-actin (McRobbie & Newell, 1985), and (iii) by use of the NBD-phallocidin assay (Condeelis et al., 1988, Hall, Schlein & Condeelis, 1988). With different species, the effect is only seen with the chemoattractant that is specific for the particular species: for Polysphondylium violaceum the dipeptide glorin (but not cyclic AMP) is able to trigger this response, and for D. lacteum the pterin monapterin is the effective compound (McRobbie & Newell, 1984b). The rise in the cytoskeletal actin is always sharp but brief and drops rapidly to values at or below the baseline within 20 s. In individual experiments further oscillations are commonly seen but these are generally not synchronous between experiments and are not seen if the mean is calculated from several experiments as in Figure 7. This association of the actin with the cytoskeleton has been visualized with immunofluorescence microscopy by Yumura, Mori & Fukui (1984) (Fig. 8).

Although the evidence that cyclic AMP can induce the actin response is very good, more uncertainty surrounds the mechanism through which it operates. The evidence that Ca^{2+} might be involved, as shown in Figure 1, comes from work with saponin-permeabilized amoebae which showed rapid polymerization of cytoskeletal actin in the presence of 5 μM

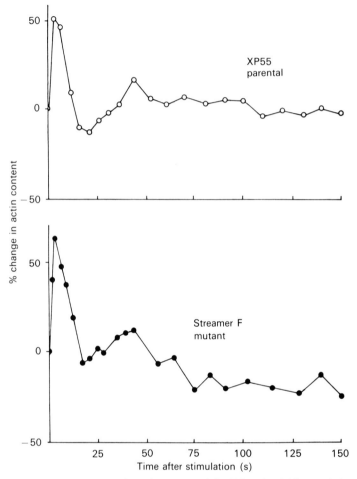

Fig. 7. Time course of changes in actin content of the Triton-insoluble cytoskeletons of 8 h amoebae of parental strain XP55 (O--O), or streamer F mutant NP368 (●--●) following stimulation with 10^{-7} M cyclic AMP. Note that the streamer F mutation has no effect on the response of actin to cAMP. Results are expresssed as percentage change over prestimulus value and are the mean of seven experiments. (Data from Liu & Newell, 1988.)

Ins(1,4,5)P_3 or 10 μM Ca^{2+} (Europe-Finner & Newell, 1986a,b). These responses were different, however, from those elicited by cyclic AMP in that, with Ins(1,4,5)P_3 and Ca^{2+}, the polymerization occurred more slowly and the actin remained polymerized, forming a plateau at about 15 s rather than a peak at 3–5 s as with cyclic AMP (Europe-Finner & Newell, 1986b; Newell, 1986a). In leucocytes, there is evidence that Ca^{2+} is not essential for the similar cytoskeletal actin polymerization that is observed in response to fMet–Leu–Phe (Sha'afi et al., 1986; Sklar, Omann & Painter, 1985;

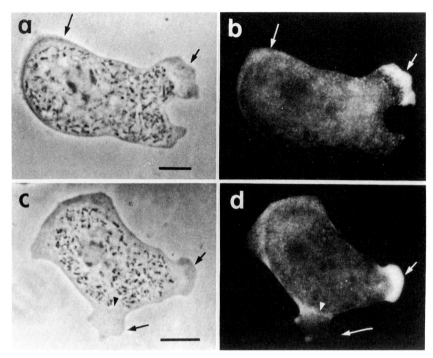

Fig. 8. Distribution of actin in fixed preparations of moving *D. discoideum* amoebae seen by phase-contrast (*a*), (*c*) and indirect immunofluorescence microscopy using anti-actin antibodies (*b*), (*d*). The samples were prepared by the 'agar-overlay' technique. Amoebae were seen to produce pseudopodia (arrows) that showed bright but transient labelling of F-actin. The bar marker represents 5 μm. (From Yumura, Mori & Fukui, 1984). Photograph kindly provided by Dr Yoshio Fukui.)

Korchak *et al.*, 1988; Zigmond *et al.*, 1988). It remains to be seen, therefore, whether the link between the cyclic AMP receptor and the actin polymerization system is as simple as shown in Figure 1.

ASSOCIATION OF MYOSIN II WITH THE CYTOSKELETON

Based on the interaction of actin and myosin in mammalian muscle tissue, it was thought likely, until recently, that the main propulsive force for the amoebal cell during chemotaxis would be derived from an interaction of actin with myosin II. That view has had to be changed by recent work from two laboratories in which amoebae have been produced which lack the myosin II heavy chain but which can still move. Strains of *Dictyostelium* have been isolated by De Lozanne & Spudich (1987) (using the technique of homologous recombination) and by Knecht & Loomis (1987, 1988) (using antisense mRNA) that do not possess any detectable normal myosin II heavy chain. Nevertheless, amoebae of such strains can still manage to move chemotactically. However, these myosin-lacking strains were found to be abnormal in their shape and less efficient in moving towards a chemotactic centre. The elongated shape of the wild-type was replaced by a more rounded shape that showed less polarity and a lowered ability to control its directional accuracy of movement (Wessels et al., 1988; Fukui, De Lozanne & Spudich, 1990). (Further details of these interesting mutants may be found in the following chapter on Behavioural Mutants by Jeffrey Segal.)

The role of myosin II may be, therefore, to regulate the shape of the cell for efficiency of chemotaxis rather than being essential for propulsion as one would originally have predicted from such a 'muscle' protein. This notion is reinforced by the work showing that myosin II associates with the cytoskeleton during the chemotactic response, and that this association is prolonged in the streamer F mutants which show a similarly prolonged period when the amoebae are elongated. The association in the wild type has been visualized using immunofluorescence microscopy by Yumura & Fukui (1985) (Fig. 9).

Work with the streamer mutants has also indicated that cyclic GMP is intimately involved in controlling this response. In the parental wild-type strain, a cyclic AMP stimulus induces the formation of a sharp peak of association of myosin II heavy chain with the Triton-insoluble cytoskeleton at 20–25 s after the stimulus (Fig. 10). In the streamer F mutants, however, although the initial rise in cytoskeletal myosin II is the same as in the parental, it does not rapidly decline but remains associated for a roughly five-fold longer period, in a similar way to the extension of the cyclic GMP peak (Liu & Newell, 1988) (Fig. 10). Such a correlation suggests that the myosin II response is, in some way, regulated by the formation of cyclic GMP. Evidence for the bifurcation of the signal pathways to actin and myosin II is provided by the finding that while the myosin II response is abnormal in the streamer F mutants, the actin response is unaffected (Fig. 7). Clearly, unlike myosin II, the actin and 190Kd proteins are not regulated by the cyclic GMP concentration (Liu & Newell, 1988; McRobbie & Newell, 1984a).

The correlation between cyclic GMP and the myosin II response is strengthened by the finding that in ras–Thr_{12} strains (which show abnormally low cyclic GMP responses) the myosin II response to cyclic AMP is similarly low (Liu & Newell, unpublished data). More recent data of Liu & Newell (unpublished) suggest that the action of cyclic GMP might not be to promote the initial association of the myosin II with the cytoskeleton as one might initially suppose, but to inhibit its removal from the cytoskeleton. This conclusion comes from a study which showed that the phosphorylation of the myosin II heavy chain (an activity that is known to be associated with chemotaxis, Berlot, Spudich & Devreotes, 1985; Berlot, Devreotes & Spudich, 1987) occurs later in the streamer F mutants

Fig. 9. Distribution of myosin II rods in fixed preparations of amoebae of *D. discoideum* during cell elongation, seen by phase-contrast microscopy (*a*) and indirect immunofluorescence microscopy (*b*), (*c*) using monoclonal anti-myosin II antibody. Micrographs were taken at two focal planes, upper (*b*) and middle (*c*). The samples were prepared by the 'agar-overlay' technique. Myosin II rods were seen to accumulate transiently in the posterior region of the amoebal cortex. (From Yumura & Fukui, 1985. Photograph kindly provided by Dr Yoshio Fukui.)

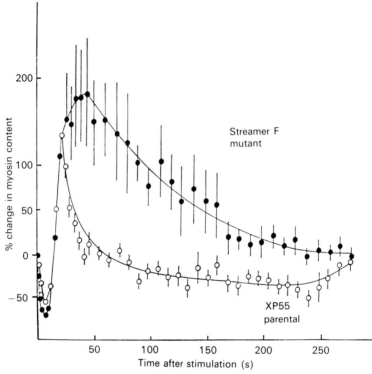

Fig. 10. Time course of changes in myosin II heavy chain content of cytoskeletons of 8 h amoebae of parental strain XP55 (O--O), or streamer F mutant NP368 (●--●), following stimulation with 10^{-7} M cyclic AMP. Results are expressed as percentage change over prestimulus value and are the mean of seven experiments. Bars represent the S.E.M. Streamer F mutant NP377 gave similar results. (From Liu & Newell, 1988.)

than in the parental strain. In both the parental strain and the mutants, this phosphorylation correlates with the period when the myosin II is being removed from the cytoskeleton. It was also found that (unlike the myosin II heavy chain in the cytosol) the myosin II heavy chain on the cytoskeleton was not detectably phosphorylated (Liu & Newell, unpublished). The significance of this finding becomes clear when considered in conjunction with the finding of Pasternak et al. (1989) that phosphorylation of myosin II heavy chain by heavy chain kinase causes bending of the molecules such that they cannot form filaments (Fig. 11). In our current working model based on this data, the myosin II heavy chain becomes associated with the cytoskeleton in an unphosphorylated state (at the same rate in the streamer F mutants as in the parental strain) and remains there until a kinase phosphorylates it and thereby removes it. If this kinase were inhibited by cyclic GMP (or a protein activated by it) then, in the wild-type strains, the brief cyclic GMP peak would lead to a brief period when more

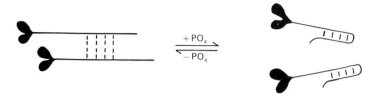

Fig 11. Schematic drawing of the proposed relationship between parallel dimers and bent monomers of myosin II. Intramolecular contacts between the myosin II molecules are prevented by the bent structures induced by the phosphorylation of the tail. (Redrawn from Pasternak et al., 1989.)

myosin II was added to the cytoskeleton than was removed. In the streamer mutants, the prolonged cyclic GMP response would induce a delay before phosphorylation of their cytoskeletal myosin II heavy chain occurred (as observed) and this delay would produce the phenotype of prolonged association of the myosin II with the cytoskeleton.

THE ROLE OF MYOSIN I

A recent report of Fukui et al. (1989) has provided evidence for a role for Myosin I in *Dictyostelium* amoebae. Myosin I proteins have been characterized in *Acanthamoeba* by Lynch et al. (1986) as a family of single-headed globular proteins with an M_r between 140000 and 150000 which are incapable of forming filaments. In *Dictyostelium* a single myosin I has been purified (Cote et al. 1985; Lynch et al. 1988) and found to be very similar to the *Acanthamoeba* proteins. Using specific anti-*Dictyostelium* myosin I antiserum and immunofluorescence techniques, Fukui et al. (1989) showed that myosin I is located in moving amoebae in the leading edge of pseudopodia and in the phagocytic cup that surrounds particles being ingested. Moreover, by double staining techniques, they found that the pattern of distribution was very similar to that of actin in the pseudopodia and in the phagocytic cups in the same cells. A substantial amount of myosin I was also found to be diffusely distributed in the cytoplasm except those areas which lay immediately behind the leading edge of the pseudopodia.

Previous to these studies it had been presumed that (because of the observed absence of myosin II from the pseudopodia) the motive force for pseudopodial extrusion was derived entirely from actin polymerization. The observations of Fukui et al. (1989), showing both actin and myosin I together in the pseudopodia, suggested another possibility, namely that this myosin isomer might be involved in an actin–myosin I interaction involved in force generation. Whether this possibility is correct or not awaits further studies. However, it is of interest that Jung et al. (1990) have recently produced mutants (lacking a functional myosin I gene) which do not pro-

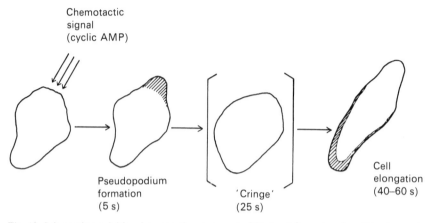

Fig. 12. Schematic model for chemotactic movement of *Dictyostelium*. A pulse of the chemoattractant cyclic AMP (represented by the arrows) elicits formation of a pseudopodium at 5 s. The shaded area represents actin and myosin I accumulation in the pseudopodium. With a 1 μM pulse or higher 'cringing' (cell rounding) may be seen at about 25 s. Cell elongation and translocation in the direction of the stimulus occur gradually at around 40–60 s and last for another 100 s in the wild-type strain. The shaded area in the elongated amoeba represents the association of myosin II with the cytoskeleton in the posterior region of the cell. (In the streamer F mutants, the elongation phase and cytoskeletal myosin II accumulation last for approximately five times longer than in the wild type.)

duce any detectable myosin I heavy chain mRNA. These cells are able to move chemotactically, although less efficiently. They are, however, impaired in their ability to phagocytose particles, as assessed by plaque formation on lawns of bacteria and their uptake of FITC-labelled *E. coli*.

FROM SIGNAL TRANSDUCTION TO CELL MOVEMENT

We propose the following model for chemotactic movement in *D. discoideum* (Fig. 12). When a cyclic AMP pulse strikes the cell, it triggers the accumulation of actin on the side initially struck and causes the protrusion at 3–5 s of an actin plus myosin I-rich pseudopodium in the direction of the stimulus. By some unknown mechanism, the rest of the cell is inhibited from responding. Somehow the cell retains a 'memory' of the pseudopodial event, even if (as sometimes happens) the cell 'cringes' or rounds up for a few seconds (Futrelle, Traut & McKee, 1982). The pseudopodium may act as a direction finder as it is in this direction that the amoeba later moves. After the pseudopodium has been withdrawn the myosin II associates with the cytoskeleton (in an unphosphorylated form) and induces the change in shape of the cell into a tube, elongated in the direction of movement, and, in this form, the cell moves toward the signal source until phosphorylation gradually removes the myosin II from the cytoskeleton.

ACKNOWLEDGEMENTS

We wish to thank Mr Ken Johnson for drawing the figures, Dr Julian Gross for helpful discussions and the Science and Engineering Research Council, the Cancer Research Campaign and the Guangxi Government of the People's Republic of China for financial support.

REFERENCES

Abe, T., Maeda, Y. & Iijima, T. (1989). Transient increase of the intracellular Ca concentration during chemotactic signal transduction in *Dictyostelium discoideum* cells. *Differentiation*, **39**, 90–6.

Barra, J., Barrand, P., Blondelet, M. H. & Brachet, P. (1980). PdsA, a gene involved in the production of active phosphodiesterase during starvation of *Dictyostelium discoideum* amoebae. *Molecular and General Genetics*, **177**, 607–13.

Berlot, C. H., Spudich, J. A. & Devreotes, P. N. (1985). Chemoattractant-elicited increases in myosin phosphorylation in *Dictyostelium*. *Cell*, **43**, 307–14.

Berlot, C. H., Devreotes, P. N. & Spudich, J. A. (1987). Chemoattractant-elicited increases in *Dictyostelium* myosin phosphorylation are due to changes in myosin localization and increases in kinase activity. *Journal of Biological Chemistry*, **262**, 3918–26.

Bonner, J. T. (1971). Aggregation and differentiation in the cellular slime molds. *Annual Review of Microbiology*, **25**, 75–92.

Condeelis, J., Hall, A., Bresnick, A., Warren, V., Hock, R., Bennett, H. & Ogihara, S. (1988). Actin polymerization and pseudopod extension during amoeboid chemotaxis. *Cell Motility & Cytoskeleton*, **10**, 77–90.

Cote, G. P., Albanesi, J. P., Ueno, T., Hammer III, J. A. & Korn, E. D. (1985). Purification from *Dictyostelium discoideum* of a low MW myosin that resembles myosin I from *Acanthamoeba castellanii*. *Journal of Biological Chemistry*, **260**, 4543–6.

Coukell, M. B. & Cameron, A. M. (1986). Characterization of revertants of *stmF* mutants of *Dictyostelium discoideum*: Evidence that stmF is the structural gene of the cGMP-specific phosphodiesterase. *Developmental Genetics*, **6**, 163–77.

Coukell, M. B., Lappano, S. & Cameron, A. M. (1983). Isolation and characterization of cAMP unresponsive (frigid) aggregation-deficient mutants of *Dictyostelium discoideum*. *Developmental Genetics*, **3**, 283–97.

De Lozanne, A. & Spudich, J. A. (1987). Disruption of the *Dictyostelium* myosin heavy chain gene by homologous recombination. *Science*, **236**, 1086–91.

Europe-Finner, G. N. & Newell, P. C. (1985). Inositol 1,4,5-trisphosphate induces cyclic GMP formation in *Dictyostelium discoideum*. *Biochemical & Biophysical Research Communications*, **130**, 1115–22.

Europe-Finner, G. N. & Newell, P. C. (1986a). Inositol 1,4,5-trisphosphate induces calcium release from a non-mitochondrial pool in amoebae of *Dictyostelium*. *Biochimica et Biophysica Acta*, **887**, 335–40.

Europe-Finner, G. N. & Newell, P. C. (1986b). Inositol 1,4,5-trisphosphate and Ca^{2+} stimulate actin polymerisation in *Dictyostelium discoideum*. *Journal of Cell Science*, **82**, 41–51.

Europe-Finner, G. N. & Newell,P. C. (1987a). Cyclic AMP stimulates accumulation of inositol trisphosphate in *Dictyostelium discoideum*. *Journal of Cell Science*, **87**, 221–9.

Europe-Finner, G. N. & Newell, P. C. (1987b). GTP analogues stimulate inositol trisphosphate formation transiently in *Dictyostelium*. *Journal of Cell Science*, **87**, 513–18.

Europe-Finner, G. N., Luderus, M. E. E., Small, N. V., Van Driel, R., Reymond, C. D., Firtel, R. A. & Newell, P. C. (1988). Mutant *ras* gene induces elevated levels of inositol tris- and hexakisphosphates in *Dictyostelium*. *Journal of Cell Science*, **89**, 13–20.

Europe-Finner, G. N., Gammon, B., Wood, C. A. & Newell, P. C. (1989). Inositol tris- and polyphosphate formation during chemotaxis of *Dictyostelium*. *Journal of Cell Science*, **93**, 585–92.

Fukui, Y., Lynch, T. J., Brzeska, H. & Korn, E. D. (1989). Myosin I is located at the leading edges of locomoting *Dictyostelium* amoebae. *Nature, London*, **341**, 328–31.

Fukui, Y., De Lozanne, A. & Spudich, J. A. (1990). Structure and function of the cytoskeleton of a *Dictyostelium* myosin-defective mutant. *Journal of Cell Biology*, **110**, 367–78.

Futrelle, R. P., Traut, J. & McKee, W. G. (1982). Cell behaviour in *Dictyostelium*: Preaggregation response to localized cyclic AMP pulses. *Journal of Cell Biology*, **92**, 807–21.

Gerisch, G. (1982). Chemotaxis in *Dictyostelium*. *Annual Review of Physiology*, **44**, 535–52.

Glazer, P. M. & Newell, P. C. (1981). Initiation of aggregation by *Dictyostelium discoideum* in mutant populations lacking pulsatile signalling. *Journal of General Microbiology*, **125**, 221–32.

Hall, A. L., Schlein, A. & Condeelis, J. (1988). Relationship of pseudopod extension to chemotactic hormone-induced actin polymerization in amoeboid cells. *Journal of Cellular Biochemistry*, **37**, 285–99.

Janssens, P. M. W., Arents, J. C., Van Haastert, P. J. M. & Van Driel, R. (1986). Forms of the chemotactic adenosine 3',5'-cyclic phosphate receptor in isolated *Dictyostelium discoideum* membranes and interconversions induced by guanine nucleotides. *Biochemistry*, **25**, 1314–20.

Janssens, P. M. W. & Van Haastert, P. J. M. (1987). Molecular basis of transmembrane signal transduction in *Dictyostelium discoideum*.*Microbiology Reviews*, **51**, 396–418.

Janssens, P. M. W. & de Jong, C. C. C. (1988). A magnesium dependent guanylate cyclase in cell free preparations of *Dictyostelium discoideum*. *Biochemical & Biophysical Research Communications*, **150**, 405–11.

Jung, G., Saxe III, C. L., Kimmel, A. R. & Hammer III, J. A. (1990). Myosin I Heavy chain gene in *Dictyostelium*. *Proceedings of the National Academy of Sciences, USA*, in press.

Kesbeke, F., Snaar-Jagalska, B. E. & Van Haastert, P. J. M. (1988). Signal transduction in *Dictyostelium fgdA* mutants with a defective interaction between cAMP receptor and a GTP-Binding regulatory protein. *Journal of Cell Biology*, **107**, 521–32.

Klein, P. S., Sun, T. J., Saxe, C. L., Kimmel, A. R., Johnson, R. L. & Devreotes, P. N. (1988). A chemotactic receptor controls development in *Dictyostelium discoideum*. *Science*, **241**, 1467–72.

Knecht, D. A. & Loomis, W. F. (1987). Antisense RNA inactivation of myosin heavy-chain gene expression in *Dictyostelium discoideum*. *Science*, **236**, 1033–48.

Knecht, D. A. & Loomis, W. F. (1988). Developmental consequences of the lack of myosin heavy chain in *Dictyostelium discoideum*. *Developmental Biology*, **128**, 178–84.

Konijn, T. M., Van de Meene., J. G. C., Bonner, J. T., & Barkley, J. T. (1967). The acrasin activity of adenosine 3',5'-cyclic phosphate. *Proceedings of the National Academy of Sciences, USA*, **58**, 1152–4.

Korchak, H. M., Vosshall, L. B., Zagon, G., Ljubich, P., Rich, A. M. & Weissmann, G. (1988). Activation of the neutrophil by calcium-mobilizing ligands. I. A chemotactic peptide and the lectin concanavalin A stimulate superoxide anion generation but elicit different calcium movements and phosphoinositide remodelling. *Journal of Biological Chemistry.* **263**, 11090–7.

Kumagai, A., Puillo, M., Gunderson, R., Miake-Lye, R., Devreotes, P. N. & Firtel, R. A. (1989). Regulation and function of G-alpha protein subunits in *Dictyostelium. Cell*, **57**, 265–75.

Liu, G. & Newell, P. C. (1988). Evidence that cyclic GMP regulates myosin interaction with the cytoskeleton during chemotaxis of *Dictyostelium. Journal of Cell Science,* **90**, 123–29.

Lynch, T. J., Albanesi, J. P., Korn, E. D., Robinson, E. A., Bowers, B. & Fujisaki, H. (1986). ATPase activities and actin-binding properties of subfragments of *Acanthamoeba* Myosin – 1A. *Journal of Biological Chemistry*, **261**, 17156–62.

Lynch, T. J., Brzeska, H., Lambooy, P. K., Fukui, Y. & Korn, E. D. (1988). Physical and enzymatic properties of *Dictyostelium* Myosin I. *Journal of Cell Biology*, **107**, 7a (abstract 25).

Mato, J. M., Krens, F. A., Van Haastert, P. J. M. & Konijn, T. M. (1977). 3',5'-cyclic AMP-dependent 3',5'-cycle GMP accumulation in *Dictyostelium discoideum. Proceedings of the National Academy of Sciences, USA*, **74**, 2348–51.

McRobbie, S. J. & Newell, P. C. (1983). Changes in actin associated with the cytoskeleton following chemotactic stimulation of *Dictyostelium discoideum. Biochemical & Biophysical Research Communications*, **115**, 351–9.

McRobbie, S. J. & Newell, P. C. (1984a). A new model for chemotactic signal transduction in *Dictyostelium discoideum. Biochemical and Biophysical Research Communications*, **123**, 1076–83.

McRobbie, S. J. & Newell, P. C. (1984b). Chemoattractant-mediated changes in cytoskeletal actin of cellular slime moulds. *Journal of Cell Science.* **68**, 139–51.

McRobbie, S. J. & Newell, P. C. (1985). Effects of cytochalasin B on cell movements and chemoattractant-elicited actin changes of *Dictyostelium discoideum. Experimental Cell Research*, **160**, 275–86.

Newell, P. C. (1981). Chemotaxis in the cellular slime moulds. In *Biology of the chemotactic response*, ed. J. M. Lackie & P. C. Wilkinson, pp. 89–114. Soc. Exp. Biol. Seminar Series vol 12, Cambridge Univ. Press.

Newell, P. C. (1986a). The role of actin polymerization in amoebal chemotaxis. *Bioessays*, **5**, 208–11.

Newell, P. C. (1986b). Receptors for cell communication in *Dictyostelium*. In *Hormones, receptors and cellular interactions in plants*, ed. C. M. Chadwick & D. R. Garrod, pp. 154–216. Cambridge Univ. Press.

Newell, P. C., Europe-Finner, G. N., Small, N. V. & Liu, G. (1988). Inositol phosphates, G-proteins and ras genes involved in chemotactic signal transduction of *Dictyostelium. Journal of Cell Science*, **89**, 123–7.

Newell, P. C., Europe-Finner, G. N., Liu, G., Gammon, B., & Wood, C. A. (1990). Signal transduction for chemotaxis in *Dictyostelium* amoebae. *Seminars in Cell Biology*, **1**, 105–13.

Padh, H. & Brenner, M. (1984). Studies of the guanylate cyclase of the social amoeba *Dictyostelium discoideum. Archives for Biochemistry & Biophysics*, **229**, 73–80.

Pasternak, C., Flicker, P. F., Ravid, S. & Spudich, J. A. (1989). Intermolecular versus intramolecular interactions of *Dictyostelium* myosin: Possible regulation by heavy chain phosphorylation. *Journal of Cell Biology*, **109**, 203–10.

Pawson, T., Amiel, T., Hinze, E., Auersperg, N., Neave, N., Sobolewski, A.

& Weeks, G. (1985). Regulation of a *ras*-related protein during development of *Dictyostelium discoideum*. *Molecular & Cellular Biology*, **5**, 33–9.

Reymond, C. D., Gomer, R. H., Mehdy, M. C. & Firtel, R. A. (1984). Developmental regulation of a *Dictyostelium* gene encoding a protein homologous to mammalian *ras* protein. *Cell*, **39**, 141–8.

Reymond, C. D., Nellen, W. & Firtel, R. A. (1985). Regulated expression of *ras* gene constructs in *Dictyostelium discoideum* transformants. *Proceedings of the National Academy of Sciences, USA*, **82**, 7005–9.

Reymond, C. D., Gomer, R. H., Nellen, W., Theibert, A., Devreotes, P. N. & Firtel, R. A. (1986). Phenotypic changes induced by a mutated *ras* gene during the development of *Dictyostelium discoideum* transformants. *Nature, London* **323**, 340–3.

Robbins, S. M., Williams, J. G., Jermyn, K. A., Spiegelman, G. B. & Weeks, G. (1989). Growing and developing *Dictyostelium* cells express different *ras* genes. *Proceedings of the National Academy of Sciences, USA*, **86**, 938–42.

Ross, F. M. & Newell, P. C. (1979). Genetics of aggregation pattern mutations in the cellular slime mould *Dictyostelium discoideum*. *Journal of General Microbiology*, **115**, 289–300.

Ross, F. M. & Newell, P. C. (1981). Streamers: chemotactic mutants of *Dictyostelium discoideum* with altered cyclic GMP metabolism. *Journal of General Microbiology*, **127**, 339–50.

Schiffmann, E., Corcoran, B. A. & Wahl, S. M. (1975). N-formylmethionyl peptides are chemotactic for leukocytes. *Proceedings of the National Academy of Sciences, USA*, **72**, 1059–62.

Sha'afi, R. I., Shefcyk, J., Yassin, R., Molski, T. F. P., Volpi, M., Naccache, P. H., White, J. R., Feinstein, M. B. & Becker, E. L. (1986). Is a rise in intracellular concentration of free calcium necessary or sufficient for stimulated cytoskeletal-associated actin? *Journal of Cell Biology*, **102**, 1459–63.

Shimomura, O., Suthers, H. L. B. & Bonner, J. T. (1982). Chemical identity of the acrasin of the cellular slime mold *Polysphondylium violaceum*. *Proceedings of the National Academy of Sciences, USA*, **79**, 7376–9.

Sklar, L. A., Omann, G. M. & Painter, R. G. (1985). Relationship of actin polymerization and depolymerization to light scattering in human neutrophils: dependence on receptor occupancy and intracellular Ca^{++}. *Journal of Cell Biology*, **101**, 1161–6.

Small, N. V., Europe-Finner, G. N. & Newell, P. C. (1986). Calcium induces cyclic GMP formation in *Dictyostelium*. *FEBS Letters*, **203**, 11–14.

Small, N. V., Europe-Finner, G. N. & Newell, P. C. (1987). Adaptation to chemotactic cyclic AMP signals in *Dictyostelium* involves the G-protein. *Journal of Cell Science*, **88**, 537–45.

Snaar-Jagalska, B. E. & Van Haastert, P. J. M. (1988). *Dictyostelium discoideum* mutant *synag7* with altered G-protein adenylate cyclase interaction. *Journal of Cell Science*, **91**, 287–94.

Van Haastert, P. J. M. & Kien, E. (1983). Binding of cAMP derivatives to *Dictyostelium discoideum* cells: Activation mechanism of the cell surface cAMP receptor. *Journal of Biological Chemistry*, **258**, 9636–42.

Van Haastert, P. J. M. & Konijn, T. M. (1982). Signal transduction in the cellular slime molds. *Molecular & Cellular Endocrinology*, **26**, 1–17.

Van Haastert, P. J. M., De Wit, R. J. W., Grijpma, Y. & Konijn, T. M. (1982a). Identification of a pterin as the acrasin of the cellular slime mold *Dictyostelium lacteum*. *Proceedings of the National Academy of Sciences, USA*, **79**, 6270–4.

Van Haastert, P. J. M., Van Lookeren Campagne, M. M. & Ross, F. M. (1982b).

Altered cGMP-phosphodiesterase activity in chemotactic mutants of *Dictyostelium discoideum*. *FEBS Letters*, **147**, 149–52.
Van Haastert, P. J. M., De Wit, R. J. W., Janssens, P. M. W., Kesbeke, F. & De Goede, J. (1986). G-protein-mediated interconversions of cell-surface cAMP receptors and their involvement in excitation and desensitization of guanylate cyclase in *Dictyostelium discoideum*. *Journal of Biological Chemistry*, **261**, 6904–11.
Van Haastert, P. J. M., Kesbeke, F., Reymond, C. D., Firtel, R. A., Luderus, M. E. E. & Van Driel, R. (1987). Aberrant transmembrane signal transduction in *Dictyostelium* cells expressing a mutated *ras* gene. *Proceedings of the National Academy of Sciences, USA*, **84**, 4905–9.
Van Haastert, P. J. M., De Vries, M. J., Penning, L. C., Roovers, E., Van der Kaay, J., Erneux, C. & Van Lookeren Campagne, M. M. (1989). Chemoattractant and guanosine 5′-[gamma-thio]triphosphate induce the accumulation of inositol 1,4,5-trisphosphate in *Dictyostelium* cells that are labelled with [3H]inositol by electroporation. *Biochemical Journal*, **258**, 577–86.
Van Lookeren Campagne, M. M., Erneux, C., Van Eijk, R. & Van Haastert, P. J. M. (1988). Two dephosphorylation pathways of inositol 1,4,5-trisphosphate in homogenates of the cellular slime mould *Dictyostelium discoideum*. *Biochemical Journal*, **254**, 343–50.
Weeks, G., Lima, A. & Pawson, T. (1987). A *ras*-encoded protein in *Dictyostelium discoideum* is acetylated and membrane-associated. *Molecular Microbiology*, **1**, 347–54.
Weeks, G. & Pawson, T. (1987). The synthesis and degradation of *ras*-related gene products during growth and differentiation in *Dictyostelium discoideum*. *Differentiation*, **33**, 207–13.
Wessels, D., Soll, D. R., Knecht, D., Loomis, W. F., De Lozanne, A. & Spudich, J. (1988). Cell motility and chemotaxis in *Dictyostelium* amebae lacking myosin heavy chain. *Developmental Biology*, **128**, 164–77.
Wurster, B., Schubiger, K., Wick, U. & Gerisch, G. (1977). Cyclic GMP in *Dictyostelium discoideum*. Oscillations and pulses in response to folic acid and cyclic AMP signals. *FEBS Letters*, **76**, 141–4.
Yumura, S. & Fukui, Y. (1985). Reversible cAMP-dependent change in distribution of myosin thick filaments in *Dictyostelium*. *Nature, London*, **314**, 194–6.
Yumura, S., Mori, H. & Fukui, Y. (1984). Localization of actin and myosin for the study of ameboid movement in *Dictyostelium* using improved immunofluorescence. *Journal of Cell Biology*, **99**, 894–9.
Zigmond, S. H., Slonczewski, J. L., Wilde, M. W. & Carson, M. (1988). Polymorphonuclear leukocyte locomotion is insensitive to lowered cytoplasmic calcium levels. *Cell Motility & Cytoskeleton*, **9**, 184–9.

CHEMOSENSORY TRANSDUCTION IN *PARAMECIUM*

JUDITH VAN HOUTEN

Department of Zoology, University of Vermont,
Burlington, VT 05405, USA

INTRODUCTION

All cells have the ability to gather information about their environment through the detection of external stimuli. Chemical stimuli, in particular, hold information about location of mates of the same species, food availability and quality, dangerous milieux, and the presence of predators. Across the eukaryotic phylogenetic spectrum, chemosensory transduction systems all appear to be initiated at the membrane surface of a receptor cell by the interaction of a stimulus with a receptor molecule (or perhaps in some cases with the membrane directly), and, subsequently, this interaction is transduced into intracellular messengers. The second and third messengers are limited in number and for the most part are recruited from cyclic nucleotides, permeant ions, phosphoinositides, diacyl glycerol, arachidonic acid and internal pH levels. The receptor cells that detect external chemical stimuli share some common transduction mechanisms, perhaps because there is a limited number of ways to accommodate the special constraints posed by external sensory systems, constraints not generally encountered by internal receptor systems: chemical stimuli are produced by other organisms, are borne in patchy plumes of air or water, and span large concentration ranges; stimuli are presented in mixtures and must be detected against a noisy chemical background; and often receptors have relatively low affinity (Atema, 1987; Moore & Atema, 1988; Van Houten *et al.*, 1981; Van Houten & Preston, 1987).

The studies of unicellular chemosensory transduction systems, such as those described in this volume in chapters on bacteria, slime moulds, and *Paramecium*, are useful for at least two reasons. First, they present examples of bona fide chemosensory receptor cells that have been successful, over evolutionary time, in dealing with external chemical stimuli and it would be insightful to determine why they are successful. Second, unicellular organisms are proving to be more amenable than metazoans to the biochemistry necessary to identify and purify receptors. Our first glimpses of chemoreceptors came from the bacteria and the first sequences of external *eukaryotic* chemoreceptors are also coming from microorganisms (Table 1). The identified eukaryotic chemoreceptors resemble some classes of

Table 1. *External eukaryotic chemoreceptors*

Arbacia spermatozoan	Resact receptor
Dictyostelium discoideum	Cyclic AMP chemotaxis receptor
Saccharomyces cerevisiae	Mating type a and *a* receptors
Paramecium tetraurelia	Cyclic AMP chemoreceptor

internal vertebrate receptors and are also likely to portend the still elusive external olfactory and gustatory receptors of vertebrates as well (Carr, 1990; Carr et al., 1989). Therefore, in this chapter I will briefly review aspects of the multicellular chemosensory systems of olfaction and taste and then describe in detail chemosensory transduction in a unicellular eukaryote, *Paramecium tetraurelia*, with an eye to features that are novel and those that are shared with other eukaryotic organisms. (Parallels with leucocyte chemotaxis, which is treated in this volume by Wilkinson, Naccache et al., and by Bacon et al., are discussed in Van Houten & Preston, 1987 and Van Houten, 1991.)

OLFACTION

Specialized cells of neuronal origin detect odorants by interaction of the ligand with the surface membrane probably at a specific receptor (Lancet & Pace, 1987). Although the definitive case for receptors will not be made until an olfactory receptor is isolated and its gene cloned, there are candidate proteins being examined for vertebrate olfaction and insect pheromone reception (Vogt et al., 1988; Price & Willey, 1987, 1988; Novoselov et al., 1988; Bignetti et al., 1987). Specific anosmias (inability to detect specific odorants) and very different odorant qualities associated with two stereoisomers of carvone argue for the existence of olfactory receptors, while hydrophobicity of odorants leaves open the possibility that the ligands interact with the lipid bilayer to indirectly affect channel function (see reviews by Anholt, 1987; Bruch, 1990; Lancet & Pace, 1987; Lancet, 1986; Dionne, 1988; Lerner et al., 1988).

In olfactory receptor neurons, the interaction with odorant generates action potentials that are communicated by axons directly to the olfactory bulb (see Kauer, 1987 for review). The mechanisms by which receptor cell channels are modulated in response to receptor cell–ligand interactions are under intense scrutiny in order to distinguish between second messenger activation of channels and direct ligand channel gating, roughly analogous to the muscarinic and nicotinic mechanisms of acetylcholine action.

The adenylate cyclase of olfactory cilia has remarkably high activity, is GTP binding protein (G protein) dependent (Lancet & Pace, 1987), and can be activated by many (but not all) odorants (Pace et al., 1985; Sklar et al., 1986). Therefore, olfactory receptors may interact with a G protein, perhaps the one that is unique to olfactory receptor neurons and that can activate adenylate cyclase in a heterologous system (G_{olf}; Jones & Reed, 1989) and the activated G protein, in turn, would stimulate the adenylate cyclase activity of the cell. The resultant cAMP would interact with channels and thereby initiate action potentials. There are reports of cyclic nucleotide-sensitive conductances in olfactory receptor neurons and correlations of receptor cell population activity with cyclase activation (Nakamura & Gold, 1987; Lowe et al., 1988). However, there are also reports of direct odorant gating of channels by the same odorants (Labarca et al., 1988), in which case the activation of adenylate cyclase might participate in the *slower* process of adaptation.

In order for a second messenger to initiate channel activity, its synthesis must precede the activity. Recent studies by Breer and colleagues have settled the important question of the time-course for second messengers by using stop flow techniques (Breer et al., 1990a,b). It appears that odorants stimulate either cAMP or inositol 1,4,5-trisphosphate (IP_3) production in rat olfactory cilia on a subsecond time-scale. In comparison, pheromone stimulates second messenger production on a time-scale sufficiently fast to be involved in conductance changes, but only IP_3 is generated in insect antennal preparations (Breer et al., 1990a,b). In all cases, the effects on second messengers are GTP dependent, reinforcing the notion that odorant receptors will eventually be isolated. Excitable cells other than olfactory receptor cells did not respond to odorants with increases in second messengers, helping to discount some of the concern that the lipophilic nature of some odorants would allow them to induce conductance changes by direct membrane interactions without benefit of receptors or specific binding sites on channels (Dionne, 1988; Lerner et al., 1988; Nomura & Kurihara, 1989).

The ionic conductances that underly the excitatory or inhibitory responses to odorants are being explored using whole cell clamp or patch techniques (Stengl et al., 1989, 1990; Michel & Ache, 1990; Dionne, 1990; Firestein & Shepherd, 1989; Pun & Gesteland, 1990; Kleene & Gesteland, 1990). The exciting possibilities include direct activation of channels by IP_3 in catfish olfactory cilia (Teeter et al., 1990), in addition to direct nucleotide gating of channels (for review, see Bruch, 1990).

Another emerging theme of metazoan olfaction is perireceptor events, that is, the processes that occur in the vicinity of the receptor to protect, carry out or degrade the stimulus (Getchell et al., 1984; Getchell & Getchell, 1987; Carr, 1990; Carr et al., 1989). Odorant binding proteins (OBP) in the mucus of vertebrate olfactory epithelium (Snyder et al., 1988; Pelosi & Dal Monte, 1990) is in great abundance and binds hydrophobic odorants non-specifically. Its function may be to protect the odorant from degradation, improve its solubilization in the aqueous mucus, carry it away from the vicinity of the receptor in preparation for the next sniff of odorant, or possibly all of the above. Insect OBP counterparts that have been known for some time (Vogt, 1987) probably function to protect the pheromone from degradation by potent esterases (Vogt et al., 1985) and to solubilize the volatile pheromone in the aqueous sensillar lymph. However, primary cultures of male moth pupal cells show less ion channel activity in response to pheromone with pheromone binding protein (PBP) than with pheromone alone (Stengl et al., 1989, 1990). These results suggest that PBP serves to clear odorant from the receptor to prevent desensitization. Unlike light and touch, chemical stimulation outlasts its transient application and must be destroyed or removed from the receptor in order for the cell to resensitize and respond to the next wave of stimulus (Atema, 1987). Another example would be lobster antennule sensillar dendrites that detect nucleotides but not adenosine at M1 and M2 purinergic receptors (Carr et al., 1989). Stimulus concentrations are controlled by enzymes that rapidly degrade the nucleotides to adenosine, which is taken up to clear the stimulus from the receptors between flicks of the antennules. Additionally, vertebrates may be utilizing cytochrome P450 enzymes and UDP glucuronosyl transferase to degrade odorants that enter receptor cells by virtue of their hydrophobicity and threaten to confound the olfaction process by diffusing freely in and out of cells (Lancet et al., 1989).

The receptor cells of the olfactory epithelium of vertebrates or the many sensilla of invertebrates are far from homogeneous. Individual receptor cells can be broadly tuned, while others are narrow in their spectrum of response (see reviews by Ache, 1987; Anholt, 1987). Therefore, when odorant stimuli from natural sources arrive at the receptor cell in mixtures, it is thought that the collective pattern of depolarized cells across the olfactory epithelium or sensillum provides the CNS with information about the quality and quantity of the odour. However, it is now clear that components of odorant mixtures can be excitatory to one cell and inhibitory to another and, while both individual cells are depolarised by the mixture, the inhibitory components can reduce the magnitude and delay the onset of the evoked depolarization (McClintock & Ache, 1989; Michel & Ache, 1990; Dionne, 1990). Therefore, the first level of integration of information is at the level of the receptor cell (Michel & Ache, 1990) and not in the CNS as previously believed.

GUSTATION

Taste receptor cells are modified neurons that synapse with interneurons and whose axons traverse to the CNS. Generally, taste stimuli elicit the calcium-dependent release of neurotransmitter by depolarizing the cell and thereby opening the voltage-sensitive Ca^{2+} channels (Kinnamon, 1988; Roper, 1989). However, the mechanisms by which internal Ca^{2+} levels are controlled are diverse and include: direct inhibition of K^+ channels by sour stimuli (H^+); passage of Na^+ through passive, amiloride-sensitive

Na$^+$ channels; inhibition of K$^+$ channels by cAMP-dependent protein kinase that is activated by the cAMP produced following the binding of sweet tastant to receptor; release of intracellular Ca^{2+} by IP$_3$ generated after bitter tastant interacts with receptor; and direct opening of channels by proline and arginine (reviewed by Kinnamon, 1988; see also Kumazawa et al., 1990; Kohbara et al., 1990).

Recently, expression of mRNA for a protein analogous to the OBP has been demonstrated in von Ebner's glands, salivary glands located directly beneath and ducting into a trough at the base of rat taste bud papillae (Schmale et al., 1990). The protein is part of the same carrier protein superfamily to which OBP belongs, suggesting that this salivary protein might function in the concentration or delivery of sapid molecules to the taste receptor cells.

To date, no gustatory receptor has been isolated, but there are some hopeful candidates for amino acid taste in catfish (Bryant et al., 1987) and for sweet taste (Persaud et al., 1988).

UNICELLULAR EUKARYOTES

The chemosensory transduction pathways of unicellular eukaryotes are basically similar to those in multicellular organisms. Stimulus binds to receptor eliciting a second messenger in the cell and eventually bringing about changes in motile behaviour, gene expression, physiology or metabolism. The responses are tailored to the organism and, therefore, differ from those of the multicellular organisms. For example, *Dictyostelium discoideum* amoebae move toward a focal cell that is pulsing cAMP as a prelude to forming a multicellular slug. The same stimulus through different receptors initiates the changes in gene expression that are required for this drastic phenotypic change. Yeast cells of complementary mating type arrest in G$_1$ and become mating reactive in response to each other's pheromones (Herskowitz & Marsh, 1987). While the responses may differ, as in metazoa, the transduction mechanisms include receptor interaction with G proteins, generation of the second messengers cAMP, cGMP, IP$_3$, or Ca^{2+} and eventual desensitization of the receptor response with prolonged exposure to stimulus.

External eukaryotic chemoreceptors, such as those of olfactory or taste receptor cells, are difficult molecules to purify because of the receptors' relatively low affinities for ligand, lack of tightly or covalently binding pharmacological agents for labelling (e.g. bungarotoxin for the acetylcholine receptor), and lack of large quantities of tissues that are highly enriched in the receptors. Success in receptor identification, isolation and cloning of receptor genes has come from unicellular eukaryotes, in part because large quantities of homogeneous cell cultures can be harvested for biochemical analysis (Table 1). The *Dictyostelium* cAMP receptor, *Arbacia* spermatozoan resact receptor, and two yeast mating type pheromone receptors have been identified and cloned (Klein et al., 1988; Singh et al., 1988; Hagen et al., 1986; Burkholder & Hartwell, 1985) and all four receptors are different in primary or secondary structure: The *D. discoideum* cAMP receptors for chemotaxis, relay and development appear to be similar to the classic G protein-mediated superfamily of receptors with seven potential transmembrane spanning regions and a cytoplasmic region for interaction with G protein; the two yeast mating type receptors likewise have these features but one receptor could not have been predicted based upon the primary DNA or amino acid sequences of the other and the interaction with G proteins may be unusual because the β and γ subunits instead of G_α appear to carry on the sensory transduction pathway (Whiteway et al., 1989); the *Arbacia* sea urchin receptor is the guanylate cyclase, apparently a member of the vertebrate atrial natriuretic factor receptor family (Garbers, 1989). Therefore, it is important to continue to isolate and sequence genes for external eukaryotic receptors because, clearly, there will be a diversity of receptor structure and receptor-mediated sensory transduction mechanisms even though there are common aspects.

Paramecium tetraurelia

Ciliated protozoa use soluble chemical cues to locate mates, detect food and avoid non-optimal environments that are high in ionic strength or at an extreme pH (Van Houten et al., 1981). *P. tetraurelia* does not appear to utilize soluble chemical stimuli to identify mates, but does accumulate in the presence of bacterial fermentation products (e.g. acetate, lactate, NH_4^+), vitamins (e.g. folate) and other indicators that bacteria, their foodstuff, are close by, and they disperse from extremes of pH and high salt. Curiously, cyclic AMP is an effective attractant for *Paramecium* (Smith et al., 1987). As with acetate and lactate, its source in the pond could be bacteria (Makman & Sutherland, 1964) or perhaps another nucleotide serves as a sign of bacterial metabolism and is the primary ligand (Zimmer-Faust & Martinez, 1988; Burt, 1977). However, we have been unable to identify such an alternative ligand (see below).

P. tetraurelia *behaviour*

The accumulation of paramecia in attractants (or dispersal from repellents) is measured in a simple T-maze assay (Van Houten et al., 1975, 1982). The T-maze is merely a modified stopcock with one arm filled with test compound in a buffer (e.g. potassium acetate, K-OAc) and the other arm filled with control compound in the same buffer (e.g. KCl). Cells loaded into the T-maze bore are free to distribute between the arms of the T-maze for 30 minutes, and they repeatedly enter and leave both arms. The cells tend to spend more time in the arm with relative attractant and, when the stopcock is closed and the cells in the arms are counted, more cells are found to be in the arm with the OAc⁻ relative to Cl⁻ (Van Houten, 1978; Van Houten *et al.*, 1982).

The behaviour that results in this accumulation or dispersal is a kinesis based on smooth, relatively fast swimming in attractants and slower swimming with more frequent turns in relative repellents (Van Houten, 1978). It is similar to the bacterial biased random walk that produces their response to stimuli (Adler, 1987), but it differs from the bacterial case in that the paramecia modulate swimming speed as well as frequency of turning whereas bacteria modulate only turns.

There are classes of stimuli that cause paramecia to be attracted and repelled by affecting swimming speed in the extreme and do not require the ability of a cell to turn (Van Houten, 1978). However, those attractants that modulate both frequency of turning and speed, e.g. KOAc, will be the focus of this chapter. These attractants are, in general, organic molecules for which the cells could have evolved receptors.

The cells respond to *changes* in attraction concentration and, as other chemoreceptor cells, adapt in the presence of uniform distributions of stimuli (Van Houten, 1978; Van Houten *et al.*, 1982). Adaptation is reflected in a return of swimming speed and frequency of turning to basal levels. As long as the attractant is present, the cells maintain their attractant-induced hyperpolarization (see below) and reset to more negative potentials their threshold for the action potentials that cause abrupt swimming turns. The membrane physiological changes prepare the cells for an immediate response to a decrease in the attractant concentration or an increase in repellent (Machemer & De Peyer, 1977). Computer simulations of swimming behaviour in the T-maze have demonstrated that adaptation is a very important component of accumulation and dispersal (Van Houten & Van Houten, 1982). Indeed, there are mutants that initially accumulate in NH_4Cl in T-mazes but cannot maintain this distribution, probably because they have defects in adaptation (Van Houten *et al.*, 1982).

The attraction behavioural responses are saturable and specific, suggesting that they are receptor mediated (Van Houten, 1976, 1978). Specificity is determined by including a potential competitor compound in both arms of the T-maze and asking whether cells can still detect a gradient of attractant, i.e. accumulate in the arm with attractant (e.g. OAc⁻) relative to the arm with control anion (e.g. Cl⁻). Acetate can be detected in the presence of lactate in concentrations that are saturating for attraction to lactate and vice versa. Likewise, cAMP is not affected by excess cGMP or 5'IMP but is inhibited by 5'AMP (Smith *et al.*, 1987; Van Houten *et al.*, 1990). Since the source of cAMP in a pond is not clear, other nucleotides such as IMP or 5'AMP that are indicators of decaying tissue, might be the primary ligand for the cAMP receptor (Zimmer-Faust & Martinez, 1988; Burt, 1977). However, IMP does not appear to compete with cAMP for receptors, based on its lack of interference with the behavioural response, and 5'AMP is a much weaker and more variable attractant than cAMP. Therefore, the natural source of nucleotide for this response remains an open question.

The threshold concentrations of attractants to which paramecia respond range from 1 μM IMP to 100 μM folate and half maximal responses are commonly in the 1 mM range. The behavioural, electrophysiological, and binding data are comparable in terms of concentrations for half maximal responses and, therefore, the assumption is that paramecia require high concentrations of stimuli before they respond and thus cells are not drawn away from clusters of cells in response to minute amounts of stimuli. The pond is a chemically noisy environment, and exquisite sensitivity would not necessarily benefit the paramecia. Instead, they move only toward high concentrations where the likelihood is great that bacteria are active. Dispersal away from clusters of paramecia and subsequent outbreeding is generally a lethal event (Anderson, 1988) and, therefore, paramecia gain an added advantage from remaining clustered, and not dispersing in response to food cues unless the likelihood of finding the source is high.

P. tetraurelia *receptors*

Paramecia have saturable, specific binding sites for several attractant stimuli: acetate, folate and cAMP (Van Houten, 1976; Schulz *et al.*, 1984; Smith *et al.*, 1987) as measured by radioligand binding. These binding sites are primarily on the cell body, not on the cilia that comprise about 50% of the surface membrane. The affinities for these attractants range from 30–250 μM, but are comparable with, although slightly lower than, the half maximal concentrations needed for attraction to these compounds (Van Houten *et al.*, 1982).

Photoaffinity labelling of whole cells by uv irradiation with N_3-cAMP renders the cells unable to respond to cAMP, but still capable of normal behavioural response to other attractants (Table 2). Likewise, affinity labelling of cells with 'activated' folate (i.e. the N-hydroxysuccinimide derivative) renders them specifically unable to detect folate (Table 3). These types of experiments support the notion that there are specific receptors, albeit of low affinity. These experiments also indicate that the covalent attachment of ligand to a small fraction of the receptors renders the cells unable to respond to the ligand since 20% is the generally accepted upper limit for affinity crosslinking of receptors (Pilch & Czech, 1984). These results are consistent with the relationship of K_Ds of binding and the concentrations for half maximal chemoresponse. An interpretation of a slightly lower K_D is that many receptors must be occupied before there is a chemoresponse. Therefore, covalent attachment of ligand to a small fraction, thus rendering them insensitive to gradients of attractant, is sufficient to reduce or eliminate chemoresponsive behaviour.

Mendelian mutants have lent credance to the receptor hypothesis (Van

Table 2.

Pretreatment	Test stimulus	Chemoresponse I_{che}	n
2.5 mM Na-cAMP	2.5 mM Na-cAMP	0.66 +/− 0.05	25
0.25 mM 8-N$_3$cAMP	5 mM Na-cAMP	0.50 +/− 0.06	14
0.25 mM 8-N$_3$cAMP	2.5 mM Na$_2$-folate	0.62 +/− 0.07	11
0.25 mM 8-N$_3$cAMP	5 mM NaOAc	0.66 +/− 0.06	12

Data are averages of n T-mazes +/− one standard deviation. Cells were UV irradiated in the presence of Na-cAMP or 8N$_3$-cAMP, washed by centrifugation, and tested in T-mazes ($I_{che} = 0.5$: no response).

Table 3. *Effect of 'activated folate' on chemoresponse*

Attractant stimulus	Chemoresponse of:	
	Normal cells	Cells treated with activated folate
Folate	0.76 +/− 0.03	0.53 +/− 0.06
Acetate	0.72 +/− 0.04	0.70 +/− 0.06
Lactate	0.80 +/− 0.04	0.78 +/− 0.05

Data are averages of 10 (for folate) or 5 T-mazes assays of chemoresponse +/− one standard deviation. Concentrations of attractants were 2.5 mM Na$_2$-folate, 5 mM Na-OAc, and 5 mM Na-lactate; 5 mM NaCl was used as control in the T-mazes.

Houten, 1990). A conditional mutant with a single site mutation (d4–534) failed to respond to folate in T-mazes, did not bind fluorescein-folate, and showed no specific binding of ^{3}H-folate (DiNallo et al., 1982; Van Houten et al., 1985). Similarly, another mutant, cyc^{-1}, showed no specific binding of cAMP and no significant attraction to this stimulus while responding normally to other stimuli (Smith et al., 1987). In the light of the binding and mutant data, a mechanism devoid of receptors, i.e. non-specific interactions of stimulus with plasma membrane, would be difficult to accept as a mechanism by which attractant stimuli initiate chemosensory transduction in *Paramecium*.

The isolation of the receptor proteins was approached very differently for folate and cAMP. Anti-folate antibodies were produced as a tool to identify purified proteins affinity labelled with folate (Sasner & Van Houten, 1989). Since 'activated' folate specifically inhibited the chemoresponse of whole cells to folate, the folate receptor should be among those cell surface proteins covalently bound by folate. ELISAs indicated that folate was bound specifically to sheets of surface membranes by the treatment with activated folate, but the subsequent solubilization of proteins for purification and identification of folate-labelled proteins proved to be technically

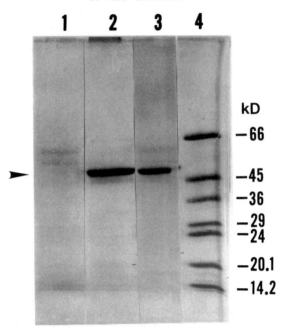

Fig. 1. Lanes from an SDS PAGE 8–12% gradient gel stained with Coomassie blue. Lane 1: proteins eluting from a cAMP affinity column with three bed volumes each of 20 mM KCl; lane 2: proteins eluting with three bed volumes of 20 mM K-cAMP in 200 mM KCl; lane 3: proteins eluting with a subsequent three bed volumes of K-cAMP; lane 4: molecular weight markers. The band of interest is at 48,000 M_r, between the top two molecular weight markers of 66,000 and 45,000.

difficult and is still on-going (Sasner & Van Houten, 1989; Yang & Van Houten, unpublished results).

In search of the cAMP chemoreceptor, affinity chromatography was used to identify a prominent doublet of proteins of 48 kD from the membrane of the cell body of *P. tetraurelia* (Van Houten et al., 1990). This doublet was specifically eluted from cAMP-agarose columns with cAMP (Fig. 1), but not with high salt or cGMP. 5'AMP, and to a lesser extent OAc^-, inhibited chemoresponse to cAMP and, therefore, were expected to elute some of the cAMP binding protein from the column, as they did. Therefore, the specificity of elution matched that expected from the specificity of behaviour and hyperpolarization by cAMP (Smith et al., 1987).

Both proteins of the doublet are most likely to be surface exposed because they are glycosylated, as shown by wheat germ agglutinin–biotin overlays of blots of the proteins and development of the blots with avidin–alkaline phosphatase (Van Houten et al., 1990). Other evidence of surface exposure from $^{125}I_2$ labelling of whole cells was negative, but this was taken to mean only that tyrosines were not available for surface labelling. While it is not possible to prevent all internal membranes from co-purifying with the sur-

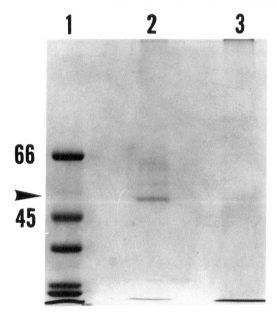

Fig 2. SDS PAGE gradient gel of cAMP affinity column fractions eluted with cAMP (see Fig. 1). Lane 2: proteins from fraction enriched in pellicle membranes were solubilized and applied to the affinity column; lane 3: proteins from fraction enriched in mitochondria and endoplasmic reticulum were solubilized and applied to the affinity column, enrichment was four- and ten-fold respectively, based on enzyme marker assays (Van Houten et al., 1990); lane 1: molecular weight markers. Note proteins of 48,000 M_r present in lane 2 but not lane 3.

face membranes, the cAMP binding protein doublet could not be from contaminating mitochondria or endoplasmic reticulum. These sources were ruled out by analysing cell fractions purposely enriched in these intracellular membranes and depleted of the surface membranes. The doublet at 48 kD is present only in Triton X-100 extracts of the fraction enriched in surface membranes (Fig. 2).

Antibodies were produced by excising gel slices containing the doublet of proteins and immunizing rabbits with the proteins in acrylamide (Van Houten et al., 1990). The resulting serum was diluted 1:40000 and used to treat whole cells prior to testing them in behavioural assays for their response to cAMP, folate and lactate. Cells with no pretreatment and those with exposure to a 1:40000 dilution of pre-immune serum were used as two different controls. There was some small loss of motility due to pre-immune serum treatment as indicated by the slight decrease in the index of chemokinesis between untreated and preimmune serum treated cells (Table 4). However, the loss of chemoresponse to cAMP after treatment of immune serum was specific for cAMP and complete. Cells showed no response to cAMP upon each replication of the experiment ($I_{che} = 0.5$) and they responded to folate and lactate within normal ranges in each

Table 4.

Pre-treatment	Test stimulus	I_{che}	n
None	2.5 mM Na-cAMP	0.73 +/− 0.07	18
Pre-immune serum		0.67 +/− 0.06	15
Immune serum		*0.50 +/− 0.04	16
None	5 mM Na-lactate	0.79 +/− 0.04	8
Pre-immune serum		0.69 +/− 0.07	8
Immune serum		0.65 +/− 0.06	8
None	5 mM NH$_4$Cl	0.94 +/− 0.02	3
Pre-immune serum		0.91 +/− 0.02	3
Immune serum		0.94 +/− 0.02	3
None	5 mM NaOAc	0.82 +/− 0.08	11
Pre-immune serum		0.72 +/− 0.10	12
Immune serum		0.69 +/− 0.12	11

Data are averages of n T-mazes +/− one standard deviation.* The T-maze responses to Na-cAMP after pretreatment with immune serum were statistically significantly different from the responses with preimmune serum or no pretreatment as determined by the Mann-Whitney U test. Within each group of data for each stimulus, all other combinations of responses showed no significant differences. T-mazes compared 2.5 mM Na-cAMP with 2.5 mM NaCl or 5 mM of each of the other stimuli with 5 mM NaCl.

replicate ($I_{che} > 0.5$). These data were taken as good evidence that the doublet of proteins identified by affinity chromatography included the receptor.

The cAMP receptor of *P. tetraurelia* is similar in molecular weight to the *D. discoideum* cAMP chemotaxis receptor (Klein *et al.*, 1989). Antibodies against the *D. discoideum* receptor (kindly provided by P. Devreotes) did not recognize the *Paramecium* receptor on western blots, nor did a cDNA clone of the *D. discoideum* chemotaxis receptor (courtesy of K. Saxe) hybridize to *Paramecium* genomic DNA under low stringency conditions (Schlichtherle & Van Houten, unpublished results). These results merely indicate that the receptors are not identical, and do not rule out similarities in tertiary structure or even in primary amino acid sequence.

The cAMP receptor of *Paramecium* has a molecular weight similar to that of the regulatory subunits of the cAMP-dependent protein kinase and proteins related to the surface antigens of *P. tetraurelia* (Van Houten *et al.*, 1990). Again, western blots ruled out identity of the cAMP receptor with these proteins (monoclonal antibodies and antisera provided by M. Hochstrasser & J. Preer).

The total amino acid compositions of the two bands of the doublet are similar although not identical (Van Houten *et al.*, 1990). The small differences between the bands in apparent molecular weights could be due to covalent modification or to proteolytic cleavage, in which case amino acid differences are expected. Therefore, it is not possible to say, based on amino acid composition, whether the bands of the doublet represent the same protein. Glycosylation and phosphorylation of the doublet proteins are currently being examined.

The cloning of the gene for this receptor is in progress and presumably with its cloning and sequencing we hope we will have identified the fifth external eukaryotic chemoreceptor.

Transduction in paramecia

Paramecia have two prominent components of their swimming behaviour: speed of forward swimming and frequency of turns. (Backward swimming in normal cells is limited to a transient part of the turn.) Both of these components are controlled by membrane potential. The level of membrane potential determines the frequency and angle of ciliary beat, hence swimming speed; the level of membrane potential determines how frequently the resting potential will reach threshold for action potentials and thereby initiate a turn in swimming direction due to the transient reversal of beat. During the calcium action potential, Ca^{2+} enters the ciliary compartment through voltage-dependent Ca^{2+} channels, and for the short period that Ca^{2+} is $>10^{-6}$M, the ciliary beat reverses causing a turn. A bias in the speed of swimming and frequency of turns can cause accumulation or repulsion of small organisms (Fraenkel & Gunn, 1961). For paramecia, the attractants discussed in this chapter hyperpolarize the cell causing a small increase in speed and a decrease in frequency of turning as the cell proceeds up a gradient of attractant. Thus paramecia accumulate by a biased random walk (Van Houten, 1978).

Repellents generally depolarize the cells and thereby disperse them. However, repellents do not tend to be large organic molecules for which the cells could have evolved receptors. The repellents that seem most likely to be physiologically relevant to *Paramecium* are low and high pH and high ionic strength. Therefore, the search for receptors has been focussed primarily on the receptors for attractant stimuli.

The cell is not evenly responsive to stimuli over its whole surface (Preston & Van Houten, 1987*b*). Cilia can be removed, thus removing 50% of the cell surface, with no decrement in the hyperpolarizing response to attractants and repellents and the anterior, ventral surface of the cell is the most responsive to pressure pulses of folate. The distribution of receptors may control the gradients of chemoresponse from anterior to posterior, just as depolarizing and hyperpolarizing mechanoreceptor channels appear to be arranged in gradients from anterior to posterior (Machemer, 1989). Therefore, the anti-receptor antisera may be useful in mapping relative chemoreceptor densities on the cell surface.

Standard electrophysiological techniques have been employed to examine the ionic conductances underlying sensory transduction (Preston & Van Houten, 1987*a,b*), but the mechanisms by which attractants cause hyperpolarization have remained elusive. There appears to be no reversal potential for the hyperpolarization induced by folate or acetate. The counterions of the anion attractants can be K^+ or Na^+, thus normal hyperpolarization can occur in the complete absence of either external K^+ or Na^+. Additionally, external K^+ and Na^+ can be manipulated to eliminate gradients of both ions across the plasma membrane with no effect on the size of the hyperpolarization. Similarly, there is little effect of external pH or agents that affect surface potential. However, there is a small decrease in resistance during acetate-induced hyperpolarization and a 0.2 nA acetate-induced current can be measured.

Mutants with specific defects in characteristic conductances (voltage-dependent Ca, voltage-dependent K, and Ca-dependent Na) and in surface charge (courtesy of C. Kung & E. Richard) showed normal stimulus-induced hyperpolarizations and helped to eliminate possible explanations of the 0.2 nA current that appeared to be the source of the hyperpolarization (Preston & Van Houten, 1987a). Of the 'borrowed' mutants, only *restless* (Richard et al., 1986) showed some alteration in the size of the acetate- and folate-induced hyperpolarizations suggesting involvement of calcium (Preston & Van Houten, 1987a). A folate chemoresponse mutant with a single Mendelian mutation (DiNallo et al., 1982) was both unable to respond to folate and to hyperpolarize in folate more than 1–2 mV compared to the 12 mV hyperpolarization of normal cells (Schulz et al., 1984), but this mutant divulged no clues to the ion carrying the hyperpolarizing current.

The many conductance and other mutants at our disposal helped to narrow our hypotheses for the stimulus-induced hyperpolarization to the stimulus activation of a calcium or proton extrusion pump that generates the hyperpolarizing current. The calcium pump was favoured because of the wide external pH range over which the hyperpolarization remained constant (Preston & Van Houten, 1987a), but these results did not rule out a proton pump that transports against a gradient. None the less, other evidence pointed to calcium: lithium inhibited both chemoresponse and calcium efflux from whole cells (Wright & Van Houten, 1990) and a mutant suspected of having abnormal calcium homeostasis was not attracted to folate or acetate (Table 5), but was normally attracted to NH_4Cl, an attractant that we believe affects internal pH and not necessarily the Ca^{2+} pump. Additionally, all of our Mendelian chemoresponse mutants were conditional, hinting that an alteration of the mechanism of hyperpolarization could be a lethal event for the cell, much as would be an alteration of the calcium homeostasis pump (Van Houten, 1990).

Effects of lithium and involvement of Ca^{2+} made it reasonable to postulate that inositol phosphates act as second messengers in *Paramecium* sensory transduction. However, assays of IP_3 and inositol phospholipids gave no hint of acetate- or folate-induced changes (Wright & Van Houten, 1990). Likewise, cAMP levels remained constant through stimulation. Internal

Table 5. *Comparison of normal and K Shy-A/K Shy-B mutant chemoresponse*

		T-maze assays of attraction			
Attractant	Control	Normal Cells	n	K Shy-A/K Shy-B	n
5 mM NaOAc	5 mM NaCl	0.76 +/− 0.05	3	0.52 +/− 0.03	5
2.5 mM K₂folate	5 mM KCl	0.75 +/− 0.08	3	0.41 +/− 0.07	5
2 mM K₂folate	4 mM KCl	0.73 +/− 0.07	6	0.51 +/− 0.09	6
5 mM NH₄Cl	5 mM NaCl	0.77 +/− 0.03	3	0.72 +/− 0.06	5

K Shy-A/K Shy-B (Evans *et al.*, 1987) is a double mutant incorporating both mutant genes that produce the *K Shy* phenotype.

Table 6. *Effects of IBMX on chemoresponse*

Test solution	Control solution	Chemokinesis response
1 mM IBMX	–	0.07 +/− 0.03
1 mM IBMX	1 mM Adenosine	0.27 +/− 0.09
1 mM Na₂folate	2 mM NaCl	
−IBMX	−IBMX	0.81 +/− 0.08
+IBMX	−IBMX	0.42 +/− 0.01
5 mM NH₄Cl	5 mM NaCl	
−IBMX	−IBMX	0.95 +/− 0.06
+IBMX	+IBMX	0.97 +/− 0.02
2 mM NaOAc	2 mM NaCl	
−IBMX	−IBMX	0.78 +/− 0.13
+IBMX	+IBMX	0.82 +/− 0.23
2 mM NacAMP	2 mM NaCl	
−IBMX	−IBMX	0.68 +/− 0.02
+IBMX	+IBMX	0.78 +/− 0.13
2.5 mM Na₂folate	5 mM NaCl	
−IBMX	−IBMX	0.96 +/− 0.05
+IBMX	+IBMX	0.98 +/− 0.01

Data are averages of three T-mazes +/− one standard deviation. 1 mM IBMX was added to one or both of the T-maze arms with test or control solutions. Index of chemokinesis >0.5 indicates attraction; <0.5 indicates repulsion. In the first three experiments, 1 mM IBMX is added only to the test solution. (In the first experiment, 1 mM IBMX in buffer is compared to buffer alone.) In the last four, 1 mM IBMX is either omitted altogether or added to both test and control solutions.

cAMP levels could be profoundly affected by treating cells with isobutyl methylxanthine (IBMX), a phosphodiesterase inhibitor. Such treatment made cells greatly increase their speed of swimming (Bonini & Nelson, 1988; Bonini *et al.*, 1986) and disperse from solutions of 1 mM IBMX by extremely fast swimming (Table 6). Despite the presence of IBMX in both arms of the T-maze, cells were able to accumulate normally in OAc, NH₄, folate and cAMP (Table 6). IBMX failed to act synergistically or additively with less than saturating amounts of Na_2-folate as attractant (Table 6, line 3). Therefore, we concluded that cAMP was probably not a second messenger in chemoresponse. These results have special significance because relatively large hyperpolarizations induced by changes in external cations are associated with increased internal cAMP and also with increased swimming speed (Hennessey et al., 1985; Bonini et al., 1986; Schultz et al., 1984) and chemoattraction likewise is associated with a hyperpolarization, albeit a smaller one. It is possible that the attractant-induced cAMP changes are beyond our ability to measure, which also calls into question their role, if any, in chemoresponse.

P. tetraurelia *Ca-ATPase*

The mechanisms of Ca homeostasis in *Paramecium* are poorly understood. The cell clearly both takes up and loses Ca^{2+} and the efflux is most likely created by a Ca^{2+}-ATPase pump (Browning & Nelson, 1976). However, the identity of this pump has remained elusive, perhaps because both the method of purifying membrane fractions and Mg^{2+} concentrations are critical for maintaining activity. Recently, we have succeeded in characterizing a high affinity ($K_m = 95$ nM for Ca^{2+}), ATP specific, Ca^{2+}-stimulated ATPase from cell surface membranes (pellicles) (Wright & Van Houten, 1990). It requires approximately 3 mM $MgCl_2$ for optimal activity and can be inhibited by vanadate, calmidazolium, trifluoperazine, and mellitin in concentrations that should have minimal nonspecific effects not associated with calmodulin inhibition (Rega & Garrahan, 1986). Na^+ and K^+ stimulate the activity while Li^+ does not, perhaps explaining the mechanism by which Li^+ inhibits chemokinesis. (Also, Li^+ treatment of cells results in significant Li^+ uptake and loss of up to 50% of the cells' K^+, possibly contributing further to inhibition of the Ca^{2+}-ATPase.) Bovine calmodulin did not stimulate the enzyme, perhaps because the enzyme is activated by proteolysis or phosphatidylserine during the purification process. (This phenomenon is seen in other Ca^{2+}-ATPases (Carafoli et al., 1982).)

A Ca^{2+}-ATPase plasma membrane pump is expected to form a phosphoenzyme intermediate that is base and hydroxylamine labile and enhanced by La^{3+} (Wuytack et al., 1982; Rega & Garrahan, 1986). Potential contaminating endoplasmic reticulum (ER) Ca^{2+}-ATPase pumps also form phosphoenzyme intermediates, but this phosphorylation is not enhanced by La^{3+} and the apparent molecular weight of the endoplasmic reticulum pump is smaller than that of the plasma membrane pumps because the ER pump lacks a calmodulin binding domain (Rega & Garrahan, 1986). Figure 3 demonstrates that the surface membrane (pellicle) preparation has a protein of approximately 133 kDa that is labelled with γ-^{32}P-ATP in the presence of Ca^{2+} (lanes A 3 and B 3) and this labelling is hydroxylamine sensitive (lanes A 3 vs A 5), and enhanced by La^{3+} (lanes B 3 vs B 5). The membrane fraction enriched in endoplasmic reticulum also

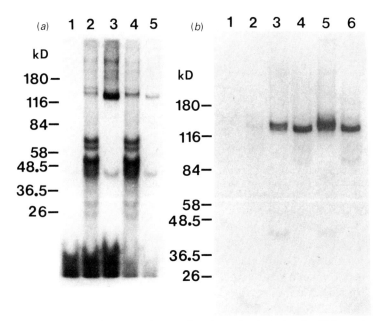

Fig 3. Analysis of phosphoenzyme intermediate formation in isolated pellicles and an ER enriched fraction. (a) Pellicles were incubated with [γ-^{32}P]-ATP, with the indicated additions to the medium: 1 mM EGTA, lane 1; 3 mM Mg^{2+}, lanes 2 and 4; 50 μM Ca^{2+}, lanes 3 and 5. TCA precipitated proteins were then treated with a control solution (lanes 1–3) or with 150 mM hydroxylamine, pH 6 (lanes 4 and 5), for 10 min at room temperature, before subjecting to SDS-PAGE. (b) Pellicles, lanes 1, 3, and 5, or ER enriched fraction, lanes 2, 4, and 6, were incubated with [γ-^{32}P]-ATP with the indicated additions to the medium, TCA precipitated, and subjected to SDS-PAGE. Additions: 1 mM EGTA, lanes 1 and 2; 50 μM Ca^{2+}, lanes 3 and 4; 50 μM Ca^{2+} plus 100 μM La^{3+}, lanes 5 and 6. The phosphoenzyme intermediates are between the 116 and 180 kD molecular mass markers.

has a phospho-protein, but of lower apparent molecular weight (115K), that is hydroxylamine labile (not shown) but not enhanced in its phosphorylation by La^{3+} (lanes B 4 vs B 6).

Currently, we are purifying this enzyme activity to clarify its relationship to the protein of the phosphoenzyme intermediate.

If the Ca^{2+}-ATPase were indeed part of the chemosensory transduction pathway, attractants would be expected to stimulate its activity. However, the increase in basal activity needed to account for the hyperpolarizing current would be very small, only a few per cent of the 30 nmoles/mg min basal activity. Stimuli added to pellicles resulted in slightly increased Ca^{2+}-ATPase activity, but too small an increase to be statistically significant. Additionally, the internal levels of free Ca^{2+} would not necessarily change with the predicted small activation of the pump.

Not all attractant stimuli are predicted to activate the Ca^{2+}-ATPase. As do OAc$^-$ and folate, NH$_4$Cl hyperpolarizes cells relative to NaCl and

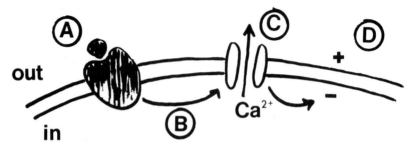

Fig. 4. Schematic picture of *Paramecium tetraurelia* chemosensory transduction pathway for some attractants. Ligand binds to receptor, which signals to the Ca^{2+}-ATPase to increase activity. The resulting increased conductance hyperpolarizes the cell and produces faster, smoother, swimming.

elicits an attractant response. However, extensive structure/activity studies have unearthed no compounds that interfere with the attraction to NH_4Cl, as might have been expected if it bound to external receptors to elicit its hyperpolarization. Also, the Ca^{2+} homeostasis mutant *K Shy-A/K Shy-B* (Evans et al., 1987) is normally attracted to NH_4Cl while it fails to respond to OAc^- and folate (Table 6). Therefore, our working hypothesis is that NH_4Cl diffuses across the cell membrane and alters internal pH as it becomes ionized. The change in internal pH would then affect a channel or perhaps a pump activity.

SUMMARY

The chemosensory system of *P. tetraurelia* responds to chemical stimuli that generally signify bacteria are in the vicinity. The sensory transduction pathway originates with stimulus binding to receptor and most likely continues with the transduction of receptor binding to the activation of a plasma membrane calcium pump. The pump activation, or increased activity of basal levels, would produce the current that hyperpolarizes the cell (Fig. 4), increases the frequency of ciliary beat, and decreases the frequency of ciliary reversals (i.e. turns). The cAMP receptor has been identified and, once its gene is cloned, is likely to be one of the growing number of external chemoreceptors being identified from eukaryotic microorganisms. The assignment to a superfamily of neurotransmitter or other internal receptors, if any, awaits the cloning and sequencing results.

The basic *Paramecium* transduction pathway of receptor, second messenger, and response is, of course, shared with other eukaryotic organisms, and even many of *P. tetraurelia*'s apparent novel aspects have parallels in metazoan systems. For example, the extremely low affinity of *P. tetraurelia* receptors is not unprecedented in chemoreception. Taste and some olfactory receptors display an even lower affinity for ligand, based on detection

CHEMOSENSORY TRANSDUCTION IN *PARAMECIUM* 315

or behavioural half maximal responses (Cagan, 1981; Hansen & Wieczorek, 1981; Price, 1981; Lancet, 1986). A stimulus-induced hyperpolarizing conductance also has precedent in the inhibitory amino acid-stimulated conductances in lobster and also rat (McClintock & Ache, 1989; Michel & Ache, 1990; Dionne, 1990).

Ca^{2+} seems to be an important second messenger in *P. tetraurelia* chemoresponse and features prominently as an important second messenger in both taste and olfaction. In these latter systems, Ca^{2+} is released from internal pools by IP_3 or IP_3 directly gates a Ca^{2+} channel (Teeter *et al.*, 1990; Akabas *et al.*, 1988; Breer *et al.*, 1990*a,b*). The activation of a Ca^{2+}-ATPase pump may seem at this time to be a novel aspect of the *P. tetraurelia* transduction system. However, the Ca^{2+}-ATPase activities of plasma membranes of taste bud cells (Barry & Savoy, 1990) and of olfactory cilia (Lo *et al.*, 1990) are high, and eventually modification of these enzymes may be found to play a part in taste and olfactory transduction.

P. tetraurelia resembles a swimming neuron because it displays so many conductances expected of metazoan excitable cells (Saimi *et al.*, 1988), and perhaps it is because it has neuronal properties that it presents some important parallels with olfactory and gustatory receptor cells. For example, *P. tetraurelia* responds to a stimulus by hyperpolarizing or depolarizing. Stimuli seem to utilize separate receptors and one of several transduction systems. We believe that there are at least three transduction mechanisms involved in attraction: stimuli such as OAc and folate stimulate a Ca^{2+}-ATPase pump and thereby hyperpolarize the cell; NH_4Cl probably diffuses into the cell without the aid of a receptor and alters pH_i to initiate hyperpolarization; some relative attractant conditions (e.g. KCl pH 7 relative to KOH pH 9) would actually relatively depolarize the cells (see Van Houten, 1977, 1978 for a discussion of this seeming paradox). The behavioural response to mixtures would therefore be the end result of stimulus-induced hyperpolarizing and depolarizing potentials. Thus in the *P. tetraurelia* chemoreceptor cell there is integration of information occurring, at the receptor cell level in lobster olfaction (Michel & Ache, 1990) and elsewhere.

REFERENCES

Ache, B. W. (1987). Chemoreception in invertebrates. In *Neurobiology of Taste and Smell*, eds. T. Finger & W. Silver, pp. 39–64. New York: J. Wiley.
Adler, J. (1987). How motile bacteria are attracted and repelled by chemicals: an approach to neurobiology. *Biological Chemistry Hoppe-Seyler*, **368**, 163–73.
Akabas, M. J. Dodd & Al-Awqati, Q. (1988). A bitter substance induces a rise in intracellular calcium in a subpopulation of rat taste cells. *Science*, **242**, 1047–50.
Anderson, O. R. (1988). *Comparative Protozoology*. Berlin: Springer-Verlag.
Anholt, R. (1987). Primary events in olfactory reception. *Trends in Biochemical Sciences*, **12**, 58–62.

Atema, J. (1985). Chemoreception in the sea: adaptations of chemoreceptors and behaviour to aquatic stimulus conditions. In *Physiological Adaptations of Marine Animals*, Society for Experimental Biology Symposium vol. 39, ed. A. S. Laverack, pp. 387–423.

Atema, J. (1987). Aquatic and terrestrial chemoreceptor organs: morphological and physiological designs for interfacing with chemical stimuli. In *Comparative Physiology: Life in Water and on Land*, eds. P. Dejours, L. Bolis, C. R. Taylow & E. R. Weibel, pp. 303–16. Fidia Research Series.

Barry, M. & Savoy, L. (1990). Elevated calcium dependent ATPase activity of plasma membranes of normal and denervated fungiform taste bud cells. *Chemical Senses*, **16**, in press.

Bignetti, E., Damiani, G., DeNegri, P., Ramoni, R., Avanzini, F., Ferrari, G. & Rossi, G. D. (1987). *Chemical Senses*, **12**, 601–8.

Bonini, N. & Nelson, D. L. (1988). Differential regulation of *Paramecium* ciliary motility by cAMP and cGMP. *Journal of Cell Biology*, **106**, 1615–23.

Bonini, N., Gustin, M. & Nelson, D. L. (1986). Regulation of ciliary motility in *Paramecium*: a role for cyclic AMP. *Cell Motility and the Cytoskeleton*, **6**, 256–72.

Breer, H., Boekhoff, I. & Tareilus, E. (1990a). Rapid kinetics of second-messenger signalling in olfaction. *Chemical Senses*, **15**, in press.

Breer, H., Boekhoff, I. & Tareilus, E. (1990b). Rapid kinetics of second-messenger formation in olfactory transduction. *Nature, London*, **345**, 65–8.

Browning, J. L. & Nelson, D. L. (1976). Biochemical studies of the excitable membrane of *Paramecium aurelia*. I. Ca^{2+} fluxes across the resting and excited membrane. *Biochemica & Biophysica Acta*, **448**, 338–51.

Bruch, R. C. (1990). Signal transduction in olfaction and taste. In *G-Proteins*, eds. R. Iyengar & L. Birnbaumer, pp. 411–28. New York: Academic Press.

Bryant, B., Brand, J. G., Kalinoski, D. L., Bruch, R. C. & Cagan, R. H. (1987). Use of monoclonal antibodies to characterize amino acid taste receptors in catfish: effects on binding and neuronal responses. In *Olfaction and taste IX*, eds. S. Roper & J. Atema, pp. 208–10.

Burkholder, A. C. & Hartwell, L. H. (1985). The yeast alpha factor receptor. *Nucleic Acids Research*, **13**, 8463–73.

Burt, J. R. (1977). Hypoxanthine: a biochemical index of fish quality. *Processes in Biochemistry*, **12**, 32–5.

Cagan, R. (1981). Recognition of taste stimuli at the initial binding interaction. In *Biochemistry of Taste and Olfaction*, eds. R. Cagan & M. Kare, pp. 175–204. New York: Academic Press.

Carafoli, E., Mauro, Z., Niggli, V. & Krebs, J. (1982). The Ca^{2+} transporting ATPase of erythrocytes. *Annals of the NY Academy of Sciences*, **402**, 304–28.

Carr, W. E. S. (1990). Chemical signaling systems in lower organisms: a prelude to the evolution of chemical communication in the nervous system. In *Evolution of the First Nervous Systems*, ed. P. A. V. Anderson. New York: Plenum Press, in press.

Carr, W. E. S., Gleeson, R. A. & Trapido-Rosenthal, H. G. (1989). Chemosensory systems in lower organisms: correlations with internal receptor systems for neurotransmitters and hormones. In *Advances in Comparative and Environmental Physiology*, vol. 5, pp. 25–49. Berlin: Springer-Verlag.

DiNallo, M., Wohlford, M. & Van Houten, J. (1982). Mutants of *Paramecium* defective in chemoreception of folate. *Genetics*, **102**, 149–58.

Dionne, V. (1988). How do you smell? Principle in question. *Trends in Neurological Sciences*, **11**, 188–9.

Dionne, V. (1990). Excitatory and inhibitory responses induced by amino acids in isolated mudpuppy olfactory receptor neurons. *Chemical Senses*, **15**, in press.

Evans, T. C., Hennessey, T. & Nelson, D. L. (1987). Electrophysiological evidence suggests a defective Ca^{2+} control mechanism in a new *Paramecium* mutant. *Journal of Membrane Biology*, **98**, 275–83.

Firestein, S. & Shepherd, G. M. (1989). Olfactory transduction is mediated by the direct action of cAMP. *Society of Neuroscience Abstracts*, **15**, 749.

Fraenkel, G. S. & Gunn, D. L. (1961). *The Orientation of Animals*. New York: Dover Publications.

Garbers, D. L. (1989). Guanylate cyclase, a cell surface receptor. *Journal of Biological Chemistry*, **264**, 9103–6.

Getchell, T. V. & Getchell, M. L. (1987). Peripheral mechanisms of olfaction: biochemistry and neurophysiology. In *Neurobiology of Taste and Smell*, eds. T. Finger & W. Silver, pp. 91–123. New York: J. Wiley & Sons.

Getchell, T. V., Margolis, F. L. & Getchell, M. L. (1984). Perireceptor and receptor events in vertebrate olfaction. *Progress in Neurobiology*, **23**, 317–45.

Hagen, D. C., McCaffrey, G. & Sprague, G. F. (1986). Evidence the yeast STE3 gene encodes a receptor for the peptide pheromone a factor: gene sequence and implications for the structure of the presumed receptor. *Proceedings of the National Academy of Sciences, USA*, **83**, 1418–22.

Hansen, K. & Wieczorek, H. (1981). Biochemical aspects of sugar reception in insects. In *Biochemistry of Taste and Olfaction*, eds. R. Cagan & M. Kare, pp. 129–62. New York: Academic Press.

Hennessey, T., Machemer, H. & Nelson, D. (1985). Injected cyclic AMP increases ciliary beat frequency in conjunction with membrane hyperpolarization. *European Journal of Cell Biology*, **36**, 153–6.

Herskowitz, I. & Marsh, L. (1987). Conservation of a receptor/signal transduction system. *Cell*, **50**, 995–6.

Jones, D. T. & Reed, R. R. (1989). G_{olf}: an olfactory neuron specific G-protein involved in odorant signal transduction. *Science*, **244**, 790–5.

Kauer, J. S. (1987). Coding in the olfactory system. In *Neurobiology of Taste and Smell*, eds. T. Finger & W. Silver, pp. 205–31. New York: J. Wiley & Sons.

Kinnamon, S. C. (1988). Taste transduction: a diversity of mechanisms. *Trends in Neurological Science*, **11**, 491–6.

Kleene, S. J. & R. C. Gesteland (1990). Transmembrane Currents in Frog Olfactory Cilia, *Journal of Membrane Biology*, in press.

Klein, P., Sun, T. J., Saxe, C. L., Kimmel, A., Johnson, R. L. & Devreotes, P. N. (1988). A chemoattractant receptor controls development in *Dictyostelium discoideum*. *Science*, **241**, 1467–72.

Kohbara, J., Wegert, S. & Caprio, J. (1990). Two types of arginine-best taste units in the channel catfish. *Chemical Senses*, **15**, in press.

Kumazawa, T., Teeter, J. H. & Brand, J. G. (1990). L-proline activates cation channels different from those activated by L-arginine in reconstituted taste epithelial membranes from channel catfish. *Chemical Senses*, **15**, in press.

Kung, C., Chang, S. Y., Satow, Y., Van Houten, J. & Hansma, H. (1975). Genetic dissection of behavior in *Paramecium*. *Science*, **188**, 898–904.

Labarca, P., Simon, S. A. & Anholt, R. H. (1988). Activation by odorants of a multistate cation channel from olfactory cilia. *Proceedings of the National Academy of Sciences, USA*, **85**, 944–7.

Lancet, D. (1986). Vertebrate olfactory reception. *Annual Reviews in Neuroscience*, **9**, 329–55.

Lancet, D. & Pace, U. (1987). The molecular basis of odor recognition. *Trends in Biochemical Science*, **12**, 63–6.

Lancet, D., Heldman, J., Khen, M., Lazard, D., Lazarovits, J., Margalit, T., Poria, Y. & Zupko, K. (1989). Molecular cloning and localization of novel olfactory-specific enzymes possibly involved in signal termination: a new role for supporting cells. *Society of Neuroscience Abstracts*, **15**, 749.

Lerner, M. R., Reagan, J., Gyorgyi, T. & Roby, A. (1988). Olfaction by melanophores: what does it mean? *Proceedings of the National Academy of Sciences, USA*, **85**, 261–4.

Lo, Y. H., Bradley, T. & Rhoads, D. E. (1990). Distribution of Ca^{2+}-ATPase activity in the olfactory rosette of Atlantic salmon: comparison with Na, K-ATPase and alanine receptors. *Chemical Senses*, **15**, in press.

Lowe, G., Nakamura, T. & Gold, G. H. (1988). EOG amplitude is correlated with odor-stimulated adenylate cyclase activity in the bullfrog olfactory epithelium. *Chemical Senses*, **13**, 710.

Machemer, H. (1976). Interactions of membrane potential and cations in regulation of ciliary activity in *Paramecium*. *Journal of Experimental Biology*, **65**, 427–48.

Machemer, H. (1989). Cellular behaviour modulated by ions: electrophysiological implications. *Journal of Protozoology*, **36**, 463–87.

Machemer, H. & De Peyer, J. E. (1977). Swimming sensory cells: electrical membrane parameters, receptor properties and motor control in ciliated protozoa. *Verhandlungen Deutschland Zoologische Gesellschaft*, **1977**, 86–110.

McClintock, T. S. & Ache, B. (1989). Hyperpolarizing receptor potentials in lobster olfactory receptor cells. *Chemical Senses*, **14**, 637–48.

Makman, R. S. & Sutherland, E. W. (1965). Adenosine-3':5'-phosphate in *Escherichia coli*. *Journal of Biological Chemistry*, **240**, 1309–13.

Michel, W. C. & Ache, B. W. (1990). Odor-activated conductance inhibits lobster olfactory receptor cells. *Chemical Senses*, **17**, in press.

Moore, P. & Atema, J. (1988). A model of a temporal filter in chemoreception to extract directional information from a turbulent odor plume. *Biological Bulletin*, **174**, 355–63.

Nakamura, T. & Gold, G. H. (1987). A cyclic nucleotide-gated conductance in olfactory receptor cilia. *Nature, London*, **325**, 442–4.

Nomura, T. & Kurihara, K. (1989). Similarity of ion dependence of odorant responses between lipid bilayer and olfactory system. *Biochimica & Biophysica Acta*, **1005**, 260–4.

Novoselov, V. I., Krapivinskaya, L. D. & Fesenko, E. E. (1988). Amino acid binding glycoproteins from the olfactory epithelium of skate (*Dasyatis partinaca*). *Chemical Senses*, **13**, 267–79.

Pace, U., Hanski, E., Salomon, Y. & Lancet, D. (1985). Odorant-sensitive adenylate cyclase may mediate olfactory reception. *Nature, London*, **316**, 255–8.

Pelosi, P. & Dal Monte, M. (1990). Purification and characterization of odorant binding proteins from nasal mucosa of pig and rabbit. *Chemical Senses*, **15**, in press.

Persaud, K. C., Chiavacci, L. & Pelosi, P. (1988). Binding proteins for sweet compounds from gustatory papillae of the cow, pig and rat. *Biochimica & Biophysica Acta*, **967**, 65–75.

Pilch, P. F. & Czeck, M. P. (1984). Affinity cross-linking of peptide hormones and their receptors. In *Membranes, Detergents, and Receptor Solubilization*, eds. J. C. Venter & L. C. Harrison, pp. 161–76. New York: Alan R. Liss.

Preston, R. R. & Van Houten, J. (1987*a*). Chemoreception in *Paramecium*: acetate-

and folate-induced membrane hyperpolarization. *Journal of Comparative Physiology*, **160**, 525–36.
Preston, R. R. & Van Houten, J. (1987*b*). Localization of chemoreceptive properties of the surface of *Paramecium*. *Journal of Comparative Physiology*, **160**, 537–41.
Price, S. (1981). Receptor proteins in vertebrate olfaction. In *Biochemistry of Taste and Olfaction*, eds. R. Cagan & M. Kare, pp. 69–84. New York: Academic Press.
Price, S. & Willey, A. (1987). Benzaldehyde binding protein from dog olfactory epithelium. In *Olfaction and Taste IX*, eds. S. Roper & J. Atema, pp. 55–60. New York: New York Academy of Sciences.
Price, S. & Willey, A. (1988). Effects of antibodies against odorant binding proteins on electrophysiological responses to odorants. *Biochimica & Biophysica Acta*, **965**, 127–29.
Pun, R. & Gesteland, R. (1990). Whole cell patch recordings show frog and salamander olfactory neurons are different. *Chemical Senses*, **15**, in press.
Rega, A. F. & Garrahan, P. J. (1986). *The Ca^{2+} Pump of Plasma Membranes*, Boca Raton: CRC Press.
Richard, E. A., Saimi, Y. & Kung, C. (1986). A mutation that increases a novel calcium-activated potassium conductance of *Paramecium tetraurelia*. *Journal of Membrane Biology*, **91**, 173–81.
Roper, S. D. (1989). The cell biology of vertebrate taste receptors. *Annual Reviews in Neuroscience*, **12**, 329–53.
Saimi, Y., Martinac, B., Gustin, M., Culbertson, M., Adler, J. & Kung, C. (1988). *Trends in Biological Sciences*, **13**, 304–9.
Sasner, J. M. & Van Houten, J. L. (1989). Evidence for a *Paramecium* folate chemoreceptor. *Chemical Senses*, **14**, 587–96.
Schiffman, S. S. (1990). The role of sodium and potassium transport pathways in taste transduction: an overview. *Chemical Senses*, **15**, 129–36.
Schmale, H., Holfgreve-Grez, H. & Heidje Christiansen, J. (1990). Possible role for salivary gland protein in taste reception indicated by homology to lipophilic-ligand carrier proteins. *Nature, London*, **343**, 366–9.
Schultz, J., Grunemund, R., Von Hirschhausen, R. & Schonefeld, U. (1984). Ionic regulation of cyclic AMP levels in *Paramecium tetraurelia* in vivo. *FEBS Letters*, **167**, 113–16.
Schulz, S., Denaro, M. & Van Houten, J. (1984). Relationship of folate binding and uptake to chemoreception in *Paramecium*. *Journal of Comparative Physiology*, **155**, 113–19.
Singh, S., Lowe, D., Thorpe, D., Rodriguez, H., Kuang, W.-J., Daggott, L., Chinkers, M., Goeddel, D. V. & Garbers, D. L. (1988). Membrane guanylate cyclase is a cell-surface receptor with homology to protein kinases. *Nature, London*, **334**, 708–12.
Sklar, P. B., Anholt, R. H. & Snyder, S. H. (1986). The odorant-sensitive adenylate cyclase of olfactory receptor cells. *Journal of Biological Chemistry*, **261**, 15538–43.
Smith, R., Preston, R. R., Schulz, S., Gagnon, M. L. & Van Houten, J. (1987). Correlation of cyclic adenosine monophosphate binding and chemoresponse in *Paramecium*. *Biochimica & Biophysica Acta*, **928**, 171–8.
Snyder, S. H., Sklar, P. B. & Pevsner, J. (1988). Molecular mechanisms of olfaction. *Journal of Biological Chemistry*, **263**, 13971–4.
Stengl, M., Zufall, F., Hatt, H., Dudel, J. & Hildebrand, J. G. (1989). Patch clamp analysis of male *Manduca sexta* olfactory receptor neurons in primary cell culture. *Soc. for Neuroscience Abstracts*, **15**, 365.

Stengl, M., Zufall, F., Hatt, H. & Hildebrand, J. G. (1990). Olfactory receptor neurons from developing male *Manduca sexta* antennae respond to species-specific sex pheromone in vitro. *Chemical Senses*, **15**, in press.
Teeter, J. H., Huque, T. & Restrepo, D. (1990). Inositol-1,4,5-triphosphate (IP_3): an alternate second messenger for olfactory transduction? *Chemical Senses*, **15**, in press.
Van Houten, J. (1976). Chemokinesis in *Paramecium*: a genetic approach. Ph.D. Dissertation, University of California, Santa Barbara.
Van Houten, J. (1977). A mutant *Paramecium* defective in chemotaxis. *Science*, **198**, 746–9.
Van Houten, J. (1978). Two mechanisms of chemotaxis in *Paramecium*. *Journal of Comparative Physiology A*, **127**, 167–74.
Van Houten, J. (1979). Membrane potential changes during chemokinesis in *Paramecium*. *Science*, **204**, 1100–3.
Van Houten, J. (1990*a*). Chemosensory transduction in unicellular eukaryotes. In *Evolution of the First Nervous Systems*, ed. P. A. V. Anderson. New York: Plenum Press, 343–56.
Van Houten, J. (1990*b*). Chemoreception in *Paramecium*: A genetic approach. In *Chemical Senses III: Genetics of Perception and Communication*, eds. C. Wysocki and M. Kare. New York: Marcel Dekker, in press.
Van Houten, J. (1991). Signal Transduction in Chemoreception. In *Modern Cell Biology*, vol. 11, ed. J. Spudich. New York: Alan R. Liss, in press.
Van Houten, J., Cote, B., Zhang, J., Baez, J. & Gagnon, M. L. (1990). Studies of the cAMP chemoreceptor of *Paramecium*. *Journal of Membrane Biology*, in press.
Van Houten, J., Hansma, H. & Kung, C. (1975). Two quantitative assays of chemotaxis in *Paramecium*. *Journal of Comparative Physiology A*, **104**, 211–23.
Van Houten, J., Hauser, D. C. R. & Levandowsky, M. (1981). Chemosensory behavior in protozoa. In *Biochemistry and Physiology of Protozoa*, vol. 4, eds. M. Levandowsky & S. H. Hutner, pp. 67–124. New York: Academic Press.
Van Houten, J., Martel, E. & Kasch, T. (1982). Kinetic analysis of chemoreception in *Paramecium*. *Journal of Protozoology*, **29**, 226–30.
Van Houten, J. & Preston, R. R. (1987). Chemoreception in unicellular organisms. In *Neurobiology of Taste and Smell*, eds. T. Finger & W. Silver, pp. 11–38. New York: J. Wiley.
Van Houten, J., Smith, R., Wymer, J., Palmer, B. & Denaro, M. (1985). Fluorescein-conjugated folate as an indicator of specific folate binding to *Paramecium*. *Journal of Protozoology*, **32**, 613–16.
Van Houten, J. & Van Houten, J. (1982). Computer analysis of *Paramecium* chemokinesis behavior. *Journal of Theoretical Biology*, **98**, 453–68.
Vogt, R. G. (1987). The molecular basis of pheromone reception: its influence on behavior. In *Pheromone Biochemistry*, ed. G. D. Prestwich & G. S. Blomquist, pp. 385–431. New York: Academic Press.
Vogt, R. G., Prestwich, G. D. & Riddiford, L. M. (1988). Sex pheromone receptor proteins. *Journal of Biological Chemistry*, **263**, 3952–9.
Vogt, R. G., Riddiford, L. M. & Prestwich, G. D. (1985). Kinetic properties of a pheromone degrading enzyme: the sensillar esterase of *Antheracea polyphemus*. *Proceedings of the National Academy of Sciences, USA*, **82**, 8827–31.
Whiteway, M., Hougan, L., Dignard, D., Thomas, D. Y., Bell, L., Saari, G. C., Grant, F. J., O'Hara, P. & MacKay, V. L. (1989). The *STE*4 and *STE*18 genes of yeast encode potential β and γ subunits of the mating factor receptor-coupled G protein. *Cell*, **56**, 467–77.

Wright, M. & Van Houten, J. (1990). Characterization of putative transporting Ca^{2+} ATPase from pellicles of *Paramecium tetraurelia*. *Biochimica & Biophysica Acta*, in press.

Wuytack, F., Raeymaekers, L., DeSchutter, G. & Casteels, R. (1982). Demonstration of the phosphorylated intermediates of the Ca^{2+}-transport ATPase in a microsomal fraction and in a $(Ca^{2+} + Mg^{2+})$-ATPase purified from smooth muscle by means of calmodulin affinity chromatography. *Biochimica & Biophysica Acta*, **693**, 45–52.

Zimmer-Faust, R. & Martinez, L. A. (1988). A proposed role of adenine nucleotides as inducers of feeding in aquatic animals. *Chemical Senses*, **13**, 751–2.

LEUCOCYTE CHEMOTAXIS: A PERSPECTIVE

PETER C. WILKINSON

Bacteriology and Immunology Department, University of Glasgow (Western Infirmary), Glasgow G11 6NT, UK

John Lackie asked me to contribute an 'overview' article to this symposium, but I have preferred to call it a 'perspective'. An overview suggests a panoramic detachment granted to few, while perspective shifts with the observer. Whatever the word chosen, the title allows the writer a latitude denied in the more conventional data-packed scientific paper. This will therefore be a personal account of some aspects of leucocyte chemotaxis which have interested me, with an emphasis on visual studies since these give so much information that cannot be obtained from the filter and other population assays. I shall discuss chemotaxis in the general context of other locomotor reactions of leucocytes since chemotaxis is only one of the many factors which determine the accumulation of cells in inflammation. This is particularly true for lymphocytes in which, as an immunologist, I have a professional interest. Since scientists have short memories, it is worth starting by recalling some of the early work which set the scene.

BEGINNINGS

During the first half of this century many beautiful studies were made of the behaviour of locomotory leucocytes and other cells (see, for example, those of Clark & Clark, 1920; Clark, Clark & Rex, 1936; McCutcheon, 1923; Lewis, 1931; 1934; 1939). Considering the inadequate equipment these workers usually had at their disposal, they gained a surprising amount of insight into the ways cells moved, which laid the foundations on which contemporary studies are based. Much of this work, alas, is little read today. Those researchers also suffered from the handicap of working before the days when the biochemists could provide pure substances for study. Progress slowed down and interest in leucocyte chemotaxis lagged until, in 1962, Stephen Boyden introduced the filter assay and promptly left the field. The filter assay produced an impetus which has not yet been exhausted, since it provided a simple way of quantifying the locomotor stimulus given to a cell by a chemical. However, it also had the drawback that it was now possible to spend a research career studying leucocyte chemotaxis without ever looking at a moving cell or, indeed, knowing its front from its back (as I discovered from a quite distinguished colleague's interpretation of his own slide at a recent meeting). The early workers used filter assays to pursue the identification of chemotactic factors of which the first to be isolated was C5a (Shin *et al.*, 1968, Snyderman *et al.*, 1970). There was not much interest in the locomotor behaviour of the responding cells. This bias still persists to some extent. Since almost everyone had abandoned visual methods, the distinction between chemotaxis, the directional reaction (*Richtungsbewegung*; Pfeffer, 1884) induced by chemical substances, and other locomotor reactions became blurred. Hansuli Keller and I attempted to revive and revise definitions that had been in use in the past, and we wrote to a number of scientists in this and adjacent fields for advice. The most useful comments came from Michael Abercrombie and Graham Dunn in Cambridge without whose advice the definitions (Keller *et al.*, 1977) would have been a good deal less tidy than they are. We distinguished forms of locomotion (random or directed) from the responses to environmental cues which led to these forms of locomotion; the cues being either chemical (chemotaxis and chemokinesis) or physical (contact guidance). These definitions have stood up reasonably well so far. The reintroduction of the term 'chemokinesis' was immediately accepted everywhere, but was unfortunately widely misunderstood since many workers then, and now, used it as a synonym for random locomotion, whereas it is defined as a reaction by which a chemical stimulus *alters* random locomotory behaviour by changing the cells' speed or their frequency of turning.

During the 1970s visual studies using time-lapse cinematography began to make a comeback. Several interesting studies using such methods deserve mention. One was the observation by Marcel Bessis (1965) that, when a red (or other) cell was destroyed by heat from a laser, nearby neutrophils migrated in by chemotaxis to phagocytose the damaged cell. This was due to a transient release of material from the damaged cell, and neutrophils failed to recognize cells that had been damaged half an hour before. This seems very likely to have been due to release of heat-denatured proteins. Laser-treated blobs of protein had an identical effect (Bessis & Boisfleury-Chevance 1984). Coincidentally, I was studying the chemotactic effect of chemically denatured haemoglobin and serum albumin at about the same time, but using filter assays (Wilkinson, 1973; Wilkinson & McKay, 1971). These observations raised interesting and unresolved questions about how phagocytic cells recognize damaged and unfolded proteins (Wilkinson, 1981) though this is not a field in which much progress has been made. Phagocytic cells are important scavengers but the study of non-specific immune recognition mechanisms is out of fashion now. Another series of studies were those of Sally Zigmond which reawakened interest in the mechanism by which leucocytes detected gradients and pointed the way to methods by which chemotactic and chemokinetic locomotion could be analysed and distinguished (Zigmond & Hirsch, 1973; Zigmond, 1974,

1977). These papers clearly demonstrated the superiority of visual methods over filter assays for such analysis. A third interesting observation was that of Keller & Bessis (1975), amplified by Malawista & Boisfleury-Chevance (1982), who heated neutrophils to 46 °C at which temperature the anterior lamellipodium broke off from the rest of the cell. These hyaline organelle- and microtubule-free, but microfilament-rich, lamellipodia continued to show locomotor and chemotactic responses, demonstrating that the nucleus and the visible organelles, left behind in the cell body, were unnecessary for these responses.

The study of leucocyte chemotaxis was opened up in the late1970s following the identification of synthetic formyl peptides (type example, formyl-Met-Leu-Phe; FMLP) as powerful, cheap, and easily synthesized chemotactic factors for neutrophils (Schiffmann et al., 1975; Showell et al., 1976) and mononuclear phagocytes. These opened the door to biochemical studies which allowed the identification of cell-surface receptors and the study of structure–function relationships using a range of related peptides (Showell et al., 1976; Toniolo et al., 1984, 1989). Later, definition of the transduction apparatus was begun and it was shown that receipt of a formyl peptide signal activated a pertussis-toxin sensitive G protein (Becker et al., 1985; Snyderman, Smith & Verghese, 1986). There has been considerable difficulty in isolating the formyl peptide receptor, or the receptor for C5a, in a suitable form to allow sequencing and genetic studies, and, at the time of writing (13 years after their description), the primary structure of these receptors is still unknown.

A difficulty that was encountered early in these biochemical studies was the unsuitability of the commonly used locomotion and chemotaxis assays as monitors for biochemical events. 'Chemotactic factors' are, in fact, general activators of neutrophil function and most investigators turned to assays of secretion or oxidative metabolism, immediate and precisely measured events, rather than to the complex and prolonged events of locomotion. The trouble with this is that the dose–response optima for secretion and metabolism are high (10^{-6} to 10^{-7} M formyl peptide) whereas those for locomotion and chemotaxis are low (10^{-8} to 10^{-9} M), that the two sets of events involve different affinity states of the receptor (Koo, Lefkowitz & Snyderman, 1982, 1983) and that therefore much of the biochemical information gleaned using secretion and metabolism assays may not apply precisely to locomotor activation (Lackie, 1988). There is an accurate assay which is suitable for this purpose, namely the polarization assay, in which the change from a spherical to a polarized shape which occurs within a few minutes after a chemoattractant is added to previously unstimulated leucocytes, is quantified. This measures the initial events of the locomotor response and is independent of secondary phenomena such as adhesion which complicate the interpretation of filming assays, filter assays, etc. The polarization assay is accurate, simple, gives good dose–response data, and has been used successfully by several groups (Smith et al., 1979; Keller, Zimmermann & Cottier, 1983; Haston & Shields, 1985).

Since the initial characterization of defined chemotaxtic factors, a substantial number of further factors have been documented in more or less detail. There is much interest at present in chemotactic cytokines released by cells which influence the locomotory behaviour of other cells, homotypic or heterotypic. These are certain to prove interesting and are dealt with elsewhere in this volume (Camp). Similarly, the biochemical pathways with specific effects on locomotor events activated by chemotactic factors are considered in another chapter (Naccache).

HOW DO LEUCOCYTES DETECT AND FOLLOW CHEMOTACTIC GRADIENTS?

For years, this has been one of the most central and demanding questions for the student of leucocyte behaviour. Leucocytes can detect gradients without translocating. A spherical immotile leucocyte forms a pseudopod accurately towards the gradient source and then moves in the direction of pseudopod formation (Zigmond, 1974; Allan & Wilkinson, 1978). The only information a leucocyte has about a gradient is about differences in receptor occupancy across its surface and Zigmond (1977) calculated that a neutrophil could detect differences as low as 1% across its length. Also, leucocytes are capable of detecting spatially fixed (e.g. substratum-bound) gradients (Wilkinson & Allan, 1978; Wilkinson & Bradley, 1981; Webster, Zanolari & Henson, 1980). These observations suggested a *spatial* detection system. However, it was not at all clear how the cell was able to make the necessary comparisons of subtle differences in receptor occupancy at different sites. Obviously, the *temporal* detection system as used by bacteria (moving from A to B and comparing concentrations at the two points), in its unmodified form, was ruled out by the fact that the stationary leucocyte detected gradients efficiently. A modified version of this was proposed (Zigmond & Sullivan, 1979; Dunn, 1981; Gerisch & Keller, 1981) in which it was not movement of the whole cell but of receptors on its surface which allowed temporal detection. Movement of receptors on an advancing pseudopod would allow comparisons of numbers of occupied receptors (n_0) at times A and B. A simple memory would allow continuation of pseudopod formation if $n_0B > n_0A$ and pseudopod retraction if $n_0B < n_0A$. This has been named '*pseudospatial*' gradient detection by Lackie (1986) and is the most plausible of the three versions (though one rarely observes formation of 'abortive' pseudopods at points other than the front of cells moving up gradients, or indeed of cells showing chemokinetic movement, and these seem to be demanded by the model) (Table 1) (Contrast the situation in *Dictyostelium* – see chapter by Segall. Eds.). A cell moving up a gradient in front of a second cell has a higher attractant concentration at its tail

Table 1. *Site of formation of pseudopods in neutrophils migrating up a formyl peptide gradient.*

	Number of pseudopods	Number of successful pseudopods
90° sector facing source	34	32
90° sector away from source	9	5

Cells were migrating in a collagen gel to a source of FMLP (10^{-8} M). Successful pseudopods were those which were not retracted and/or whose formation was followed by a contraction moving down the cell body. Pseudopods formed in other sectors were ignored. Control cells moving randomly showed no preferred direction of pseudopod formation.

than the second cell has at its head. This is explicable on the 'pseudospatial' mechanism, provided that the cell can adapt to (become unresponsive to) an ambient concentration but responds as the concentration is raised. Adaptation is a feature of many sensory systems (including chemotaxis in bacteria); and evidence that it occurs in leucocytes was presented by Zigmond & Sullivan (1979).

HOW DO LEUCOCYTES POLARIZE WHEN THERE IS NO GRADIENT?

None of the gradient detection systems proposed above explains the polarization and chemokinetic responses induced by chemotactic factors present in uniform concentration. Under such conditions, leucocytes move in a 'persistent random walk' at about the same speed as they would if moving up a gradient within the same approximate concentration range (Allan & Wilkinson 1978). Shields & Haston (1985) studied the ability of *uniform* concentrations of attractant to induce polarization of human blood neutrophils. These cells in suspension are spherical when first purified from blood, but once a chemoattractant is added, the cells rapidly (<5 min) acquire an anteroposterior polarity essentially similar to that of cells exposed to a gradient. This polarity was evident only if the attractant concentration was sufficiently low ($<10^{-8}$ M FMLP). At higher concentrations, the cells either failed to polarize or assumed multipolar shapes without a defined front and back, since none of the pseudopods achieved dominance. Likewise, if neutrophils attached to a surface were exposed to uniform concentrations of FMLP at $<10^{-8}$ M, they showed a typical persistent random walk, starting in the direction in which the anterior pseudopod was first formed, but, at higher concentrations, they showed little persistence and poor displacement because different pseudopods competed with one another.

A concomitant of morphological polarization is redistribution of membrane proteins towards the anterior pole of the cell as the cell polarizes. This occurs on addition of an attractant either in uniform concentration

Table 2. *Forward distribution of membrane proteins in polarized leucocytes.*

Protein	Cell-type	Method	Reference
Fc receptor	Neutrophil	Fc rosetting	Walter et al., 1980
			Wilkinson et al., 1980
FcRII but not FcRIII	Neutrophil	Immunofluorescence	Petty et al., 1989
Complement receptor (undefined)	Neutrophil	C3b rosetting	Shields & Haston, 1985
Not CR3	Neutrophil	Immunofluorescence	Pytowski et al., 1990
		Immunogold TEM	Haston & Maggs, 1990
Formyl peptide receptor	Neutrophil	Haemocyanin-linked peptide	Sullivan et al., 1984
Formyl peptide receptor	Neutrophil	Autoradiography	Walter & Marasco, 1987
Thy-1	Lymphocyte	Immunofluorescence	Shields & Haston, 1985
CD15	Neutrophil	Immunogold TEM	Haston & Maggs, 1990
CD45	Mononuclear	Immunogold TEM	Haston & Maggs, 1990

or as a gradient. It has been observed for several proteins including chemotactic factor receptors (Table 2) and is independent of the polarizing ligand. Proteins redistribute in the absence of their own ligand. The time course of redistribution has not been determined exactly. Redistribution is by no means absolute, polarized cells vary considerably in the degree to which proteins can be shown to have moved to the anterior pole (Sullivan, Daukas & Zigmond, 1984; Haston & Maggs, 1990) and, in some reports, no redistribution was seen (Pytowski, Maxfield & Michl, 1990). These proteins may become linked to submembranous microfilaments (Jesaitis et al., 1984) (although at least one, Thy-1, has a phosphatidyl–inositol tail and therefore the link must be indirect). Newly formed pseudopods are rich in filamentous actin (Fechheimer & Zigmond, 1983; Haston, 1987; Zimmermann, Gehr & Keller, 1988) which may move rapidly down the cell, possibly sweeping membrane proteins with it. In dictyostelium and acanthamoebae, and presumably also in leucocytes, myosin I is concentrated at the leading edge (Adams & Pollard, 1989; Fukui et al., 1989) and may link membrane proteins to the cytoskeleton. Kucik, Elson & Sheetz (1989) have made direct observations of forward movement of membrane glycoproteins. Their observations suggest that forward distribution is achieved by lateral movement in the plane of the membrane rather than by membrane recycling through the cytoplasm following backward lipid flow.

STOCHASTIC MODELS OF LEUCOCYTE MOVEMENT

Observations of leucocyte behaviour in uniform attractant concentrations, and of its similarities with behaviour in gradients, stimulated the attempt to formulate unifying hypotheses to explain how attractants regulated both chemokinetic and chemotactic locomotion. We suggested (Haston & Wilkinson, 1987; Wilkinson & Haston, 1988) that a leucocyte might respond by polarizing in the direction of the first perceived signal, under conditions where such a signal could be distinguished from subsequent signals. This would be a stochastic process. In uniform attractant concentrations, the site of activation would be random, so the cells would polarize in directions random in relation to one another. In gradients, the probability of signal reception would be greatest on the side facing the gradient source, and most cells would polarize in that direction (Fig. 1). Following receipt of a signal, receptor redistribution to the newly-forming pseudopod would reinforce the probability of subsequent signals being received at the head of the cell and therefore of persistence of locomotion in that direction. However, this model does not fit with the observed behaviour of spherical leucocytes in suspension immediately on exposure to a uniform and optimal concentration of attractant. As shown in Table 3 and Figure 2, these cells often do not start shape-change by forming a pseudopod at one pole. Rather (within a few seconds) they form a large ruffle which may occupy over 50%, or even the whole circumference, of the cell image, a behaviour similar to that of neutrophils exposed to sudden jumps in attractant concentration (Zigmond & Sullivan, 1979). At optimal concentrations, this ruffle rapidly becomes restricted to one pole which becomes the anterior end of the cell and may remain as such for as long as an hour. At supraoptimal concentrations, however, the cells remain ruffled over most of their surface indefinitely.

Another difficulty with a 'first-signal' model is that, at optimal attractant concentrations, the number of ligand molecules available for receptor binding seems much too high for such a signal to be distinguished from later signals. Much depends on what proportion of bound ligand actually activates a signal. Activation of transduction proteins following ligand binding, and subsequent activation of effector function, are stochastic events which probably require random diffusion for close contact of the relevant proteins. The receptor itself may require assembly (like the interleukin-2 receptor) or clustering (like CR3; Detmers et al., 1987). So little is known about the properties of chemotactic receptors at present that any of these processes is possible. However, integration of a response resulting from activation at multiple receptor sites may fit better with the observed initial behaviour of stimulated cells than a 'first signal' model. Such integration would only be possible in the presence of ligand at optimal or suboptimal, not at supraoptimal concentrations.

A stochastic model to explain both chemotactic and chemokinetic responses has been proposed by Tranquillo and colleagues (Tranquillo, Lauffenburger & Zigmond, 1988), based on leucocyte responses to perceived fluctuations in attractant concentrations due to real fluctuations in receptor binding ('noise'). Essentially this model (which is discussed else-

(a) Chemokinesis

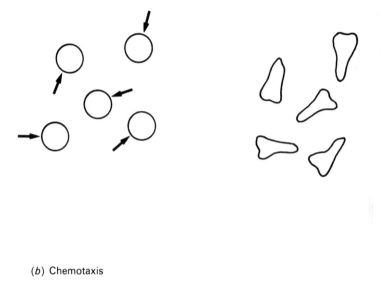

(b) Chemotaxis

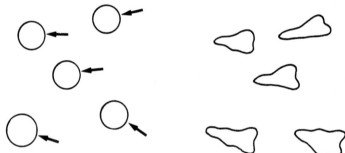

Fig. 1. Schematic sketch of how stochastic events might account for polarization and locomotion in both uniform concentrations and gradients of attractant. (a) Chemokinesis is initiated and maintained by anisotropies or 'fluctuations' in an isotropic environment. The arrows on the left-hand side might represent either local, transient rises in concentration above a mean, or sites where a signal is first received. In either case, the cell polarizes and moves in the direction of the arrow. (b) In gradients, the concentration of attractant is highest on the side of the gradient source so that, by essentially the same mechanisms as for chemokinesis, polarization and locomotion to that side is favoured statistically.

where in this volume: Tranquillo) proposed that a polarized cell moving in a uniform attractant concentration would perceive fluctuations in concentration. These would be interpreted by the cell as fluctuating gradients in the absence of a mean gradient and the cell would respond by turning, leading to a persistent random walk. Likewise, cells moving chemotactically in a mean gradient would perceive fluctuations in the gradient which would

Table 3. *Shape-change in human blood neutrophils at different times after adding formyl-Met-Leu-Phe to the cells.*

Time after adding FMLP (10^{-8} M)	Shape category: % of population			
	Spherical	Front–tail polarity	Incipient polarity	Irregular ruffling
Zero time	96.5	3	0.5	0
30 secs	9	16	40	35
90 secs	11.5	42.5	35	11
10 min	4	77	12	7

Front–tail polarity: Elongated cells with a definable head and tail.
Incipient polarity: Cells with a large ruffle occupying > 50% of the cell circumference, but with a sharp margin at the opposite pole.
Irregular ruffling: Overall ruffling with no discernible polarity.

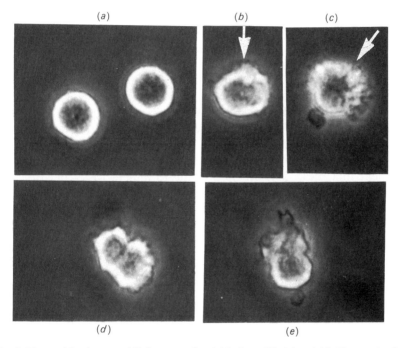

Fig. 2. Human blood neutrophils in suspension (a) before, (b), (c) and (d) 30 seconds after, and (e) 5 minutes after addition of FMLP at 10^{-8} M. The cells in (b) and (c) show 'incipient' polarity (Table 3) with a very broad head ruffle (arrowed), and a sharp margin at the opposite pole of the cell. The cell in (d) is irregular in shape without definable polarity. Within a few minutes most of these cells develop clear polarity as in (e). *Note* that in supraoptimal peptide concentrations (10^{-7} M and 10^{-6} M) many cells never develop clear polarity as in (e). Also, that at suboptimal concentrations (10^{-9} and 10^{-10} M), most cells respond by an immediate change from a spherical to a well-polarized shape by protrusion of a localized pseudopod, without an intermediate stage of transient, widespread, ruffling.

account for the turning behaviour seen in cells moving directionally up gradients. The model considered fluctuations in receptor binding, but could also apply to fluctuations in signal transduction, etc. As originally proposed it was restricted to locomotor behaviour of already-polarized cells and provided an account of how stochastic events might control this behaviour. It did not address the question of how a pseudopod is formed at one side of the spherical cell (but see Tranquillo, this volume). Possibly this also is a response to a fluctuation in attractant concentration or in receptor binding. It is not clear how either this or a first signal model can explain how extensive initial ruffles become localized to a dominant pole.

An attractive feature of stochastic models is that they provide an explanation for leucocyte movement in both uniform attractant concentrations and gradients, whereas the earlier models were concerned solely with chemotaxis. All chemotactic factors so far studied cause both an increase in cell speed in uniform concentrations and a directional response in gradients, and it seems probable that the biochemical events are identical in both situations.

One unexplained phenomenon that any theory of attractant-induced polarization and locomotion will have to accommodate is the effect of microtubule-disaggregating drugs such as colchicine and nocodazole. These drugs, added to spherical leucocytes, induce polarization and locomotion in the absence of any external signal. The locomotor response is slow to develop (15–30 min, Keller *et al.*, 1984) presumably because the drugs enter the cytoplasm slowly. The effect appears to by-pass cell surface receptors altogether yet colchicine-treated neutrophils and lymphocytes show a random walk similar to that induced by conventional attractants. The argument that these cells are responding to minute traces of attractant seems implausible, since *untreated* leucocytes can probably respond in the presence of less than ten molecules of chemotactic peptide per cell anyway. We do not know enough about interactions between microtubules and microfilaments to know how these drugs are acting yet.

MECHANISM FOR LEUCOCYTE ACCUMULATION

How important is chemotaxis as a mechanism for mobilizing leucocytes at sites of inflammation or at other sites where leucocytes accumulate? Leucocytes move at speeds of 5–20 μm per minute *in vitro* and, obviously, if this movement is in near-straight lines towards a gradient source, this will allow rapid accumulation of a large number of cells at a site, e.g. where a microorganism has gained entry. However, the tissues that cells must traverse *in vivo* are a great deal more complex than their *in vitro* models, and there are a number of other determinants of accumulation to be taken into consideration. These include locomotor responses to environmental cues other than chemotaxis; adhesive behaviour; and the locomotor capacity of the cells.

LOCOMOTOR RESPONSES OTHER THAN CHEMOTAXIS

I considered these in some detail in an earlier review (Wilkinson, 1987) and there is also a good general discussion of their characteristics and implications (Dunn, 1981), so they will only be considered briefly here. As stated above, factors which are chemotactic when presented in a concentration gradient also, in all cases so far studied, stimulate a concentration-dependent increase in the speed of leucocyte locomotion (chemokinesis), when presented in uniform concentration. Thus the presence of chemotactic factors in a tissue, even when there is no gradient, will cause leucocytes to accelerate. Leucocytes entering a lesion will migrate at random, and some of them will reach the centre of the lesion. This is a very inefficient way to accumulate cells compared to chemotaxis (Dunn, 1981). However, accumulation by chemokinesis alone may be made much more efficient if some mechanism is available for trapping cells which enter the lesion. Neutrophils migrating randomly at a boundary between a protein-coated surface (BSA) and an immune complex-coated surface (BSA-anti BSA) show increased adhesion on the latter surface by binding to the substratum-linked antibody by their Fc receptors (Wilkinson *et al.*, 1984). They thus become tethered and gradually accumulate on the immune complexes by slowing of locomotion (negative chemokinesis). This is seen in the absence of serum. In the presence of complement, the immune complexes generate gradients of C5a, and the combination of the chemotactic response to C5a with the adhesive trap causes more efficient cell accumulation than either alone.

Another form of locomotor behaviour which may modify chemotactic responses is *contact guidance*. The direction of cell locomotion may be determined by the curvature, shape or patterning of tissues. This is not a response to a chemical agent and may not depend on ligand-receptor interactions. It has been demonstrated in many cell types (reviewed by Dunn, 1982) and its mechanisms, which may be multifold, are not well understood. If gels of collagen or fibrin are aligned so that the fibres of the gel run parallel, then leucocytes (of all types) placed in the gel will follow the axis of alignment of the gel rather than migrating in other directions (Wilkinson, Shields & Haston, 1982). Since leucocyte locomotion through such gels does not depend on adhesion to the fibres of collagen or fibrin (Haston, Shields & Wilkinson, 1982; Schmalstieg *et al.*, 1986), this is not due to preferential adhesion. If cells in such a gel are presented with a chemotactic gradient, the cells for which the axis of the gradient is parallel to the axis of gel alignment show much better chemotactic responses than those for which the two axes are perpendicular to one another and in which the tactic and guidance cues conflict (Wilkinson &

Lackie, 1983). The complex patterning of tissues is therefore likely to be an important determinant of the success or failure of chemotactic responses of leucocytes migrating into those tissues.

CELL ADHESION TO ENDOTHELIUM

The first event in an inflammatory response, as in lymphocyte recirculation, is adherence to vascular endothelium. For many years this event was not understood, but recently its molecular basis has begun to be unravelled. Vascular endothelial cells respond to inflammatory events (endotoxin, cytokines) by new expression of adhesion molecules (ICAM 1 and 2, ELAM-1 and certainly others) (Dustin *et al.*, 1986; Bevilacqua *et al.*, 1989; Springer *et al.*, 1987), which are recognized by leucocyte surface ligand molecules (of the integrin family and certainly others) (Springer, *et al.*, 1987; Hynes, 1987) so that, at the inflammation site, there is increased adhesion of leucocytes to endothelium. The leucocytes then migrate out of the vessel and detach from endothelium, the latter by mechanisms still be to determined. This adhesion event is essential for an effective defence against infection as has been strikingly demonstrated in patients with leucocyte adhesion deficiency (LAD). The leucocytes of these patients lack surface molecules of the CD11/CD18 family and are defective in many functions which require adhesion, including adhesion to endothelium (Pohlman *et al.*, 1986). There is no defect of locomotion *per se* in these patients or in normal neutrophils treated with anti CD11/CD18 antibodies. Their cells show normal polarization responses to chemotactic factors (Anderson *et al.*, 1984) and invade normally in assays in which locomotion is not dependent on adhesion. Thus they invade collagen gels normally (Schmalstieg *et al.*, 1986) and enter micropore filters though they do not migrate as far as normal cells into filters, suggesting that adhesion contributes to locomotion in this assay. Likewise, they respond badly to chemotactic factors when moving over planar surfaces because they do not adhere to these surfaces (Anderson *et al.*, 1984, 1986). The clinical problem with recurrent and severe infections in LAD is much more serious than in the ill-assorted and poorly characterized groups of patients with 'chemotaxis' defects, which pinpoints the importance of seeing chemotaxis in the context of the whole cellular response in inflammation.

The migratory pathways followed by lymphocytes are more complex than those of neutrophils or monocytes. Lymphocytes recirculate from blood to lymphoid tissue by crossing the high endothelial venules (HEV), which are composed of tall, columnar endothelial cells which line the postcapillary venules of lymphoid tissue. Again, progress has been considerable in defining the basis of the adhesion step by which lymphocytes bind to HEV, and lymphocytes have been shown to carry 'homing receptors' which bind to 'vascular addressins' present on HEV, and differing in different lymphoid tissues. These were well reviewed by Duijvestijn & Hamann (1989). This allows specific attachment of different lymphocytes to HEV in their appropriate lymphoid tissues, but absolutely nothing is known about what happens once the lymphocytes have crossed the HEV and entered the lymphoid tissue. These cells distribute themselves within the lymphoid tissue in complex patterns and eventually leave the tissue by the efferent lymphatic from which they reach the bloodstream again. This must involve complex locomotory behaviour, and possibly chemotactic events, about which nothing is known.

Lymphocytes also migrate into inflammatory sites. Early in inflammation, the endothelium is non-specialized and is, of course, the same endothelium that is traversed by neutrophils and monocytes. Later, in chronic inflammation, the vascular endothelium may acquire the characteristic columnar morphology of HEV in lymphoid tissues. The phenotype of T-lymphocytes in chronic inflammatory lesions has been studied and shown to differ from that of lymphocytes found in the bloodstream. Of particular interest are two variants of the CD45 surface molecule, the extracellular domains of which contain different sequences due to alternative gene splicing. CD45RA is a marker found on 'naïve' T-lymphocytes which have not encountered antigen. After priming with antigen, 'memory' T-lymphocytes show a different version, CD45RO. Both versions are found in blood, but CD45RO+ cells have been shown to adhere better to vascular endothelium than CD45RA+ cells (Pitzalis et al., 1988). Also CD45RO+ cells are found selectively in a variety of inflammatory sites. In unpublished studies using collagen gel assays, Ian Newman in the author's laboratory examined whether there were differences in locomotor behaviour between these maturation subsets under circumstances where adhesion was not involved. He found that the proportion of locomotor (polarized) cells in both populations was not greatly different, and that both migrated well into gels. Thus the accumulation of these cells in inflammatory sites may be explained, not by differences in their locomotory or chemotactic properties, but by differences in their adhesion to vascular endothelium. However, this is not the end of the story as far as lymphocytic locomotion and accumulation is concerned, and, in the next section, I shall discuss evidence that the locomotor properties of lymphocytes are markedly altered by growth activators.

LYMPHOCYTE ACTIVATION AND LOCOMOTOR CAPACITY

Among the definitions listed by Keller et al. (1977) was included 'intrinsic locomotor capacity: An intrinsic capacity of the cell to perform active locomotion'. This definition seemed self-evident and, at the time, not particularly important as far as leucocytes were concerned. There was no question but that the vast majority of blood neutrophils and monocytes had locomotor capacity. However, locomotor capacity is a property that varies with the state of the cell: witness the cessation of movement and the rounding-up of cells which are about to divide: and, even though blood leucocytes of the myeloid series are motile, their bone-marrow precursors are not.

Lymphocytes and their precursors are cells whose locomotor capacity varies at different stages of maturation or activation. This has been documented early in embryonic development in the chick in which haemopoietic stem cells which are to populate the thymus can be seen, during a brief period of development, to show chemotactic responses to factors derived from thymic epithelium. Shortly afterwards this capacity is lost (Slimane et al., 1983; Champion et al., 1986; Savagner et al., 1986).

We have investigated the locomotor properties of the lymphocytes of human blood for several years. Locomotor capacity of these cells is determined by whether they are in cell cycle or not. The majority of lymphocytes from normal blood are in a resting stage (G_0) and it is usually not possible to induce locomotor responses in more than about 20–30% of lymphocytes direct from blood. Culture with appropriate mitogens causes the cells to enter cell cycle. We have used anti-CD3 antibody as a T cell-specific mitogen (Wilkinson & Higgins, 1987a) and interleukin 4 or the Cowan strain of *Staphylococcus aureus* as B cell-specific mitogens for studying locomotion (Wilkinson & Islam, 1989). Within the first 24 h of culture with these mitogens, a high proportion of the stimulated population acquires locomotor capacity, shows head–tail polarity and invades collagen gels (Wilkinson, 1986). These cells are larger than the non-motile cells and show increased RNA and protein synthesis, but not DNA synthesis, so they are in the G_1 stage of cell cycle. It should be noted that this activation, which eventually leads to cell division, is not the same thing as the maturation of naïve cells, following contact with antigen, to become memory cells, mentioned above, since the latter represents a permanent alteration of the lymphocyte's maturation status, whether or not it is resting or cycling at any particular time. These *in vitro* observations, that activated blood lymphocytes acquire locomotor capacity, are in accord with many observations of the migration of lymphocytes into chronic inflammatory lesions *in vivo*, reviewed by Parrott & Wilkinson (1981). Large, activated, lymphocytes migrate from the blood into such lesions in greater numbers than small, resting, lymphocytes. In contrast, the latter cells cross specialized HEV, but it is not obvious how they traverse lymphoid tissues, since they appear to be non-motile when taken from blood. Possibly they respond to contact signals for locomotion – from HEV or from the lymphoid stroma – rather than to classical chemotactic factors. Activated lymphocytes no longer recirculate but, instead, behave more like neutrophils and monocytes in crossing non-specialized endothelia in inflammatory sites. However, they do not recognise the same range of chemotactic factors (formyl peptides, C5a, leukotriene B_4) as those cells. These findings support the idea that the selection of cells to accumulate in a particular site can be achieved not

Table 4. *Lymphocytes activated with anti-CD3 polarize in the presence of a material in their own supernatant which is not itself anti-CD3.*
Cells cultured in anti CD3 (10 ng/ml), washed, and retested in a 30 min polarization assay

	% polarized cells ± SEM (3 experiments)
Hanks-HSA (negative control)	20.9 ± 2.1
Anti-CD3 (25 ng/ml) in Hanks-HSA	20.2 ± 2.0
Cell supernatant (contains anti-CD3 at 25 ng/ml)	48.6 ± 3.0
Foetal calf serum 25% (positive control)	53.4 ± 2.3

Lymphocytes were cultured for 24 h in aCD3 then washed and the supernatants retained. Lymphocytes were then tested in a polarization assay against their own supernatants and appropriate controls.

only by chemical or physical cues to locomotion, but also by the fact that a population may contain locomotor and non-locomotor cells, only the former of which can respond to inflammatory signals.

The development of polarized morphologies in lymphocytes after culture overnight with an activator of growth suggests that these cells have not only acquired locomotor capacity, but have also responded to an attractant by changing from a spherical to a polarized shape. Lymphocytes washed and retested with the growth activator in pure form in a short-term polarization assay do not show a locomotor response to it; however, when the original supernatant is added back to the cells, they show rapid polarization (Table 4). This suggests the presence in the supernatant of an attractant, other than the growth activator, which is probably a product released by the cells themselves during culture. Studies of lymphocyte locomotion have been hampered by the lack of purified attractants. They respond to mixtures such as cell supernatants or serum: – autologous or foetal calf serum: – though not plasma. They also respond to inflammatory exudates, e.g. glycogen-induced peritoneal exudate fluids. Various cytokines have been suggested as lymphocyte attractants. The most recently described of these, interleukin-8 (Larsen et al., 1989), is released by mononuclear phagocytes, and is the most active of the cytokines we have tested, though none of them, in purified form, is as active as 20% serum. Still, the availability of at least one defined lymphocyte attractant in a recombinant form will be a help. Lymphocytes themselves release lymphocyte attractants, (LCF, Potter & Van Epps, 1987) though the relation of these to the interleukins remains to be investigated.

The two stages of lymphocyte locomotor activation can thus be seen as follows. (a) Acquisition of locomotor capacity which is growth-determined, occupies a period of hours and may need expression of new genes; and (b) response by polarization and locomotion to an attractant, similar

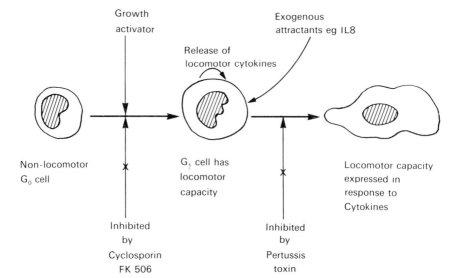

Fig. 3. A two-stage model for activation of lymphocyte locomotion in which the first stage is growth-driven activation of locomotor capacity and the second stage is a locomotor response to environmental chemicals. Each stage can be blocked by specific inhibitors. From Wilkinson & Watson (1990).

to the response of neutrophils and taking only minutes. These two stages can be distinguished pharmacologically (Fig. 3). Two *immunosuppressant drugs*, cyclosporin and FK506, specifically inhibit mitogen-activated lymphocyte growth, acting early in G_1. Each has a separate cellular binding protein, both of which are peptidyl-prolyl isomerases believed to play a role in Ca^{2+}-mediated signal transduction (Takahashi, Hayano & Suzuki, 1989; Fischer et al., 1989; Siekierka et al., 1989; Harding et al., 1989). These drugs inhibit the cell cycle-related acquisition of locomotor capacity by lymphocytes (Wilkinson & Higgins, 1987b; Wilkinson & Watson, 1990), but have no effect on the locomotor responses of already motile lymphocytes. Conversely, *pertussis toxin* has no effect on acquisition of locomotor capacity but does inhibit the immediate response of lymphocytes to IL-8 and FCS (Wilkinson & Watson, 1990), their locomotion in filter assays (Spangrude et al., 1985) and their entry into lymphoid tissues (but not by blocking adhesion to endothelium; Spangrude, Braaten & Davies, 1984). These observations suggest separate transduction pathways, one mediated by a pertussis toxin-sensitive G protein for attractant-induced lymphocyte locomotion; the other for growth activation and locomotor activation, the pathway for which is under intensive study and is probably not directly mediated by a pertussis toxin-sensitive G protein.

Table 5. *Locomotor and adhesive behaviour of T cell maturation subsets.*

	Naïve T cells CD45RA+	Memory T cells CD45RO+
Resting	Non-adhesive Non-motile	Adhesive Non-motile
Mitogen- or antigen-activated	Non-adhesive Motile	Adhesive Motile

Since both adhesion to endothelium and locomotion are necessary for entry into an inflammatory site, only the cells in the bottom right-hand category will enter such sites.

Coming back to the 'naïve' and 'memory' T-lymphocyte populations, it is possible using mitogens to separate each of these into a resting and an activated population only the latter of which is motile. Thus, based on the data of Pitzalis *et al.* (1988) and our own data, there may be four potential T-lymphocyte subpopulations, only one of which is recruited into inflammatory sites (Table 5).

These studies show the diversity of lymphocyte patterning that may be seen before one even begins to consider the chemotactic responses of these cells. Another factor to consider is that in immune activation, lymphocytes must cluster with MHC Class II+ accessory cells to become activated. This clustering is necessary *in vitro* for lymphocytes cultured with anti CD3 antibody and other mitogens to begin growth and to acquire locomotor capacity (McFarland, Heilman & Moorhead, 1966; Wilkinson & Higgins, 1987*a*). Time-lapse filming shows no evidence that this clustering involves chemotactic attraction; rather heterotypic contacts occur randomly and slowly and are followed by preferential adhesion (Wilkinson, 1990). Later, the activated and newly motile lymphocytes break away from such clusters (McFarland *et al.*, 1966). Lymphocyte chemotaxis towards macrophages can, however, be seen in a different situation (Wilkinson, 1990), using cells from the peritoneal cavities of mice after immunological challenge with *Corynebacterium parvum*. Lymphocytes from these exudates are sufficiently adherent for their locomotion to be studied on planar surfaces. These lymphocytes show chemotactic responses to macrophages (derived from the same exudates) for a brief period after the macrophages have phagocytosed a spore of *Candida albicans* (Fig. 4). Following phagocytosis, a brief pulse of attractant appears to be released within about 15 min which attracts lymphocytes to cluster round the macrophage. Thereafter, no further attraction is seen. In immunological terms, this release may be a useful device by which a macrophage which is presenting newly ingested antigen can focus lymphocytes around itself and maximize the possibility of antigen recognition.

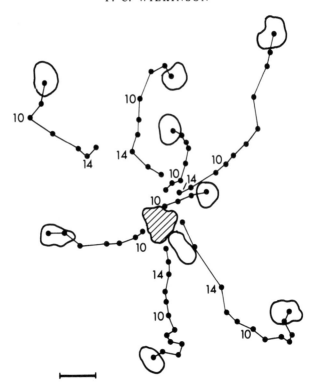

Fig. 4. Chemotaxis of mouse lymphocytes towards a macrophage (hatched) following phagocytosis of *Candida albicans* by the latter. The paths of nearby lymphocytes were tracked by dotting in the cell centres at 80-second intervals and joining the dots. The numerals are times in minutes after phagocytosis. From Wilkinson (1990) by courtesy of the editor of *Immunology*.

CONCLUSION

Chemotaxis is a locomotor response of major importance in recruitment of leucocytes into inflammatory lesions. Cells may move through three-dimensional tissues by non-adhesive mechanisms, but adhesion is absolutely necessary for the first event, binding to endothelium. Adhesion within tissues and the patterning of tissues are factors which also contribute to the accumulation of inflammatory cells. The diversity of processes necessary for successful invasion of tissues is exemplified by the diversity of behaviour of lymphocytes which require to be activated to acquire full locomotor capacity. Understanding of the molecular biology of the chemotactic response is still very incomplete and it is still difficult to draw firm conclusions about how cells detect chemotactic gradients. Hypotheses which unify chemotactic and chemokinetic locomotion are now being considered. It seems likely that these two responses differ at the behavioural, but not at the molecular, level.

ACKNOWLEDGEMENTS

I wish to thank the Medical Research Council, the Leukaemia Research Fund and the Arthritis and Rheumatism Council who have at various times supported my work quoted here. At least as much, and probably more, I should thank my present and former colleagues, especially Wendy Haston, John Lackie and Jim Shields, discussions and arguments with whom set me thinking and sometimes showed me the error of my ways. FK506 was a gift of the Fujisawa Pharmaceutical Co. Ltd and of Dr Angus Thomson (University of Aberdeen). Dr Roger Parton (Microbiology Dept., University of Glasgow) kindly donated the pertussis toxin.

REFERENCES

Adams, R. J. & Pollard, T. D. (1989). Binding of myosin I to membrane lipids. *Nature, London*, **340**, 565–8.

Allan, R. B. & Wilkinson, P.C. (1978). A visual analysis of chemotactic and chemokinetic locomotion of human neutrophil leucocytes. *Experimental Cell Research*, **111**, 191–203.

Anderson, D. C., Miller, L. J., Schmalstieg, F. C., Rothlein, R. & Springer, T.A. (1986). Contributions of the Mac-1 glycoproteins to adherence-dependent granulocyte functions: structure–function assessments employing subunit-specific monoclonal antibodies. *Journal of Immunology*, **137**, 15–27.

Anderson, D. C., Schmalstieg, F. C., Kohl, S., Arnaout, M. A., Hughes, B. J., Tosi, M.F., Buffone, G. J., Brinkley, B. R., Dickey, W. D., Abramson, J.S., Springer, T. A., Boxer, L. A., Hollers, J. M. & Smith, C. W. (1984). Abnormalities of polymorphonuclear leukocyte function associated with a heritable deficiency of a high molecular weight surface glycoprotein (GP 138): common relationship to diminished cell adherence. *Journal of Clinical Investigation*, **74**, 536–51.

Becker, E. L., Kermode, J. C., Naccache, P. H., Yassin, R., Marsh, M. L., Munoz, J. J. & Sha'afi, R. I. (1985). The inhibition of neutrophil granule enzyme secretion and chemotaxis by pertussis toxin. *Journal of Cell Biology*, **100**, 1641–6.

Bessis, M. & Burte, B. (1965). Positive and negative chemotaxis as observed after the destruction of a cell by U. V. or laser microbeams. *Texas Reports in Biology and Medicine*, **23**, 204–12.

Bessis, M. & de Boisfleury-Chevance, A. (1984). Facts and speculation about necrotaxis. *Blood Cells*, **10**, 5–22.

Bevilacqua, M. P., Stengelin, S., Gimbrone, M. A. & Seed, B. (1989). Endothelial leukocyte adhesion molecule I: an inducible receptor for neutrophils related to complement regulatory proteins and lectins. *Science*, **243**, 1160–5.

Boyden, S. V. (1962). The chemotactic effect of mixtures of antibody and antigen on polymorphonuclear leucocytes. *Journal of Experimental Medicine*, **115**, 453–66.

Champion, S., Imhof, B. A., Savagner, P. & Thiery, J-P (1986). The embryonic thymus produces chemotactic peptides involved in the homing of hemopoietic precursors. *Cell*, **44**, 781–90.

Clark, E. R. & Clark, E. L. (1920). Reactions of cells in the tail of amphibian larvae to injected croton oil (aseptic inflammation). *American Journal of Anatomy*, **27**, 221–54.

Clark, E. R., Clark, E. L. & Rex, R. O. (1936). Observations on polymorphonuclear leukocytes in the living animal. *American Journal of Anatomy*, **59**, 123–73.

Detmers, P.A., Wright, S. D., Olsen, E., Kimball, B. & Cohn, Z. A. (1987).

Aggregation of complement receptors on human neutrophils in the absence of ligand. *Journal of Cell Biology*, **105**, 1137–45.
Duijvestijn, A. & Hamann, A. (1989). Mechanisms and regulation of lymphocyte migration. *Immunology Today*, **10**, 23–8.
Dunn, G. A. (1981). Chemotaxis as a form of directed cell behaviour: some theoretical considerations. In *Biology of the Chemotactic Response*, ed. J. M. Lackie and P. C. Wilkinson, pp. 1–26, Cambridge, Cambridge University Press.
Dunn, G. A. (1982). Contact guidance of cultured tissue cells: a survey of potentially relevant properties of the substratum. In *Cell Behaviour*, ed. R. Bellairs, A. S. G. Curtis & G. A. Dunn, pp. 247–280, Cambridge, Cambridge University Press.
Dustin, M. L., Rothlein, R., Bhan, A. K., Dinarello, C. A. & Springer, T. A. (1986). Induction by IL-1 and interferon-γ: tissue distribution, biochemistry, and function of a natural adherence molecule (ICAM-1). *Journal of Immunology*, **137**, 245–54.
Fechheimer, M. & Zigmond, S. H. (1983). Changes in cytoskeletal proteins of polymorphonuclear leukocytes induced by chemotactic peptides. *Cell Motility and Cytoskeleton*, **3**, 349–61.
Fischer, G., Wittmann-Liebold, B., Lang, K., Kiefhaber, T. & Schmid, F. X. (1989). Cyclophilin and peptidyl–prolyl *cis–trans* isomerase are probably identical proteins. *Nature, London*, **337**, 476–8.
Fukui, Y., Lynch, T. J., Brzeska, H. & Korn, E. D. (1989). Myosin I is located at the leading edge of locomoting Dictyostelium amoebae. *Nature, London*, **341**, 328–31.
Gerisch, G. & Keller, H. U. (1981). Chemotactic reorientation of granulocytes stimulated with fMet-Leu-Phe. *Journal of Cell Science*, **51**, 1–10.
Harding, M. W., Galat, A., Uehling, D. E. & Schreiber, S. L. (1989). A receptor for the immunosuppressant FK506 is a *cis–trans* peptidyl–prolyl isomerase. *Nature, London*, **341**, 758–60.
Haston, W. S. (1987). F-actin distribution in polymorphonuclear leucocytes. *Journal of Cell Science*, **88**, 495–51.
Haston, W. S. & Maggs, A. F. (1990). Evidence for membrane differentiation in polarised leucocytes: the distribution of surface antigens analysed with Ig–gold labelling. *Journal of Cell Science*, **95**, 471–9.
Haston, W. S. & Shields, J. M. (1984). Contraction waves in lymphocyte locomotion. *Journal of Cell Science*, **68**, 227–41.
Haston, W. S. & Shields, J. M. (1985). Neutrophil leucocyte chemotaxis: a simplified assay for measuring polarising responses to chemotactic factors. *Journal of Immunological Methods*, **81**, 229–37.
Haston, W. S., Shields, J. M. & Wilkinson, P. C. (1982). Lymphocyte locomotion and attachment on 2-dimensional surfaces and in 3-dimensional matrices. *Journal of Cell Biology*, **92**, 747–52.
Haston, W. S. & Wilkinson, P. C. (1987). Gradient perception by neutrophil leucocytes. *Journal of Cell Science*, **87**, 373–4.
Hynes, R. O. (1987) Integrins: a family of cell surface receptors. *Cell*, **48**, 549–54.
Jesaitis, A. J., Naemura, J. R., Sklar, L. A., Cochrane, C. G. & Painter, R. G. (1984). Rapid modulation of *N*-formyl chemotactic peptide receptors on the surface of human granulocytes: formation of high-affinity ligand–receptor complexes in transient association with cytoskeleton. *Journal of Cell Biology*, **98**, 1378–87.
Keller, H. U. & Bessis, M. (1975). Migration and chemotaxis of anucleate cytoplasmic leucocyte fragments. *Nature, London*, **258**, 723–4.
Keller, H. U., Naef, A. & Zimmermann, A. (1984). Effects of colchicine, vinblas-

tine and nocodazole on polarity, motility, chemotaxis and cAMP levels of human polymorphonuclear leukocytes. *Experimental Cell Research*, **153**, 173–85.

Keller, H. U., Wilkinson, P. C., Abercrombie, M., Becker, E. L., Hirsch, J. G., Miller, M. E., Ramsay, W. S. & Zigmond, S. H. (1977). A proposal for the definition of terms related to locomotion of leucocytes and other cells. *Clinical and Experimental Immunology*, **27**, 377–80.

Keller, H. U., Zimmermann, A. & Cottier, H. (1983). Crawling-like movements, adhesion to solid substrata and chemokinesis of neutrophil granulocytes. *Journal of Cell Science*, **64**, 89–106.

Koo, C., Lefkowitz, R. J. & Snyderman, R. (1982). The oligopeptide chemotactic factor receptor on human polymorphonuclear leukocyte membranes exists in two affinity states. *Biochemical and Biophysical Research Communications*, **106**, 442–9.

Koo, C., Lefkowitz, R. J. & Snyderman, R. (1983). Guanine nucleotides modulate the binding affinity of the oligopeptide chemoattractant receptor on human polymorphonuclear leukocytes. *Journal of Clinical Investigation*, **72**, 748–53.

Kucik, D. F., Elson, E. L. & Sheetz, M. P. (1989). Forward transport of glycoproteins on leading lamellipodia in locomoting cells. *Nature, London*, **340**, 315–17.

Lackie, J. M. (1986). *Cell Movement and Cell Behaviour*, p. 224, London, Allen & Unwin.

Lackie, J. M. (1988). The behavioural repertoire of neutrophils requires multiple signal transduction pathways. *Journal of Cell Science*, **89**, 449–52.

Larsen, C. G., Anderson, A. O., Appella, E., Oppenheim, J. J. & Matsushima, K. (1989). The neutrophil-activating protein (NAP-1) is also chemotactic for T lymphocytes. *Science*, **243**, 1464–6.

Lewis, W. H. (1931). Locomotion of lymphocytes. *Bulletin of the Johns Hopkins Hospital*, **49**, 29–36.

Lewis, W. H. (1934). On the locomotion of the polymorphonuclear neutrophils of the rat in autoplasma cultures. *Bulletin of the Johns Hopkins Hospital*, **55**, 273–9.

Lewis, W. H. (1939). The role of a superficial plasmagel layer in changes in form, locomotion and division of cells in tissue cultures. *Archiv für experimentelle Zellforschung*, **23**, 1–7.

McCutcheon, M. (1946). Chemotaxis in leukocytes. *Physiological Reviews*, **29**, 319–36.

McFarland, W., Heilman, D. H. & Moorhead, J. F. (1966). Functional anatomy of the lymphocyte in immunological reactions *in vitro*. *Journal of Experimental Medicine*, **124**, 851–8.

Malawista, S. E. & de Boisfleury-Chevance, A. (1982). The cytokineplast: purified, stable and functional motile machinery from human blood polymorphonuclear leukocytes. *Journal of Cell Biology*, **95**, 960–73.

Parrott, D. M. V. & Wilkinson, P. C. (1981). Lymphocyte locomotion and migration. *Progress in Allergy*, **28**, 193–284.

Petty, H. R., Francis, J. W. & Anderson, C. L. (1989). Cell surface distribution of Fc receptors II and III on living neutrophils before and during antibody dependent cellular cytotoxicity. *Journal of Cellular Physiology*, **141**, 598–605.

Pfeffer, W. (1884). Locomotorische Richtungsbewegung durch chemische Reize. *Untersuchungen aus der Botanischen Institut zu Tübingen*, **1**, 363.

Pitzalis, C., Kingsley, G., Haskard, D. & Panayi, G. (1988). The preferential accumulation of helper-inducer T lymphocytes in inflammatory lesions: evidence for regulation by selective endothelial and homotypic adhesion. *European Journal of Immunology*, **18**, 1397–404.

Pohlman, T. H., Stanness, K. A., Beatty, P. G., Ochs, H. D. & Harlan, J. M. (1986). An endothelial surface factor(s) induced in vitro by lipopolysaccharide, interleukin 1 and tumor necrosis factor-α increases neutrophil adherence by a CDw18-dependent mechanism. *Journal of Immunology*, **136**, 4548–53.

Potter, J. W. & Van Epps, D. E. (1987). Separation and purification of lymphocyte chemotactic factor (LCF) and interleukin 2 produced by human peripheral blood mononuclear cells. *Cellular Immunology*, **105**, 9–22.

Pytowski, B., Maxfield, F. R. & Michl, J. (1990). Fc and C3bi receptors and the differentiation antigen BH2-Ag are randomly distributed in the plasma membrane of locomoting neutrophils. *Journal of Cell Biology*, **110**, 661–8.

Savagner, P., Imhof, B. A., Yamada, K. M. & Thiery, J-P (1986). Homing of hemopoietic precursor cells to the embryonic thymus: characterization of an invasive mechanism induced by chemotactic peptides. *Journal of Cell Biology*, **103**, 2715–27.

Schiffmann, E., Corcoran, B. A., & Wahl, S. A. (1975). *N*-formylmethionyl peptides as chemoattractants for leukocytes. *Proceedings of the National Academy of Sciences of the USA*, **72**, 1059–62.

Schmalstieg, F. C., Rudloff, H. E., Hillman, G. R. & Anderson, D. C. (1986). Two-dimensional and three-dimensional movement of human polymorphonuclear leukocytes: two fundamentally different mechanisms of locomotion. *Journal of Leukocyte Biology*, **40**, 677–91.

Shields, J. M. & Haston, W. S. (1985). Behaviour of neutrophil leucocytes in uniform concentrations of chemotactic factors: contraction waves, cell polarity and persistence. *Journal of Cell Science*, **74**, 75–93.

Shin, H. S., Snyderman, R., Friedman, E., Mellors, A. & Mayer, M. M. (1968). Chemotactic and anaphylatoxic fragment cleaved from the fifth component of guinea pig complement. *Science*, **162**, 361–3.

Showell, H. J., Freer, R. J., Zigmond, S. H., Schiffmann, E., Aswanikumar, S., Corcoran, B. & Becker, E. L. (1976). The structure-activity relations of synthetic peptides as chemotactic factors and inducers of lysosomal enzyme secretion for neutrophils. *Journal of Experimental Medicine*, **143**, 1154–69.

Siekierka, J. J., Hung, S. H. Y., Poe, M., Lin, C. S. & Sigal, N. H. (1989). A cytosolic binding protein for the immunosuppressant FK506 has peptidyl–prolyl isomerase activity but is distinct from cyclophilin. *Nature, London*, **341**, 755–7.

Slimane, S. B., Houillier, F., Tucker, G. & Thiery, J-P (1983). In vitro migration of avian hemopoietic cells to the thymus: preliminary characterization of a chemotactic mechanism. *Cell Differentiation*, **13**, 1–24.

Smith, C. W., Hollers, J. C., Patrick, R. A. & Hassett, C. (1979). Motility and adhesiveness in human neutrophils: Effects of chemotactic factors. *Journal of Clinical Investigation*, **63**, 221–9.

Snyderman, R., Phillips, J. & Mergenhagen, S. W. (1970). Polymorphonuclear leukocyte chemotactic activity in rabbit serum and guinea pig serum treated with immune complexes. Evidence for C5a as the major chemotactic factor. *Infection and Immunity*, **1**, 521–5.

Snyderman, R., Smith, C. D. & Verghese, M. W. (1986). Model of leukocyte regulation by chemotactic receptors: roles of a guanine nucleotide regulatory protein and phosphoinositide metabolism. *Journal of Leukocyte Biology*, **40**, 785–800.

Spangrude, G. J., Braaten, B. A. & Davies, R. A. (1984). Molecular mechanisms of lymphocyte extravasation I. Studies of two selective inhibitors of lymphocyte recirculation. *Journal of Immunology*, **132**, 354–62.

Spangrude, G. J., Sacchi, F., Hill, H. R., Van Epps, D. E. & Daynes, R. A. (1985). Inhibition of lymphocyte and neutrophil chemotaxis by pertussis toxin. *Journal of Immunology*, **135**, 4135–43.

Springer, T. A., Dustin, M. L., Kishimoto, T. K. & Marlin, S. D. (1987). The lymphocyte function-associated LFA-1, CD2 and LFA-3 molecules: cell adhesion receptors of the immune system. *Annual Review of Immunology*, **5**, 223–52.

Sullivan, S. J., Daukas, G. & Zigmond, S. H. (1984). Asymmetric distribution of the chemotactic peptide receptor on polymorphonuclear leukocytes. *Journal of Cell Biology*, **99**, 1461–7.

Takahashi, N., Hayano, T & Suzuki, M. (1989). Peptidyl–prolyl *cis–trans* isomerase is the cyclosporin A-binding protein cyclophilin. *Nature, London*, **337**, 473–5.

Toniolo, C., Bonora, G. M., Showell, H., Freer, R. J. & Becker, E. L. (1984). Structural requirements for formyl homooligopeptide chemoattractants. *Biochemistry*, **23**, 698–704.

Toniolo, C., Crisma, M., Valle, G., Bonora, G. M., Polinelli, S., Becker, E. L., Freer, R. J., Sudhanand, R., Rao, B., Balaram, P. & Sukumar, M. (1989). Conformationally restricted formyl methionyl tripeptide chemoattractants: a three-dimensional structure–activity study of analogs incorporating a $C^{\alpha \alpha}$-dialkylated glycine at position 2. *Peptide Research*, **2**, 275–81.

Tranquillo, R. T., Lauffenburger, D. A. & Zigmond, S. H. (1988). A stochastic model for leukocyte random motility and chemotaxis based on receptor binding fluctuations. *Journal of Cell Biology*, **106**, 303–10.

Walter, R. J., Berlin, R. D. & Oliver, J. M. (1980). Asymmetric Fc receptor distribution on human PMN oriented in a chemotactic gradient. *Nature, London*, **286**, 724–5.

Walter, R. J. & Marasco, W. A. (1987). Direct visualization of formylpeptide receptor binding on rounded and polarized human neutrophils: cellular and receptor heterogeneity. *Journal of Leukocyte Biology*, **41**, 377–91.

Webster, R. O., Zanolari, B. & Henson, P. M. (1980). Neutrophil chemotaxis in response to surface-bound C5a. *Experimental Cell Research*, **129**, 55–62.

Wilkinson, P. C. (1973). Recognition of protein structure in leukocyte chemotaxis. *Nature, London*, **244**, 512–13.

Wilkinson, P. C. (1981). Peptide and protein chemotactic factors and their recognition by neutrophil leucocytes. In *Biology of the Chemotactic Response*, ed. J. M. Lackie & P. C. Wilkinson, pp. 53–72, Cambridge, Cambridge University Press.

Wilkinson, P. C. (1986). The locomotor capacity of human lymphocytes and its enhancement by cell growth. *Immunology*, **57**, 281–9.

Wilkinson, P. C. (1987). Leucocyte locomotion: behavioural mechanisms for accumulation. *Journal of Cell Science, Supplement*, **8**, 103–19.

Wilkinson, P. C. (1990). Relation between locomotion, chemotaxis and clustering of immune cells. *Immunology*, **69**, 127–33.

Wilkinson, P. C. & Allan, R. B. (1978). Chemotaxis of neutrophil leucocytes towards substratum-bound protein attractants. *Experimental Cell Research*, **111**, 403–12.

Wilkinson, P. C. & Bradley, G. (1981). Chemotactic and enzyme-releasing activity of amphipathic proteins for neutrophils. A possible role for proteases in chemotaxis on substratum-bound protein gradients. *Immunology*, **42**, 637–48.

Wilkinson, P. C. & Haston, W. S. (1988). Chemotaxis: an overview. *Methods in Enzymology*, **162**, 3–16.

Wilkinson, P. C. & Higgins, A. (1987*a*). OKT3-activated locomotion of human blood lymphocytes: a phenomenon requiring contact of T cells with Fc-receptor bearing cells. *Immunology*, **60**, 445–51.

Wilkinson, P. C. & Higgins, A. (1987b) Cyclosporin A inhibits mitogen-activated but not phorbol ester-activated locomotion of human lymphocytes. *Immunology*, **61**, 311–16.

Wilkinson, P. C. & Islam, L. N. (1989). Recombinant interleukin 4 and gamma interferon activate locomotor capacity in human B cells. *Immunology* **76**, 237–43.

Wilkinson, P. C. & Lackie, J. M. (1983). The influence of contact guidance on chemotaxis of human neutrophil leukocytes. *Experimental Cell Research*, **146**, 117–26.

Wilkinson, P. C., Lackie, J. M., Forrester, J. V. & Dunn, G. A. (1984). Chemokinetic accumulation of human neutrophils on immune-complex-coated substrata: analysis at a boundary. *Journal of Cell Biology*, **99**, 1761–8.

Wilkinson, P. C. & McKay, I. C. (1971). The chemotactic activity of native and denatured serum albumin. *International Archives of Allergy and applied Immunology*, **41**, 237–47.

Wilkinson, P. C., Michl, J. & Silverstein, C. (1980) Receptor distribution in locomoting neutrophils. *Cell Biology International Reports*, **4**, 736.

Wilkinson, P. C., Shields, J. M. & Haston, W. S. (1982). Contact guidance of human neutrophil leucocytes. *Experimental Cell Research*, **140**, 55–62.

Wilkinson, P. C. & Watson, E. A. (1990) FK506 and pertussis toxin distinguish growth-induced locomotor activation from attractant-stimulated locomotion in human blood lymphocytes. *Immunology*, **71**, in press.

Zigmond, S. H. (1974). Mechanisms of sensing chemical gradients by polymorphonuclear leukocytes. *Nature, London*, **249**, 450–2.

Zigmond, S. H. (1977). Ability of polymorphonuclear leukocytes to orient in gradients of chemotactic factors. *Journal of Cell Biology*, **75**, 606–16.

Zigmond, S. H. & Hirsch, J. G. (1973). Leukocyte locomotion and chemotaxis. New methods for evaluation, and demonstration of a cell-derived chemotactic factor. *Journal of Experimental Medicine*, **137**, 387–410.

Zigmond, S. H. & Sullivan, S. J. (1979). Sensory adaptation of leukocytes to chemotactic peptides. *Journal of Cell Biology*, **82**, 517–27.

Zimmermann, A., Gehr, P. & Keller, H. U. (1988). Diacylglycerol-induced shape changes, movements and altered F-actin distribution in human neutrophils. *Journal of Cell Science*, **90**, 657–66.

SIGNAL TRANSDUCTION IN LEUCOCYTES: THE LINKAGE BETWEEN RECEPTOR AND MOTOR

P. H. NACCACHE, M. GAUDRY, S. BOURGOIN AND S. R. McCOLL

Unité de recherche 'Inflammation et Immunologie-Rhumatologie', Centre de recherche du CHUL, 2705 Boulevard Laurier, Ste Foy, Québec G1V 4G2, Canada

The past few years have witnessed major advances in our understanding of the signalling pathways implicated in the responses of leucocytes. A consolidation of previous conceptual and experimental paradigms together with an expansion towards new transduction mechanisms has redefined the framework of this field. Recent data have also provided evidence for the multiplicity and inter-connectedness of the signalling pathways controlling leucocyte behaviour. This review will summarize the salient features of the signalling pathways operative in leucocytes with a view towards understanding the linkage between receptor and motor in leucocytes in general, and in polymorphonuclear neutrophils (neutrophils) in particular.

LEUCOCYTE AGONISTS AND THEIR RECEPTORS

Plasma membrane-located receptors on leucocytes are for the most part glycoproteins, the structures of which range from a single polypeptide (the formyl peptide receptor), to dynamic multimeric entities (e.g. the T-cell receptor (TCR)). The nature and basic properties of these receptors have been described in detail (e.g. Synderman & Uhing, 1988; Gordon, 1988) and will not be dealt with further in the present review. Recent evidence has demonstrated the presence on neutrophils of glycosyl phosphatidyl-inositol-linked and independent-Fc (and transmembrane-spanning) receptors that may function in part co-operatively (Selvaraj *et al.*, 1988) and in part independently through distinct signalling pathways (Kimberly *et al.*, 1990).

Phagocyte-subtype specific agonists have recently been described. For example, neutrophils store and can be induced to release a family of monocyte-specific chemoattractants, the defensins, which only exhibit weak chemokinetic activity towards neutrophils (Territo et al., 1989), as well as a 37 kD granule-associated cationic protein that exhibits monocyte-specific chemotactic activity (Pereira et al., 1990). Monocytes, on the other hand, secrete stimuli (NAP/IL 8) with relative neutrophil specificity (Baggiolini et al., 1989) but which have also been shown to cause the release of histamine and sulphidoleukotrienes from interleukin 3-primed basophils (Dahinden et al., 1989) and to induce human T lymphocyte chemotaxis *in vivo* as well as *in vitro* (Leonard et al., 1990; Bacon et al., this volume). Other well-known neutrophil (and monocyte) chemotactic factors include the complement fragment C5a, the lipid mediator leucotriene B_4, and the bacterial-derived formylated oligopeptides (reviewed in Becker, 1987).

ACTIVATION PATHWAYS

The activation of two major transduction pathways has been documented, albeit to different extent, in leucocytes upon the occupancy of surface receptors: the polyphosphoinositide- and the tyrosine kinase-dependent pathways. In addition, recent studies have demonstrated the presence of an activatable phospholipase D (PLD) in leucocytes. As will be summarized in the next few sections, guanine nucleotide-binding proteins (G proteins) play a central role in the mediation of the activation of most, if not all, of these pathways (Fig.1). Cyclic nucleotides, on the other hand, apparently only serve to modulate (particularly attenuate) responses elicited upon the activation of one or the other of the above signalling pathways (Simchowitz et al., 1983).

The phosphoinositide cycle

The major signal transduction pathway operating in the responses of leucocytes to chemotactic factors is the phosphatidylinositol cycle originally hypothesized by Michell (1975) to be related to calcium-mobilizing mechanisms and more recently characterized in a large variety of cell types (Michell & Putney, 1987). Occupancy of receptors coupled to this pathway leads to the activation, in a G protein-dependent manner, of a phospholipase C (phosphoinositidase, or PIC) specific for phosphatidylinositol, 4,5-bis-phosphate (PtdIns4,5-P_2). The hydrolysis of PtdIns4,5-P_2 results in the generation of inositol 1,4,5-trisphosphate (Ins1,4,5-P_3), and diglyceride (DG). The former compound binds to specific receptors on intracellular organelles and induces the liberation of sequestered calcium, while the latter, in conjunction with calcium and phosphatidyl serine, activates protein kinase C. These two second messengers are thought to act co-ordinately to initiate and control cell functions. The involvement of this signalling pathway in leucocytes has been documented and its essential elements found to be similar to those described in other non-muscle cells (for recent reviews, see Snyderman & Uhing, 1988 and Sha'afi & Molski, 1988).

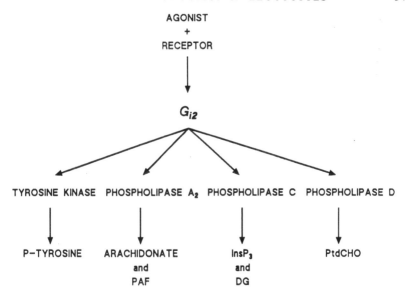

Fig. 1. Schematic representation of the central role of G proteins in the signal transduction pathways operative in leucocytes.

The presence of phosphatidylinositol trisphosphate (PIP$_3$) in human neutrophils has recently been demonstrated (Traynor-Kaplan et al., 1989). It is presumably formed through the action of a tyrosine kinase-linked PI-3-kinase. The levels of PIP$_3$ are up-modulated in a pertussis toxin-sensitive, calcium-insensitive fashion by formyl-methionyl-leucyl-phenylalanine (fMet-Leu-Phe) and to a smaller extent by leucotriene B$_4$. A correlation between the levels of this phospholipid and the degree of neutrophil activation has been observed, suggesting that it participates to the signal transduction pathways of granulocytes. PIP$_3$ may be the precursor for Ins(1,3,4,5)P$_4$ and Ins(1,3,4)P$_3$, the levels of which also increase upon stimulation of neutrophils with chemotactic factors (Traynor-Kaplan et al., 1989).

Guanine nucleotide-binding proteins

Guanine nucleotide-binding proteins (G proteins) play critical roles in the coupling of leucocyte receptors to PIC. Supporting evidence for this conclusion comes from the demonstration of the potentiating role of GTP in the activation of PIC in isolated plasma membranes (Cockcroft & Gomperts, 1985; Smith et al., 1985; Haslam & Davidson, 1984), the sensitivity of the activation of PIC to inhibition by pertussis toxin (Smith et al., 1987) and the ability of purified G proteins to restore the chemotactic factor-stimulated high-affinity GTPase of membranes isolated from pertussis toxin-treated cells (Okajima et al., 1985).

The presence of various G proteins has been demonstrated in leucocytes.

This includes 'classical' G proteins such as G_s and G_i, as well as more recently characterized ras-related small molecular weight GTP binding proteins. The G protein involved in the coupling of the leucocyte receptors belong to the G_i family (Itoh et al., 1988; Kim et al., 1988). Three members of the latter have been identified on the basis of molecular cloning (Murphy et al., 1987; Didsbury et al., 1987; Didsbury & Snyderman, 1987) and immunological techniques (Gierschik et al., 1986, 1987). The major pertussis toxin substrate in neutrophils is a 40 kD protein that has been identified as the α-subunit of G_{i2} (Murphy et al., 1987; Didsbury et al., 1987; Didsbury & Snyderman, 1987). Evidence has been obtained indicating that the α-subunit of G_{i2} is present in excess of the $\beta\gamma$-subunits in neutrophils (Bokoch et al., 1988a) and may contribute to the inhibition of adenylate cyclase in these cells (Bokoch, 1987). The α-subunit of G_{i2} has been found to be primarily associated with the plasma membrane, and to a minor extent with the cytosol (Rudolph et al., 1989a, 1989b) or the specific granule fractions (Bokoch et al., 1988a; Rotrosen et al., 1988). Rotrosen et al., (1988) have also observed that the addition of fMet-Leu-Phe to neutrophils induced a redistribution of the G proteins such that greater amounts were detected in the plasma membrane. Furthermore, stimulation by GTP[γ]S of isolated plasma membranes (Rudolph et al., 1989b), or by fMet-Leu-Phe of intact human neutrophils (Naccache et al., manuscript in preparation) leads to the release of α-subunit to the supernatants of the membranes or to the cytosol of the cells. The potential functions of non-membranous (cytosolic or granule-associated) G proteins are presently unknown.

Small molecular weight, botulinum toxin-substrate G proteins (rac1, rac2, and rap1) have recently been characterized in myeloid cells including neutrophils, HL-60 cells, U937 cells, Jurkat T cell line, and platelets (Polakis et al., 1989; Didsbury et al., 1989; Bhullar & Haslam, 1988; Bokoch et al., 1988b; Quinn et al., 1989; Nagata et al., 1989). The function of the small molecular weight G proteins in neutrophil activation mechanisms are presently unclear. Mege et al (1988) reported that botulinum D toxin did not affect chemotactic factor- and/or PMA-stimulated calcium mobilization, superoxide production, degranulation or actin polymerization in rabbit peritoneal neutrophils, despite the concomitant ADP-ribosylation of up to 70% of the 22 kD substrate. On the other hand, Norgauer et al (1988), have observed an enhancement of superoxide anion production and of degranulation, and an inhibition of migration in botulinum C2 toxin-treated human neutrophils. Only 20% of cellular actin was ADP-ribosylated under the latter conditions, implying that the modification of this relatively minor pool has pronounced effects on the locomotion of neutrophils. The reasons underlying the above differences are not known. A potential link between small G proteins and the activation of the NADPH oxidase has recently been provided by Quinn et al., (1989) who co-purified and co-immunoprecipitated cytochrome b and a small G protein related to rap1. The possibility

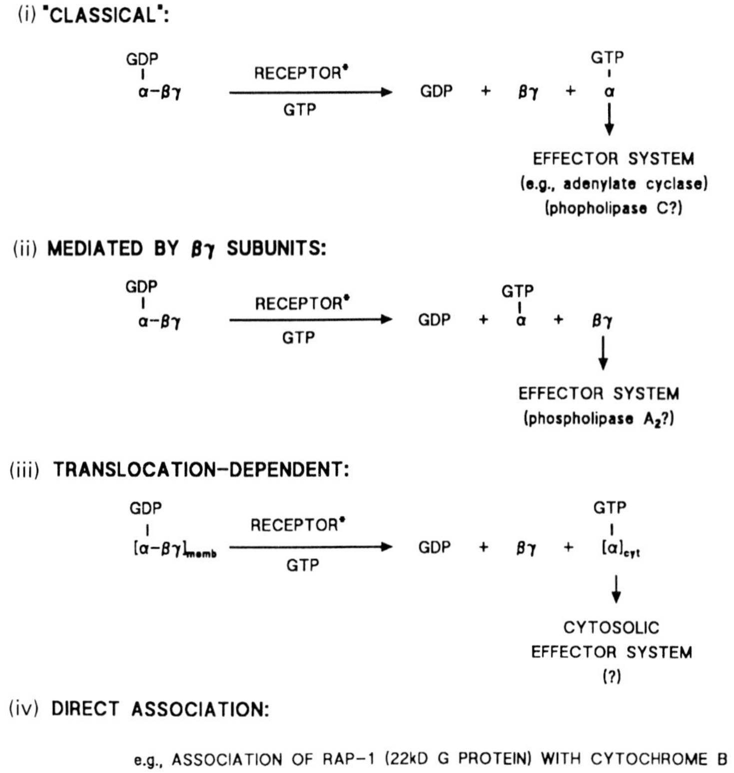

(i) "CLASSICAL":

$$\text{GDP-}\alpha\text{-}\beta\gamma \xrightarrow[\text{GTP}]{\text{RECEPTOR}^*} \text{GDP} + \beta\gamma + \alpha\text{-GTP}$$

↓

EFFECTOR SYSTEM
(e.g., adenylate cyclase)
(phopholipase C?)

(ii) MEDIATED BY βγ SUBUNITS:

$$\text{GDP-}\alpha\text{-}\beta\gamma \xrightarrow[\text{GTP}]{\text{RECEPTOR}^*} \text{GDP} + \alpha\text{-GTP} + \beta\gamma$$

↓

EFFECTOR SYSTEM
(phospholipase A_2?)

(iii) TRANSLOCATION-DEPENDENT:

$$[\alpha\text{-}\beta\gamma]_{\text{memb}}\text{-GDP} \xrightarrow[\text{GTP}]{\text{RECEPTOR}^*} \text{GDP} + \beta\gamma + [\alpha]_{\text{cyt}}\text{-GTP}$$

↓

CYTOSOLIC
EFFECTOR SYSTEM
(?)

(iv) DIRECT ASSOCIATION:

e.g., ASSOCIATION OF RAP-1 (22kD G PROTEIN) WITH CYTOCHROME B

Fig. 2. Summary of the various potential mechanisms of action of G proteins.

that ras, or ras-like, G proteins may be involved in the activation of a PC-specific phospholipase C or of a PLD remains to be tested.

Thus, the results of recent studies have suggested several modes of action for G proteins, some of which have been indirectly implicated in the mechanism of activation of leucocytes. These are schematically represented in Fig.2.

Phospholipase C

The biochemical characteristics of the leucocyte phospholipase C specific for the polyphosphoinositides (PIC) include sensitivity to guanine nucleotides, sub-micromolar concentrations of calcium, and optimal pH dependence around 7.0 (Cockcroft & Gomperts, 1985; Smith et al., 1985; Haslam & Davidson, 1984; Mackin & Stevens, 1988). The leucocyte PIC(s) has not been cloned as yet and its relationship to the various (7) isozymes presently described (Rhee et al., 1989) is unknown. On the other hand, undifferentiated HL60 cells contain large amounts of PLC type IV but no type I, II or III (Homma et al., 1989).

Phospholipase A2

The evidence linking G proteins to the activation of phospholipase A_2 in leucocytes includes the demonstration of the inhibition of fMet-Leu-Phe-induced arachidonic acid release in pertussis toxin-treated cells (Bokoch & Gilman, 1984), as well as the GTPγ S-(Cockcroft & Stutchfield, 1989) and NaF- (Brom et al., 1989) stimulation of arachidonic acid release and production of 5-lipoxygenase products in HL60 cells, neutrophils, monocytes and platelets. Although Kajiyama et al., (1989) have presented evidence that guanine nucleotides lower the calcium requirement of the platelet membrane phospholipase A_2, the generality of this finding remains to be established. Reconstitution experiments in transducin-depleted rod outer segments as well as in heart membranes point to the $\beta\gamma$-subunits as being responsible for coupling receptors to phospholipase A_2 (for a review, see Burch, 1990). Interrelationships among the PLA_2 and PIC (and PLD?) pathways are suggested by the findings that the activity of PLA_2 may be modulated by PIC (or PLD?)-produced diglycerides (Dawson et al., 1984; Burch, 1988). Furthermore, while both acyl-and alkyl-glycerols prime neutrophils for enhanced arachidonic acid release, only the former stimulate the further metabolism of the fatty acid through the 5-lipoxygenase pathway (Bauldry et al., 1988). The activation of phospholipase A_2 in leucocytes may also be dependent on accessory proteins such as phospholipase A_2-activating protein (Bomalaski et al., 1989), and the inhibitory lipocortins, the activity of which may be modulated by tyrosine kinase as well as protein kinase C (Flower, 1988).

The liberated arachidonic acid can serve as substrate for lipoxygenase- and cyclooxygenase-type enzymes which transform it into a variety of bioactive molecules including the potent chemotactic factor leucotriene B_4. Arachidonic acid also activates neutrophils directly, in part by interacting with the leucotriene B_4 receptors (Naccache et al., 1989) and in part by inducing a release of calcium from the $InsP_3$-sensitive intracellular pools of calcium (Beaumier et al., 1987). The metabolic fate of arachidonic acid and the biological activities of its metabolites (leucotriene B_4 in particular) in granulocytes have been recently described (Naccache et al., 1989). By acting in an autocoid fashion, arachidonic acid and its metabolites amplify and modulate neutrophil responses to primary stimuli. The activation of the 5-lipoxygenase in neutrophils relies on a calcium and ATP-dependent translocation of the enzyme to the plasma membrane (Kargman & Rouzer, 1989). The translocation of the 5-lipoxygenase is mediated by a small molecular weight protein (FLAP) that has been purified, cloned, and sequenced (Miller et al., 1990).

Phospholipase D

The results of various lines of investigation have implicated a PLD in the regulation of leucocyte function. Phosphatidylcholine-specific PLD activities have been detected in eosinophils (Kater et al., 1976), HL-60 cells (Anthes et al., 1989), neutrophils, and monocytes (Balsinde et al., 1989). Neutrophils contains two PLD activities, one that is present in the cytosol and the other in the azurophil granules, with varying calcium and pH dependence (Balsinde et al., 1989). The PLD from HL-60 cells was, in addition, found to be guanine nucleotide-dependent (Anthes et al., 1989). A calcium ionophore-stimulated phosphatidylinositol-specific PLD has also been described in the post-nuclear fraction of human neutrophils (Balsinde et al., 1988).

The stimulation of neutrophils and of HL-60 cells by fMet-Leu-Phe and C5a results in the generation of phosphatidylcholine-derived phosphatidic acid due to the activation of a PLD (Pai et al., 1988; Billah et al., 1989; Mullmann et al., 1990; Gelas et al., 1989; Reinhold et al., 1990). The stimulation of PLD by fMet-Leu-Phe is pertussis toxin-sensitive and thus presumably G protein dependent (Bourgoin et al., unpublished observations). Phosphatidic acid is acted upon by a phosphohydrolase leading to the generation of diglyceride. A functional link between the activation of PLD and that of the NADPH oxidase has been proposed on the basis of the parallel inhibition of (PC-derived) diradylglycerol and superoxide anion production by alcohols (Bonser et al., 1989) and wortmannin (Reinhold et al., 1990). The physiological roles and mechanism of action of phosphatidic acid remain to be elucidated. It is noteworthy to recall that phosphatidic acid (as well as lysophosphatidic acid) has been shown to mobilize intracellular calcium and have growth factor-like activity in human A431 carcinoma cells (Moolenaar et al., 1986) and to activate human platelets (Kroll et al., 1989).

Tyrosine phosphorylation

Several proto-oncogene-related tyrosine kinases including $p60^{c-src}$, $p55^{c-fgr}$, $p93^{c-fes}$, $p59^{c-hck}$ and $p56^{c-lck}$ are present in high amounts in blood cells (Huang, 1989; Perlmutter et al., 1988). Their specific functions are not known at present with the exception of $p56^{c-lck}$ which is part of the TCR-CD4 complex, the activation of which leads to the tyrosine phosphorylation of the ζ-subunit of the TCR (Veillette et al., 1989). Tyrosine phosphorylation following ligation of the T-cell receptor precedes the activation of PIC and may involve tyrosine kinases other than $p56^{c-lck}$ (June et al., 1990). Increased expression of $p93^{c-fes}$ and of $p55^{c-fgr}$ during the DMSO-induced differentiation of HL-60 cells has been reported (Glazer et al., 1986; Frank & Sartorelli, 1988). Furthermore, $p55^{c-fgr}$ has been shown to be translocated to the plasma membrane during neutrophil degranulation (Gutkind & Robbins, 1989).

The presence of phosphotyrosine-containing proteins on the surface of human neutrophils has been detected (Skubitz et al., 1988). The stimulation of neutrophils and HL-60 cells by chemotactic factors as well as by cytokines such as GM-CSF, G-CSF, TNF-α and γ-interferon has, in addition, been shown to lead to the tyrosine phosphorylation of several proteins of molecular weight varying between 40 and 118 kD (Evans et al., 1990; Huang et al., 1988; Gomez-Cambronero et al., 1989a, 1989b). The nature and function of the agonist-sensitive tyrosine kinase(s) as well as of its substrates are presently unknown. The stimulation of tyrosine phosphorylation is an indirect event initiated by the occupation of non-tyrosine kinase receptors and is mediated by G proteins. Guanine nucleotides have thus been shown to stimulate tyrosine phosphorylation in electropermeabilized neutrophils (Nasmith et al., 1989). In accord with the latter results, GM-CSF-stimulated tyrosine phosphorylation in human neutrophils has been found to be inhibited by pertussis toxin (Gomez-Cambronero et al., 1989a).

Tyrosine kinase inhibitors such as ST638 and erbstatin inhibit the stimulation of the oxidative burst by chemotactic factors in human neutrophils (Berkow et al., 1989; Gomez-Cambronero et al., 1989b; Naccache et al., submitted for publication). The chemotactic factor-stimulated polymerization of actin, mobilization of calcium or degranulation are not affected under the same conditions. Erbstatin has also been found to inhibit the locomotory responses (chemotaxis as well as chemokinesis) of human neutrophils to a variety of chemotactic factors including fMet-Leu-Phe and leucotriene B_4. This effect may be related to a decrease in adherence due to an altered expression of adhesion proteins (CD11b) on the neutrophils' surface (Gaudry & Naccache, manuscript in preparation). These latter results indicate that cell motility and oxidative metabolism in human neutrophils are both dependent on tyrosine kinases. Tyrosine kinases and their substrates have been involved in other type of cell movements including ligand-induced membrane ruffling of the human epidermoid carcinoma KB cells (Izumi et al., 1988) and this association is supported by the linkage of $p60^{src}$ with the cytoskeleton of chicken embryo fibroblasts (Hamaguchi & Hanafusa, 1987).

Additional evidence for the involvement of tyrosine kinases can be deduced from the activating effects of platelet-derived growth factor (PDGF) on granulocytes. The latter stimulates, among others, monocyte and neutrophil chemotaxis (Deuel et al., 1982), neutrophil adherence (Tzeng et al., 1984), and CR3 expression (Lanser et al., 1988). We have recently observed that PDGF elicited a rapid (and pertussis toxin-sensitive) mobilization of calcium in human neutrophils, and that the latter response was dose-dependently inhibited by erbstatin (Fig.3). These results indicate that a tyrosine kinase-dependent pathway may, in addition to the well-

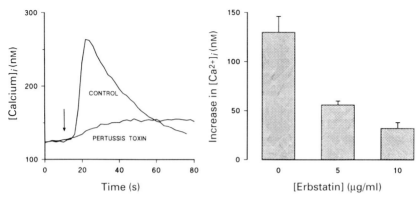

Fig. 3. The effect of pertussis toxin and erbstatin on the PDGF-induced mobilization of calcium in human neutrophils. Calcium mobilization was measured using the fluorescent calcium indicator fura-2. The left panel represent the time course of the increases in cytoplasmic-free calcium elicited by the addition of 1 nM PDGF (at the arrow) in control and pertussis toxin-treated cells (0.5 μg/ml, 2 hours at 37°C. The data in the right panel illustrate the peak increases in cytoplasmic-free calcium (mean ± sem) induced by 1 nM PDGF in control cells and in cells pre-treated for 1 hour at 37°C with the indicated concentrations of erbstatin.

described PLC-dependent pathway, lead to calcium mobilization in neutrophils. The relationship of these observations to the PI-3 kinase (Traynor-Kaplan et al., 1989) which is known to associate with the PDGF receptor (Kazlauskas & Cooper, 1989) and to be tyrosine-phosphorylated in PDGF-treated cells (Kaplan et al., 1987) remains to be investigated.

Membrane-associated phosphotyrosine phosphatases have also been described in lymphocytes (Charbonneau et al., 1988), neutrophils (Kraft & Berkow, 1987), and HL-60 cells (Frank & Sartorelli, 1988). The lymphocyte enzyme, termed CD45, constitutes up to 10% of the lymphocyte cell surface proteins, is variably glycosylated, and can associate with the cytoskeletal protein fodrin. A linkage between CD45 and the TCR has been demonstrated resulting in activation of CD3- and inhibition of CD4-dependent signal transduction (reviewed in Tonks & Charbonneau, 1989).

Protein phosphotyrosine phosphatase activity has been shown to increase during the differentiation of HL-60 cells (Frank & Sartorelli, 1986) and decrease in fMet-Leu-Phe- and phorbol ester-stimulated human neutrophils (Kraft & Berkow, 1987). The phosphotyrosine phosphatase inhibitor vanadate increases tyrosine phosphorylation and oxygen consumption and secretion in electropermeabilized human neutrophils (Grinstein et al., 1990) and platelets (Lerea et al., 1989), respectively. However, the mechanism of action of vanadate remains subject to interpretation as Yang et al., (1989) have suggested that vanadate also stimulates phosphatidylinositol kinase.

The results of these various studies have strongly implicated a role for tyrosine protein kinases in the excitation–response coupling sequence activated by chemotactic factors in human neutrophils. It should be noted

Function	Sensitivity to tyrosine kinase inhibitors
Chemotaxis	Yes
Adherence	Yes
Superoxide production	Yes
Degranulation	No
Calcium mobilization	No
Actin polymerization	No

Fig. 4. Summary of the effects of the tyrosine kinase inhibitor erbstatin on chemotactic factor-stimulated functions in human neutrophils.

that the receptors for these agonists do not themselves possess intrinsic kinase activity. This functional implication is selective for some, but not all, neutrophil functions (Fig.4).

SECOND MESSENGERS AND EFFECTOR SYSTEMS

Calcium

Calcium mobilization in neutrophils, as in most other non-muscle cells, results from interplays between intracellular and extracellular components (Sha'afi & Naccache, 1981).

$InsP_3$ and the calciosomes

The predominant role of $InsP_3$ in the intracellular mobilization of calcium in leucocytes is well established (Feinstein, 1989). Specific receptors for IP_3 in permeabilized neutrophils have been demonstrated, some, but perhaps not all of which may be coupled to calcium release (Spät et al., 1986). An $InsP_3$ binding protein has recently been isolated from brain and shown to be able to mediate calcium fluxes when reconstituted in lipid vesicles (Ferris et al., 1989). Specific calcium storage organelles have also been described (Pozzan et al., 1988; Volpe et al., 1988). These were termed calciosomes in part to indicate their being distinct from other known cell organelles as judged by a variety of biochemical, immunological and morphological criteria. The calciosomes contain a calsequestrin-related calcium binding protein which is presumably responsible for the high-capacity binding of calcium within their lumen. The potential role of other inositol phosphates in calcium release and/or influx has only been sketchily examined. Neither $Ins(1,3,4,5)P_4$ nor inositol 1:2-cyclic 4,5-trisphosphate elicited a release of calcium in saponin-permeabilized rat basophilic leukaemia cells under conditions where InsP$_3$ was able to induce a rapid (<4 s) and highly co-operative opening of intracellular calcium channels (Meyer et al., 1988). Arachidonic acid liberated upon cell stimulation may also cause a release of calcium from the InsP$_3$-sensitive pool both in permeabilized (Beaumier et al., 1987) and intact (Naccache et al., 1989) human neutrophils.

Calcium influx

The stimulated influx of calcium was first detected using radio-isotopic methods (reviewed in Sha'afi & Naccache, 1981), and the results more recently confirmed using fluorescent calcium indicators such as quin2, fura-2 and indo-1 (reviewed in Sha'afi & Molski, 1988). The mechanisms mediating the increased permeability to calcium are still unclear. Although calcium-activated calcium (and sodium) channels have been detected in patch-clamping experiments (Tscharner et al., 1986), Nasmith & Grinstein (1987) have obtained evidence that Mn^{2+} entry, an index of calcium permeability, was preserved in neutrophils loaded with calcium chelators. In addition, the agonist-activated calcium channels in neutrophils are not voltage-dependent (Sha'afi & Naccache, 1981; Merritt et al., 1989; Krause & Welsh, 1990). Furthermore, while calcium mobilization induced by fMet-Leu-Phe is clearly associated with polyphosphoinositide hydrolysis, that induced by leucotriene B$_4$ and PAF is significantly less so. For example, the mobilization of calcium stimulated by PAF in neutrophils (Naccache et al., 1985) and peripheral blood mononuclear cells (Ng & Wong, 1989) is insensitive to inhibition by pertussis toxin. Similarly, calcium mobilization in rabbit and human neutrophils in response to leucotriene B$_4$ can be demonstrated in the absence of a concomitant accumulation of phosphatidic acid, a consequence of the phosphorylation of PLC-generated diglyceride (Volpi et al., 1984; Rossi et al., 1988). It is thus presently unclear whether the increase in the membrane permeability elicited by chemoattractants in neutrophils requires the activation of the PI cycle.

Studies on individual cells

Stimulated calcium mobilization has recently been measured in individual cells using various imaging techniques. The latter studies have uncovered cyclic fluctuations or oscillations in the level of cytosolic free calcium in leukaemic T cells (Lewis & Cahalan, 1989) and adherent neutrophils (Jaconi et al., 1988; Marks & Maxfield, 1990). These oscillatory phenomena may be related to interplays between InsP$_3$-sensitive and insensitive pools and the stimulated influx of calcium from the extracellular medium (Berridge & Taylor, 1988; Harootunian et al., 1988), or to the levels of adenine nucleotides (Ferris et al., 1990). It is intriguing to postulate that the InsP$_3$-insensitive pool may be involved in the mediation of the mobilization of calcium induced by leucotriene B_4 and PAF and that is relatively independent of the activation of PIC (see preceding section).

Protein kinase

Protein kinase C

The calcium and phospholipid-dependent protein kinase C is the major protein kinase in neutrophils. Of the seven known isoenzymes of protein kinase C, two (α and β) (Pontremoli et al., 1990) or three (α, β, and γ) (Fujiki et al., 1988; Huang, 1989) have been detected in neutrophils. Protein kinase C activation by chemotactic factors as well as by phorbol esters has been documented. The functional implication of the activation of protein kinase C has been deduced on the basis of several lines of evidence. Firstly, exogenous activators of protein kinase C such as phorbol ester and diglycerides stimulate several neutrophil functions including the production of superoxide anions, specific granule exocytosis and activation of PLD. Secondly, synergistic effects of phorbol esters and calcium mobilizing agonists have been documented. Of particular relevance to the present discussion are the findings that low concentrations of PMA stimulate chemotactic factor-induced neutrophil locomotion and degranulation (Gaudry et al., 1990) as well as calcium mobilization induced by chemotactic factors (Gaudry & Naccache, manuscript in preparation). Thirdly, protein kinase C inhibitors depress some (but not all, see below) neutrophil responses to chemotactic factors including locomotion (Gaudry et al., 1988).

The effects of the activation of protein kinase C are quite complex. Indeed, inhibitors of protein kinase C have modest effects on chemotactic factor-stimulated neutrophil functional responsiveness such as degranulation, oxidative burst and calcium mobilization despite marked inhibition of protein phosphorylation and of phorbol ester-induced effects. However, a recent report indicates that low concentrations of staurosporine potentiate the initial rate and the total amount of oxidase activation in fMet-Leu-Phe-stimulated neutrophils (Combadiere et al., 1990). Furthermore, pre-incubation with high concentrations of phorbol esters results in reduced binding of PAF and leucotriene B_4 (Yamazaki et al., 1989; O'Flaherty et al., 1989) and in an inhibition of degranulation, chemotaxis, and calcium mobilization in human neutrophils (Naccache et al., 1985; Yamazaki et al., 1989). These two sets of results indicate that protein kinase C exerts bidirectional effects on leucocyte functions. The relative magnitude of the stimulatory and inhibitory facets of protein kinase C remain to be determined under each experimental conditions.

Activation of protein kinase C leads to the phosphorylation of multiple substrates, some of which have been identified. One of the earliest events to follow the stimulation of leucocytes and platelets by phorbol esters is the phosphorylation of a 47 kD protein recently cloned and expressed in a λgt11 expression library (Tyers et al., 1988). Several locomotion-related proteins are phosphorylated by activated protein kinase C including myosin light chain (Inagaki et al., 1984), profilin (Hansson et al., 1988), vimentin, and lipocortin (for a recent review see Huang, 1989) and the microtubule-related τ-proteins (Baudier et al., 1987) and MAP-2 (Hoski et al., 1988). Actin can also be phosphorylated by protein kinase C, G-actin being a better substrate than F-actin; furthermore, phosphorylated G-actin polymerizes more efficiently than its unphosphorylated counterpart (Ohta et al., 1987). These results may underly the small, but significant, polymerization of actin induced by phorbol esters in neutrophils (Yassin et al., 1985).

Other kinases

Several other protein kinases are known to be present in leucocytes. This includes cyclic AMP- and calcium-calmodulin-dependent protein kinases (for a recent review, see Huang, 1989). Whereas the activation of the former is most probably related to the termination or down-regulation of leucocyte functions, the (likely) role(s) of the latter, in addition to regular 'housekeeping' functions such as the regulation of the plasma membrane-located Ca^{2+}-pump (Volpi et al., 1983), remains unknown. Of relevance to the motile functions of leucocytes are the findings that actin and the light chain of myosin are substrates of cyclic AMP-dependent and calmodulin-dependent kinases, respectively. The functional consequences of the modifications of these cytoskeletal proteins are still to be determined. The presence and role of MAP kinase (Ray & Sturgill, 1987) in leucocytes and its possible identity with the 40–42 kD tyrosine kinase substrate in neutrophils remain to be investigated.

THE CYTOSKELETON

The leucocyte cytoskeleton responsible for mechanochemical transduction consists primarily of actin and actin-associated proteins, the presence of a large number of which has been demonstrated in various phagocytes (reviewed in Stossel, 1989; Yin & Hartwig, 1989).

The spatial arrangement of the leucocyte cytoskeleton, in contrast to that of striated and cardiac muscles, is morphologically quite amorphous and displays a great amount of dynamic remodelling upon cell stimulation (Stossel, 1989; Yin & Hartwig, 1988). Neutrophil stimulation by chemotactic factors leads to rapid cycles of actin polymerization and depolymerization (Roa & Varani, 1982; White et al., 1983; Fechheimer & Zigmond, 1983; Howard & Meyer, 1984; Yassin et al., 1985) and exhibits oscillatory behaviour (Wyman et al., 1990; Omann et al., 1989). The main characteristics of the actin polymerization response are consistent with the locomotory responses elicited by the respective agonists (for reviews, see Omann et al., 1987 and Sha'afi & Molski, 1988).

In addition to actin, evidence has also been obtained indicating increased amounts of a 65 kD protein (possibly profilin) (Yassin et al., 1985) and of α-actinin (Nigli & Jenni, 1989) in the Triton X-100-insoluble pellet (the cytoskeleton) of chemotactic peptide-stimulated neutrophils. The latter protein may play a role in the bundling of newly formed actin filaments. On the other hand, the presence of immuno-reactive talin, a putative actin-membrane linker protein associated with surface integrins (see section below), could not be demonstrated in human neutrophils (Nigli & Jenni, 1989).

Several lines of evidence suggest that a guanine nucleotide binding protein (G protein) plays a mediatory role in the stimulation of the polymerization of actin in neutrophils. Exogeneously added GTP augments the actin nucleating activity of plasma membrane preparations from neutrophils (Carson et al., 1986), solubilizes actin and myosin associated with these cell fractions (Huang & Devanney, 1986), and induces the polymerization of actin in permeabilized neutrophil preparations (Therrien & Naccache, 1989; Downey et al., 1989a). An association between the cytoskeleton and the α-subunit of G_{i2} has been noted in lymphocytes where it is apparently required for the GTP-induced activation of PIC as well as receptor patching and capping (Bourguignon et al., 1990), as well as in neutrophils (Caon & Naccache, manuscript in preparation). Furthermore, the amounts of G_{i2} associated with the neutrophils' cytoskeleton increase upon stimulation with fMet-Leu-Phe with a time course that does not, however, correspond with that of the polymerization of actin (Caon & Naccache, manuscript in preparation). These results indicate that the association of α_{i2} with the cytoskeleton may be involved not in the initiation of the actin polymerization response but with its subsequent regulation including receptor recycling. Taken together, the above-summarized data strongly support the hypothesis that the activation of a pertussis toxin-sensitive G protein is involved in the initiation, and possibly also of the subsequent modulation of actin polymerization in neutrophils.

It is presently unclear whether the pathway leading to the polymerization of actin shares elements other than G proteins with those leading to the other neutrophil responses. The stimulated polymerization of actin has been dissociated from the increases in the levels of cytoplasmic-free calcium (Sha'afi et al., 1986; Naccache et al., 1989), the changes in internal pH (Naccache et al., 1989; Downey et al., 1989b) and from the activation of the NADPH oxidase (Naccache et al., 1989). The cytoskeletal responses of neutrophils are similarly insensitive to tyrosine kinase inhibitors (Gomez-Cambronero et al., 1989b; Naccache et al., submitted for publication). However, the inhibition of the chemotactic response induced by the latter compounds indicate that more subtle, although critical, steps of the functional responsiveness of the cytoskeleton are under the control of tyrosine phosphorylation.

Familial classification	Structure	Ligand
β1 Integrins		
(Very Late Antigens)		
VLA-1	$\alpha 1 \beta 1$	Cl, CIV, LM
VLA-2	$\alpha 2 \beta 1$	Cl, CIV, LM
VLA-3	$\alpha 3 \beta 1$	LM, Cl, FN
VLA-4	$\alpha 4 \beta 1$	FN alt
VLA-5	$\alpha 5 \beta 1$	FN
VLA-6	$\alpha 6 \beta 1$	LM
β2 Integrins		
(Leukocyte Cell-Adhesion Molecules)		
CD11a/CD18 (LFA-1)	$\alpha_L \beta 2$	ICAM-1, ICAM-2,
CD11b/CD18 (Mo-1, MAC-1, CR3)	$\alpha_M \beta 2$	iC3b, ICAM-1, LPS, FB, FX
CD11c/CD18 (p150,95, LeuM5)	$\alpha_X \beta 2$	LPS, iC3b
β3 Integrins		
(Cytoadhesins)		
IIb/IIIa	$\alpha_{IIb} \beta 3$	FB, FN, TSP, VN, vWF
Vitronectin receptor	$\alpha_V \beta 3$	FB, VN, vWF
Leukocyte response integrin	$\alpha_R \beta 3$	FB, FN, VN, vWF

Abbreviations; Cl, collagen I; CIV, collagen IV; LM, laminin; FN, fibronectin; FN alt, alternatively spliced CS1 region of fibronectin; ICAM, intercellular adhesion molecule; iC3b, proteolytic fragment of C3; FB, fibrinogen; FX, factor X; LPS, lipopolysaccharide; TSP, thrombospondin; VN, vitronectin; vWF, von Willebrand's factor. (Adapted from Ruoslahti & Giancotti, 1989 and Arnaout, 1990)

Fig. 5. Summary of the various leucocyte integrin receptors, their subunit structure and their ligands.

MODULATION OF CELL RESPONSIVENESS BY EXTRACELLULAR MATRIX COMPONENTS

The ability of leucocytes to emigrate from the bloodstream in response to chemotactic signals generated at sites of inflammation is critically dependent on their ability to adhere to, and migrate through, the endothelium and connective tissue. In recent years, various aspects of the nature of the receptor–ligand interactions controlling these events have been unravelled.

The receptors involved in the adhesion of leucocytes to endothelium are collectively referred to as integrins and have recently been classified into several distinct but closely related families (Fig.5). The three best characterized integrin families are the β1 or very late antigen (VLA), β2 or leucocyte cell-adhesion molecules (Leu-CAM) and the β3 or cytoadhesin families (for reviews see Hynes, 1987, Arnaout, 1990, Ruoslahti & Giancotti, 1989). Although the Leu-CAM integrins are expressed exclusively on leucocytes, members of the β1 and β3 families are also expressed on leucocytes and all three families play an important role in leucocyte adhesion in particular and function in general.

Integrin receptors are composed of two subunits, the α and β subunits. The α subunits typically have a molecular weight of between 120 and 180 kD and are antigenically specific for a given receptor (Hynes, 1987). In contrast the β subunits are approximately 95–130 kD and are antigenically identical within a given integrin family (Hynes, 1987). Thus the Leu-CAM family of receptors consists of three distinct α subunits (CD11a, CD11b and CD11c) with one common β subunit (Arnaout, 1990). The principal ligands for many of these receptors are now known (see Fig. 5).

Leucocyte adhesion to vascular endothelium can be upregulated in at least two ways. Firstly, chemotactic factors upregulate the expression of integrins on the cell surface (Tonnesen et al., 1989; Detmers et al., 1990; Carveth et al., 1989). This effect is rapid, occurring within minutes of stimulation and depends on translocation of pre-synthesized receptors from specific granules to the membrane (Singer et al., 1989). There is, however, substantial uncertainty as to whether the increased binding of leucocytes mediated by chemotactic factor to endothelium is simply due to an increased expression of $\beta2$ integrins on the cell surface. For instance, the PMA-induced increase in expression of CD11b does not correlate quantitatively or in time with the binding of C3bi (Wright & Meyer, 1986). Therefore, it seems likely that integrin activity must be regulated in other ways in addition to cell surface expression. One possibility is that the activity of integrins is modulated by phosphorylation reactions (Wright & Detmers, 1988).

The second mechanism by which leucocyte adhesion is upregulated occurs more slowly (over a period of hours) and is dependent on *de novo* protein synthesis of and expression of ICAMs by cells of the vascular endothelium. This effect is mediated by cytokines such as IL-1, IFNγ and TNF as well as LPS (Pober et al., 1986; Dustin et al., 1986; Rothlein et al., 1986; Pohlman et al., 1986).

In addition to enabling adherence of leucocytes to endothelium, integrins and their ligands may also modify leucocyte functions other than adherence. For example, C3bi-mediated phagocytosis by (chemotactic factor-treated) neutrophils and monocytes is enhanced following prior exposure of these cells to fibronectin (Wright et al., 1983; Pommier et al., 1984). In addition, integrins can affect the neutrophil oxidative burst by at least two mechanisms. Incubation of neutrophils in suspension with laminin stimulates degranulation of specific granules, increases expression of fMet-Leu-Phe receptors on the cell surface and enhances superoxide production in response to fMet-Leu-Phe (Pike et al., 1989). In addition, exposure of neutrophils (but not monocytes) to plastic microtitre plates coated with a variety of different extracellular matrix proteins, fetal calf serum or confluent human umbilical vein endothelial cells dramatically altered the responsiveness of neutrophils not only to fMet-Leu-Phe but also GM-CSF, G-CSF and TNFα (Nathan, 1987, 1989). This effect was not observed with cells in suspension. Furthermore, it appears to be principally mediated by $\beta2$ integrins since specific antibodies directed against members of the β2 integrin family inhibited the ability of the cytokines to stimulate the oxidative burst and the effect was not observed using cells from patients suffering from leucocyte adhesion deficiency (a genetic disease manifested by a lack of expression of β2 integrins) (Nathan et al., 1989).

These studies, although in a relatively preliminary stage, have already provided information that the functional responsiveness of leucocytes is likely to be profoundly affected by both cellular interactions and the occupation of integrin receptors. Most current progress is still fairly restricted to identification of both receptors and their respective ligands on various cell types. At present little is known about the mechanism of signal transduction following receptor occupancy. For example, amongst other questions, it is still not known whether G proteins or the cytoskeleton play a role in the modification of leucocyte response following integrin–ligand interactions or whether there is a requirement for intracellular phosphorylation reactions to occur. At the very least, it appears that the integrins themselves are phosphorylated following activation (Buyon et al., 1990) although the precise role of this event in modulation of leucocyte activation awaits detailed clarification. In addition, the effect of adherence of leucocytes to specific integrin ligands on the critical function of these cells, including chemotaxis, degranulation, cytotoxicity and release of inflammatory mediators such as PAF, leucotriene B_4 and lymphokines requires further investigation.

MODULATION OF CELL RESPONSIVENESS BY CYTOKINES

The functional responsiveness of leucocytes *in vivo* is modulated by interactions with various cytokines. Recent *in vitro* data have described the profound effects of some of these cytokines, and of GM-CSF and TNFα in particular, on the responses of neutrophils and monocytes/macrophages. GM-CSF has several direct effects on neutrophils including changes in light scatter, membrane ruffling and surface expression of chemotactic receptors and adherence proteins (Lopez et al., 1986; Arnaout et al., 1986) and of c-fos mRNA (Colotta et al., 1987). Various neutrophil functions are enhanced following a preincubation with GM-CSF, a phenomenon referred to as priming. These include phagocytosis (Fleischmann et al., 1986; Lopez et al., 1986), antibody-dependent cytotoxicity (Lopez et al., 1986), superoxide radical production (Weisbart et al., 1987), chemotaxis (Gasson et al., 1984) and degranulation both *in vivo* (Devereux et al., 1990) and *in vitro* (Smith et al., 1990).

Recent studies have begun to explore the biochemical basis for the effects of GM-CSF in neutrophils. Treatment of neutrophils with GM-CSF does not directly activate PIC (Naccache et al., 1988; Sullivan et al., 1987; Corey & Rosoff, 1989) or alter the subsequent activation of PIC by chemotactic factors (Bourgoin et al., submitted for publication). Evidence for the involvement of G proteins (McColl et al., 1989; Gomez-Cambronero et al., 1989a; Corey & Rosoff, 1989), and for the indirect activation of tyrosine kinase (the GM-CSF receptor does not possess intrinsic tyrosine kinase activity (Gearing et al., 1989) in the direct and priming effects of GM-CSF has also been obtained (Gomez-Cambronero et al., 1989a; McColl & Naccache, manuscript in preparation). The pertussis toxin sensitivity of the tyrosine phosphorylation induced by GM-CSF (Gomez-Cambronero et al., 1989a) indicates that the activation of the latter kinase is a secondary event mediated by G proteins. By itself, GM-CSF activates the Na^+/H^+ antiport in human neutrophils and induces a pertussis toxin-sensitive cytoplasmic alkalinization (Gomez-Cambronero et al., 1989a). Furthermore, preincubation of human neutrophils with GM-CSF leads to an enhancement of the mobilization of calcium induced by chemotactic factors (Naccache et al., 1987). The latter effect of GM-CSF is also pertussis toxin-sensitive (McColl et al., 1989) and is due to increased intracellular mobilization of calcium without a concomitant increase in the production of $InsP_3$ (Naccache, manuscript in preparation).

Although the mechanism of action of GM-CSF remains to be established, recent evidence suggests that up-regulation of a phosphatidylcholine-specific PLD plays an important role in this process. While GM-CSF did not directly activate PLD, pretreatment of human neutrophils with the cytokine led to a marked enhancement of the stimulation of PLD by chemotactic factors (Bourgoin et al., submitted for publication). A tight correlation between the production of phosphatidic acid and of superoxide anions has been noted (Bonser et al., 1989; Rossi et al., 1990). The calcium dependence of partially purified PLD indicates that a relationship may exist between the stimulation of the chemotactic factor-induced activation of PLD in GM-CSF-treated neutrophils and the enhanced mobilization of calcium (Naccache et al., 1987). The implication of tyrosine kinase in the modulation of the activity of PLD is an intriguing possibility that remains to be investigated. The potential sequence of events involved in priming by GM-CSF that is suggested by the presently available data is illustrated schematically in Fig.6.

CONCLUDING REMARKS

Significant advances have taken place in the last few years regarding the understanding of the signalling pathways involved in leucocyte activation. The data supporting the multiplicity, and interdependence, of the signalling pathways illustrate the complexity of the regulatory mechanisms that may be called into play in the modulation of the responses of leucocytes. A schematic representation of some of the relationships of some of these pathways is presented in Fig.7. The central role of G proteins in the transduction of the signals initiated upon receptor occupancy has been firmly

SIGNAL TRANSDUCTION IN LEUCOCYTES

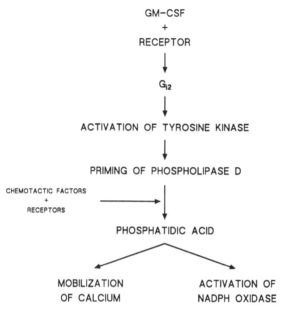

Fig. 6. Schematic representation of the postulated mechanism of action of GM-CSF.

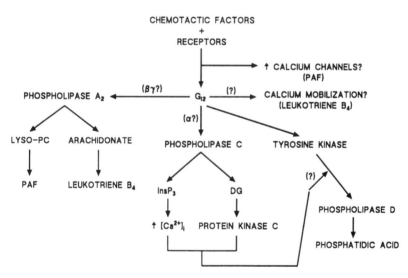

Fig. 7. Summary flow diagram of the major signalling pathways involved in neutrophil activation by chemotactic factors.

established. The phosphoinositide bifurcating pathway is similarly tightly associated with the mechanism of cell activation and in particular with regard to the mobilization of internal calcium and the stimulation of protein kinase C. Recent developments include the realization of the multiplicity of signalling pathways and their relationships to G proteins. Thus, non-receptor-linked tyrosine kinases and PLD and PLA_2 are assuming important roles in the initiation as well as the modulation of cell responsiveness.

In spite of the very real progress recently accomplished, several outstanding questions remain. Current investigations in several cell types including leucocytes are thus focused, among others, on: (i) the elucidation of the respective roles of, and interrelationships (cross-talk) among the various pathways, (ii) the elucidation of the physical linkage between extracellular matrix binding proteins (integrins) and intracellular effector systems, and (iii) the further characterization of the modulation of cell responsiveness by single and multiple cytokines. A conceptually and experimentally more difficult shortcoming of our present knowledge concerns the integration of the biochemical events outlined in the preceding discussion into a working knowledge of the behaviour of the cells. It remains to be established, for example, how these various responses are related to the directed movement of leucocytes, which are responsible for the deactivation and desensitization of the cells, and which are the nuclear sites of action of the various second messengers generated upon cell stimulation.

ACKNOWLEDGMENTS

Supported in parts by grants and fellowships from the Medical Research Council of Canada, the National Cancer Institute of Canada, the Arthritis Society of Canada and the Fonds de la Recherche en Santé du Québec.

REFERENCES

Anthes, J. C., Eckel, S., Siegel, M. I., Egan, R. W. & Billah, M. M. (1989). Phospholipase D in homogenates from HL-60 granulocytes: implications of calcium and G protein control. *Biochemical Biophysical Research Communications*, **163**, 657–64.

Arnaout, M. (1990). Structure and function of the leucocyte adhesion molecules CD11/CD18. *Blood* **75**, 1037–50.

Arnaout, M. A., Wang, E. A., Clark, S. C. & Sieff, C. A. (1986). Human recombinant granulocyte-macrophage colony-stimulating factor increases cell-to-cell adhesion and surface expression of adhesion-promoting surface glycoproteins on mature granulocytes. *Journal of Clinical Investigation*, **78**, 597–601.

Baggiolini, M., Walz, A. & Kunkel, S. L. (1989). Neutrophil-activating peptide-1/interleukin 8, a novel cytokine that activates neutrophils. *Journal of Clinical Investigation*, **84**, 1045–9.

Balsinde, J., Diez, E., Fernandez, B. & Mollinedo, F. (1989). Biochemical characterization of phospholipase D activity from human neutrophils. *European Journal of Biochemistry*, **186**, 717–24.

Balsinde, J. Diez, E. & Mollinedo, F. (1988). Phosphatidylinositol-specific phospholipase D: a pathway for generation of a second messenger. *Biochemical & Biophysical Research Communications*, **154**, 502–8.

Baudier, J., Lee, S. H. & Cole, R. D. (1987). Separation of the different microtubule-associated tau protein species from bovine brain and their mode of phosphorylation by Ca^{2+}/phospholipid-dependent protein kinase C. *Journal of Biological Chemistry*, **262**, 17584–90.

Bauldry, S. A., Wykle, R. L. & Bass, D. A. (1988). Phospholipase A2 activation in human neutrophils. Differential actions of diacylglycerols and alkylglycerols in priming cells for stimulation by N-formyl-met-leu-phe. *Journal of Biological Chemistry*, **263**, 16787–95.

Beaumier, L., Faucher, N. & Naccache, P. H. (1987). Arachidonic acid-induced release of calcium in permeabilized human neutrophils. *FEBS Letters*, **221**, 289–92.

Becker, E. L. (1987). The formylpeptide receptor of the neutrophil: a search and conserve operation. *American Journal of Pathology*, **129**, 16–24.

Berkow, R. L., Dodson, D. W. & Kraft, A. S. (1989). Human neutrophils contain distinct cytosolic and particulate tyrosine kinase activities: possible role in neutrophil activation. *Biochimica & Biophysica Acta*, **997**, 292–301.

Berridge, M. J. & Taylor, C. W. (1988). Inositol trisphosphate and calcium signaling. *Cold Spring Harbor Symposia on Quantitative Biology*, **53**, 927–33.

Bhullar, R. P. & Haslam, R. J. (1988). Gn-proteins are distinct from ras p21 and other known low molecular mass GTP-binding proteins in the platelet. *FEBS Letters*, **237**, 168–72.

Billah, M. M., Eckel, S., Mullman, T. J., Egan, R. W. & Siegel, M. I. (1989). Phosphatidylcholine hydrolysis by phospholipase D determines phosphatidate and diglyceride levels in chemotactic peptide-stimulated human neutrophils. Involvement of phosphatidate phosphohydrolase in signal transduction. *Journal of Biological Chemistry*, **264**, 17069–77.

Bokoch, G. M. (1987). The presence of free G protein β/γ subunits in human neutrophils results in suppression of adenylate cyclase activity. *Journal of Biological Chemistry*, **262**, 589–94.

Bokoch, G. M. & Gilman, A. G. (1984). Inhibition of receptor mediated release of arachidonic acid by pertussis toxin. *Cell*, **39**, 301–8.

Bokoch, G. M., Bickford, K. & Bohl, B. P. (1988a). Subcellular localization and quantitation of the major neutrophil pertussis toxin substrate, Gn. *Journal of Cell Biology*, **106**, 1927–36.

Bokoch, G. M., Parkos, C. A. & Mumby, S. M. (1988b). Purification and characterization of the 22,000-dalton GTP-binding protein substrate for ADP-ribosylation by botulinum toxin, G22K. *Journal of Biological Chemistry*, **263**, 16744–9.

Bomalaski, J. M., Baker, D. G., Brophy, L., Resurreccion, N. V., Spilberg, I., Munian, M. & Clark, M. A. (1989). A phospholipase A_2-activating protein (PLAP) stimulates human neutrophil aggregation and release of lysosomal enzymes, superoxide and eicosanoids. *Journal of Immunology*, **142**, 3957–62.

Bonser, R. W., Thompson, N. T., Randall, R. W. & Garland, L. G. (1989). Phospholipase D activation is functionally linked to superoxide generation in the human neutrophil. *Biochemical Journal*, **264**, 617–20.

Bourguignon, L. Y. W., Walker, G. & Huang, H. S. (1990). Interactions between a lymphoma membrane-associated guanosine 5'-triphosphate-binding protein and the cytoskeleton during receptor patching and capping. *Journal of Immunology*, **144**, 2242–52.

Brom, C., Koller, M., Brom, J. & Konig, W. (1989). Effect of sodium fluoride on the generation of lipoxygenase products from human polymorphonuclear granulocytes, mononuclear cells and platelets – indication for the involvement of G proteins. *Immunology*, **68**, 240–6.

Burch, R. M. (1988). Diacylglycerol stimulates phospholipase A_2 from Swiss 3T3 fibroblasts. *FEBS Letters*, **234**, 283–6.

Burch, R. M. (1990). Regulation of phospholipase A_2 by G proteins. In *G Proteins and Calcium Signalling*, pp. 48–58, ed. P. H. Naccache. CRC Press, Boca Raton, Florida.

Buyon, J. P., Slade, S. G., Reibman, J., Abramson, S. B., Phillips, M. R., Weissmann, G. & Winchester, R. (1990). Constitutive and induced phosphorylation of the α- and β-chains of the Cd11/CD18 leukocyte integrin family. Relationship to adhesion-dependent functions. *Journal of Immunology*, **144**, 191–7.

Carson, M., Wilde, M. W. & Zigmond, S. H. (1986). Chemotactic peptides and GTPγ-S increase actin nucleation on PMN membranes. *Journal of Cell Biology*, **108**, 548a.

Carveth, H. J., Bohnsack, J. F., McIntyre, T. M., Baggiolini, M., Prescott, S. M. & Zimmerman, G. A. (1989). Neutrophil activating factor (NAF) induces polymorphonuclear leukocyte adherence to endothelial cells and to subendothelial matrix proteins. *Biochemical & Biophysical Research Communications*, **162**, 387–93.

Charbonneau, H., Tonks, N. K., Walsh, K. A. & Fischer, E. H. (1988). The leukocyte common antigen (CD45): a putative receptor-linked protein tyrosine phosphatase. *Proceedings of the National Academy of Sciences, USA*, **85**, 7182–6.

Cockcroft, S. & Gomperts, B. (1985). Role of guanine nucleotide binding proteins in the activation of polyphosphoinositide phosphodiesterase. *Nature, London*, **314**, 328–30.

Cockcroft, S. & Stutchfield, J. (1989). The receptors for ATP and fMet-Leu-Phe are independently coupled to phospholipases C and A_2 via G-protein(s). Relationship between phospholipase C and A_2 activation and exocytosis in HL60 cells and human neutrophils. *Biochemical Journal*, **263**, 715–23.

Colotta, F., Wang, J. M., Polentarutti, N. & Mantovani, A. (1987). Expression of c-fos protooncogene in normal human peripheral blood granulocytes. *Journal of Experimental Medicine*, **165**, 1224–9.

Combadiere, C., Hakim, J., Giroud, J.-P. & Perianin, A. (1990). Staurosporine, a protein kinase C inhibitor, up-regulates the stimulation of human neutrophil respiratory burst by N-formyl peptides and platelet activating factor. *Biochemical & Biophysical Research Communications*, **168**, 65–70.

Corey, S. J. & Rosoff, P. M. (1989). Granulocyte-macrophage colony-stimulating factor primes neutrophils by activating a pertussis toxin-sensitive G protein not associated with phosphatidylinositol turnover. *Journal of Biological Chemistry*, **264**, 14165–71.

Dahinden, C. A., Kurimoto, Y., DeWeck, A. L., Lindley, I., Dewald, B. & Baggiolini, M. (1989). The neutrophil-activating peptide NAF/NAP-1 induces histamine and leukotriene release by interleukin 3-primed basophils. *Journal of Experimental Medicine*, **170**, 1787–92.

Dawson, R. M. C., Irvine, R. F., Bray, J. & Quinn, P. J. (1984). Long-chain unsaturated diacylglycerols cause a perturbation in the structure of phospholipid bilayers rendering them susceptible to phospholipase attack. *Biochemical & Biophysical Research Communications*, **125**, 836–42.

Detmers, P. A., Lo, S. K., Olsen-Egbert, E., Walz, A., Baggiolini, M. & Cohn, Z. A. (1990). Neutrophil-activating protein 1/interleukin 8 stimulates the binding activity of the leukocyte adhesion receptor CD11b/CD18 on human neutrophils. *Journal of Experimental Medicine*, **171**, 1155–62.

Deuel, T. F., Senior, R. M., Huang, J. S. & Griffin, G. L. (1982). Chemotaxis of monocytes and neutrophils to platelet-derived growth factor. *Journal of Clinical Investigation*, **69**, 1046–9.

Devereux, S., Porter, J. B., Hoyes, K. P., Abeysinghe, R. D., Saib, R. & Linch, C. D. (1990). Secretion of neutrophil secondary granules occurs during granulocyte-macrophage colony-stimulating factor induced margination. *British Journal of Haematology*, **74**, 17–23.

Didsbury, J. R. & Snyderman, R. (1987). Molecular cloning of a new human G protein. Evidence for two Giα-like protein families. *FEBS Letters*, **219**, 259–63.

Didsbury, J. R., Ho, Y.-.S. & Snyderman, R. (1987). Human Gi protein α-subunit: deduction of amino acid structure from a cloned cDNA. *FEBS Letters*, **211**, 160–4.

Didsbury, J., Weber, R. F., Bokoch, G. M., Evans, T. & Snyderman, R. (1989). rac, a novel ras-related family of proteins that are botulinum toxin substrates. *Journal of Biological Chemistry*, **264**, 16378–82.

Downey, G. P., Chan, C. K. & Grinstein, S. (1989a). Actin assembly in electropermeabilized neutrophils: role of G-proteins. *Biochemical & Biophysical Research Communications*, **164**, 700–5.

Downey, G. P. & Grinstein, S. (1989b). Receptor-mediated actin assembly in electropermeabilized neutrophils: role of intracellular pH. *Biochemical & Biophysical Research Communications*, **160**, 18–24.

Dustin, M. L., Rothlein, R., Bhan, A. K., Dinarello, C. A. & Springer, T. A. (1986). Induction by IL 1 and interferon gamma: tissue distribution, biochemistry, and function of a natural adherence molecule (ICAM-1). *Journal of Immunology*, **137**, 245–54.

Evans, J. P. M., Mire-Sluis, A. R., Hoffbrand, A. V. & Wickremasinghe, R. G. (1990). Binding of G-CSF, GM-CSF, tumor necrosis factor-α, and γ-interferon to cell surface receptors on human myeloid leukaemia cells triggers rapid tyrosine and serine phosphorylation of a 75-kD protein. *Blood*, **75**, 88–95.

Fechheimer, M. & Zigmond, S. (1983). Changes in cytoskeletal proteins of polymorphonuclear leukocytes induced by chemotactic peptides. *Cell Motility*, **3**, 349–61.

Feinstein, M. B. (1989). Platelets, macrophages and neutrophils. In *Inositol Lipids in Cell Signalling*, pp. 247–82, ed. R. H. Michell, A. H. Drummond and C. P. Downes. Academic Press Ltd., London.

Ferris, C. D., Huganir, R. L., Supattapone, S. & Snyder, S. H. (1989). Purified inositol 1,4,5-trisphosphate receptor mediates calcium flux in reconstituted lipid vesicles. *Nature, London*, **342**, 87–9.

Ferris, C. D., Huganir, R. L. & Snyder, S. H. (1990). Calcium flux mediated by purified inositol 1,4,5-trisphosphate receptor in reconstituted lipid vesicles is allosterically regulated by adenine nucleotides. *Proceedings of the National Academy of Sciences, USA*, **87**, 2147–51.

Fleischmann, J., Golde, D. W., Wiesbart, D. W. & Gasson, J. C. (1986). Granulocyte-macrophage colony-stimulating factor enhances phagocytosis of bacteria by human neutrophils. *Blood*, **68**, 708–11.

Flower, R. J. (1988). Lipocortins and the mechanism of action of the glucocorticoids. *British Journal of Pharmacology*, **94**, 987–1015.

Frank, D. A. & Sartorelli, A. C. (1986). Regulation of protein phosphotyrosine content by changes in tyrosine kinase and protein phosphotyrosine phosphatase activities during induced granulocytic and monocytic differentiation of HL-60

leukaemia cells. *Biochemical & Biophysical Research Communications*, **440**, 440–7.
Frank, D. A. & Sartorelli, A. C. (1988). Biochemical characterization of tyrosine kinase and phosphotyrosine phosphatase activities of HL-60 cells. *Cancer Research*, **48**, 4299–306.
Frank, D. A. & Sartorelli, A. C. (1988). Alterations in tyrosine phosphorylation during the granulocytic maturation of HL-60 leukemia cells. *Cancer Research*, **48**, 52–8.
Fujiki, T., Philips, W. A., Johnston, R. B. & Korchak, H. M. (1988). Phorbol ester and calcium elicit selective translocation of the isotype of protein kinase C in human neutrophils. *Journal of Cell Biology*, **107**, 57a.
Gasson, J. C., Weisbart, R. H., Kaufman, S. E., Clark, S. C., Hewick, R. M. & Wong, G. G. (1984). Purified human granulocyte-macrophage colony-stimulating factor: direct effects on neutrophils. *Science*, **226**, 1339–42.
Gaudry, M., Perianin, A., Marquetty, C. & Hakim, J. (1988). Negative effect of a protein kinase C inhibitor (H-7) on human polymorphonuclear neutrophil locomotion. *Immunology*, **63**, 715–19.
Gaudry, M., Marquetty, C. & Hakim, J. (1990). Dual effect of phorbol myristate acetate on chemoattractant-induced locomotion of human neutrophils. *Nouvelle Revue Française Hematology*, (in press).
Gearing, D. P., King, J. A., Gough, N. M. & Nicola, N. A. (1989). Expression cloning of a receptor for human granulocyte-macrophage colony-stimulating factor. *EMBO Journal*, **8**, 3667–76.
Gelas, P., Ribbes, G., Record, M., Terce, F. & Chap, H. (1989). Differential activation by fMet-Leu-Phe and phorbol ester of a plasma membrane phosphatidylcholine-specific phospholipase D in human neutrophil. *FEBS Letters*, **251**, 213–18.
Gierschik, P., Falloon, J., Milligan, G., Pines, M., Gallin, J. I. & Spiegel, A. (1986). Immunochemical evidence for a novel pertussis toxin substrate in human neutrophils. *Journal of Biological Chemistry*, **261**, 8058–62.
Gierschik, P., Sideropoulos, D., Spiegel, A. & Jakobs, K. H. (1987). Purification and immunochemical characterization of the major pertussis toxin-sensitive guanine nucleotide-binding protein of bovine neutrophil membranes. *European Journal of Biochemistry*, **165**, 185–94.
Glazer, R. I., Chapekar, M. S., Hartman, K. D. & Knode, M. C. (1986). Appearance of membrane-bound tyrosine kinase during differentiation of HL-60 leukemia cells by immune interferon and tumor necrosis factor. *Biochemical & Biophysical Research Communications*, **140**, 908–15.
Gomez-Cambronero, J., Yamazaki, M., Metwally, F., Molski, T. F. P., Bonak, V. A., Huang, C. K., Becker, E. L. & Sha'afi, R. I. (1989a). Granulocyte–macrophage colony-stimulating factor and human neutrophils: role of guanine nucleotide regulatory proteins. *Proceedings of the National Academy of Sciences, USA*. **86**, 3569–73.
Gomez-Cambronero, J., Huang, C. K., Bonak, V. A., Wang, E., Casnellie, J. E., Shiraishi, T. & Sha'afi, R. I. (1989b). Tyrosine phosphorylation in human neutrophils. *Journal of Biological Chemistry*, **162**, 1478–85.
Gordon, S. (1988). *Macrophage Plasma Membrane Receptors: Structure and Function*. The Company of Biologists Limited, Cambridge, UK.
Grinstein, S., Furuya, W., Lu, D. J. & Mills, G. B. (1990). Vanadate stimulates oxygen consumption and tyrosine phosphorylation in electropermeabilized human neutrophils. *Journal of Biological Chemistry*, **265**, 318–27.
Gutkind, J. S. & Robbins, K. C. (1989). Translocation of the FGR protein-tyrosine

kinase as a consequence of neutrophil activation. *Proceedings of the National Academy of Sciences, USA*, **86**, 8783–7.

Hamaguchi, M. & Hanafusa, H. (1987). Association of p60src with triton X-100-resistant cellular structure correlates with morphological transformation. *Proceedings of the National Academy of Sciences, USA*, **84**, 2312–16.

Hansson, A., Skoglund, G., Lassing, I., Lindberg, U. & Ingelman-Sundberg, M. (1988). Protein kinase C-dependent phosphorylation of profilin is specifically stimulated by phosphatidylinositol bisphosphate (PIP_2). *Biochemical & Biophysical Research Communications*, **150**, 526–31.

Harootunian, A. T., Kao, J. P. Y. & Tsien, R. Y. (1988). Agonist-induced calcium oscillations in depolarized fibroblasts and their manipulation by photoreleased $Ins(1,4,5)P_3$, Ca^{++}, and Ca^{++} buffer. *Cold Spring Harbor Symposia in Quantitative Biology*, **53**, 935–43.

Haslam, R. J. & Davidson, M. M. L. (1984). Receptor-induced diacylglycerol formation in permeabilized platelets: possible role for a GTP-binding protein. *Journal of Receptor Research*, **4**, 605–29.

Homma, Y., Takenawa, T., Emori, Y., Sorimachi, H. & Suzuki, K. (1989). Tissue- and cell type-specific expression of mRNAs for four types of inositol phospholipid phospholipase C. *Biochemical & Biophysical Research Communications*, **164**, 406–12.

Hoski, M., Akiyama, T., Shinohara, Y., Miyata, Y., Ogawara, H. & Nishida, E. (1988). Protein kinase C-catalyzed phosphorylation of the microtubule-binding domain of microtubule-associated protein 2 inhibits its ability to induce tubulin polymerization. *European Journal of Biochemistry*, **174**, 225–32.

Howard, T. H. & Meyer, W. H. (1984). Chemotactic peptide modulation of actin assembly and locomotion in neutrophils. *Journal of Cell Biology*, **98**, 1265–71.

Huang, C.-K. (1989). Protein kinases in neutrophils: a review. *Membrane Biochemistry*, **8**, 61–79.

Huang, C. K. & Devanney, J. F. (1986). GTP-γ-S-induced solubilization of actin and myosin from rabbit peritoneal neutrophil membranes. *FEBS Letters*, **202**, 41–4.

Huang, C. K., Laramee, G. R. & Casnellie, J. E. (1988). Chemotactic factor induced tyrosine phosphorylation of membrane associated proteins in rabbit peritoneal neutrophils. *Biochemical & Biophysical Research Communications*, **151**, 794–801.

Hynes, R. O. (1987). Integrins: a family of cell surface receptors. *Cell*, **48**, 549–54.

Inagaki, M., Kawamoto, S. & Hidaka, H. (1984). Serotonin secretion from human platelets may be modified by the Ca^{2+}-activated, phospholipid-dependent myosin phosphorylation. *Journal of Biological Chemistry*, **259**, 14321–9.

Itoh, H., Toyama, R., Kozasa, T., Tsukamoto, T., Matsuoka, M. & Kaziro, Y. (1988). Presence of three distinct molecular species of G_i protein α subunit. Structure of rat cDNAs and human genomic DNAs. *Journal of Biological Chemistry*, **263**, 6656–64.

Izumi, T., Saeki, Y., Akanuma, Y., Takaku, F. & Kasuga, M. (1988). Requirement for receptor-intrinsic tyrosine kinase activities during ligand-induced membrane ruffling of KB cells. Essential sites of src-related growth factor receptor kinases. *Journal of Biological Chemistry*, **263**, 10386–93.

Jaconi, M. E. E., Rivest, R. W., Schlegel, W., Wollheim, C. B., Pittet, D. & Lew, P. D. (1988). Spontaneous and chemoattractant-induced oscillations of cytosolic free calcium in single adherent human neutrophils. *Journal of Biological Chemistry*, **263**, 10557–60.

June, C. H., Ledbetter, J. A. & Samelson, L. E. (1990). Increases in tyrosine phosphorylation are detectable before phospholipase C activation after T cell receptor stimulation. *Journal of Immunology*, **144**, 1591–9.

Kajiyama, Y., Murayama, T. & Nomura, Y. (1989). Pertussis toxin-sensitive GTP-binding proteins may regulate phospholipase A_2 in response to thrombin in rabbit platelets. *Archives in Biochemistry & Biophysics*, **274**, 200–8.

Kaplan, D. R., Whitman, M., Schaffhausen, B., Pallas, D. C., White, M., Cantley, L. & Roberts, T. M. (1987). Common elements in growth factor stimulation and oncogenic transformation: 85 kD phosphoprotein and phosphatidylinositol kinase activity. *Cell*, **50**, 1021–9.

Kargman, S. & Rouzer, C. A. (1989). Studies on the regulation, biosynthesis, and activation of 5-lipoxygenase in differentiated HL60 cells. *Journal of Biological Chemistry*, **264**, 13313–20.

Kater, L. A., Goetzl, E. J. & Austen, K. F. (1976). Isolation of human eosinophil phospholipase D. *Journal of Clinical Investigation*, **57**, 1173–80.

Kazlauskas, A. & Cooper, J. A. (1989). Autophosphorylation of the PDGF receptor in the kinase region regulates interactions with cell proteins. *Cell*, **58**, 1121–33.

Kim, S., Ang, S.-L., Bloch, D. B., Kawahara, Y., Tolman, C., Lee, R., Seidman, J. G. & Neer, E. J. (1988). Identification of cDNA encoding an additional α subunit of a human GTP-binding protein: expression of three (α)i subtypes in human tissues and cell lines. *Proceedings of the National Academy of Sciences, USA*, **85**, 4153–7.

Kimberly, R. P., Ahlstrom, J. W., Click, M. E. & Edberg, J. C. (1990). The glycosyl phosphatidylinositol-linked FcγRIII$_{pmn}$ mediates transmembrane signaling events distinct from FcγRIII. *Journal of Experimental Medicine*, **171**, 1239–55.

Kraft, A. S. & Berkow, R. L. (1987). Tyrosine kinase and phosphotyrosine phosphatase activity in human promyelocytic leukaemia cells and human polymorphonuclear leukocytes. *Blood*, **70**, 356–62.

Krause, K.-H. & Welsh, M. J. (1990). Voltage-dependent and Ca^{2+}-activated ion channels in human neutrophils. *Journal of Clinical Investigation*, **85**, 491–8.

Kroll, M. H., Zavoico, G. B. & Schafer, A. I. (1989). Second messenger function of phosphatidic acid in platelet activation. *Journal of Cellular Physiology*, **139**, 558–64.

Lanser, M. E., Brown, G. E. & Garcia-Aguilar, J. (1988). Neutrophil CR3 induction by platelet supernatants is due to platelet-derived growth factor. *Surgery*, **104**, 287–91.

Leonard, E. J., Skeel, A., Yoshimura, T., Noer, K., Kutvirt, S. & Epps, D. van (1990). Leukocyte specificity and binding of human neutrophil attractant/activation protein-1. *Journal of Immunology*, **144**, 1323–30.

Lerea, K. M., Tonks, N. K., Krebs, E. G., Fischer, E. H. & Glomset, J. A. (1989). Vanadate and molybdate increase tyrosine phosphorylation in a 50-kilodalton protein and stimulate secretion in electropermeabilized platelets. *Biochemistry*, **28**, 9286–92.

Lewis, R. S. & Cahalan, M. D. (1989). Mitogen-induced oscillations of cytosolic Ca^{2+} and transmembrane Ca^{2+} current in human leukemic T cells. *Cell Regulation*, **1**, 99–112.

Lopez, A. F., Williamson, D. J., Gamble, J. R., Begley, C. G., Harlan, J. M., Klebanoff, S. J., Waltersdorph, A. (1986). Recombinant human granulocyte-macrophage colony-stimulating factor stimulates in vitro mature human neutrophil and eosinophil function, surface receptor expression and survival. *Journal of Clinical Investigation*, **78**, 1220–8.

Mackin, W. M. & Stevens, T. M. (1988). Biochemical and pharmacologic characterization of a phosphatidylinositol-specific phospholipase C in rat neutrophils. *Journal of Leukocyte Biology*, **44**, 8–16.

Marks, P. W. & Maxfield, F. R. (1990). Transient increases in cytosolic free calcium appear to be required for the migration of adherent human neutrophils. *Journal of Cell Biology*, **110**, 43–52.

McColl, S. R., Kreis, C., DiPersio, J. F., Borgeat, P. & Naccache, P. H. (1989). Involvement of guanine nucleotide binding proteins in neutrophil activation and priming by GM-CSF. *Blood*, **73**, 588–91.

Mege, J. L., Volpi, M., Becker, E. L. & Sha'afi, R. I. (1988). Effect of botulinum D toxin on neutrophils. *Biochemical & Biophysical Research Communications*, **152**, 926–32.

Merritt, J. E., Jacob, R. & Hallam, T. J. (1989). Use of manganese to discriminate between calcium influx and mobilization from internal stores in stimulated human neutrophils. *Journal of Biological Chemistry*, **264**, 1522–7.

Meyer, T., Holowka, D. & Stryer, L. (1988). Highly cooperative opening of calcium channels by inositol 1,4,5-trisphosphate. *Science*, **240**, 653–5.

Michell, R. H. (1975). Inositol phospholipids and cell surface receptor function. *Biochimica & Biophysica Acta*, **415**, 81–147.

Michell, R. H. & Putney, J. W. (1987). *Inositol Lipids in Cellular Signaling*. Cold Spring Harbor Laboratory, Cold Spring Harbor.

Miller, D. K., Gillard, J. W., Vickers, P. J., Sadowski, S., Leveille, C., Mancini, J. A., Charleson, P., Dixon, R. A. F., Ford-Hutchinson, A. W., Fortin, R., Gauthier, J. Y., Rodkey, J., Rosen, R., Rouzer, C., Sigal, I. S., Strader, C. D. & Evans, J. F. (1990). Identification and isolation of a membrane protein necessary for leukotriene production. *Nature, London*, **343**, 278–81.

Moolenaar, W. H., Kruijer, W., Tilly, B. C., Verlaan, I., Bierman, A. J. & Laat, S.W.d.e. (1986). Growth factor-like action of phosphatidic acid. *Nature, London*, **323**, 171–3.

Mullmann, T. J., Siegel, M. I., Egan, R. W. & Billah, M. M. (1990). Complement C5a activation of phospholipase D in human neutrophils. A major route to the production of phosphatidates and diglycerides. *Journal of Immunology*, **144**, 1901–8.

Murphy, P. M., Eide, B., Goldsmith, P., Braun, M., Gierschik, P., Spiegel, A. & Malech, H. (1987). Detection of multiple forms of Giα in HL-60 cells. *FEBS Letters*, **221**, 81–6.

Naccache, P. H., Molski, T. F. P., Borgeat, P. & Sha'afi, R. I. (1985). Phorbol esters inhibit the fMet-Leu-Phe and leukotriene B$_4$ stimulated calcium mobilization and enzyme secretion in rabbit neutrophils. *Journal of Biological Chemistry*, **260**, 2125–32.

Naccache, P. H., Molski, T. F. P., Volpi, M., Becker, E. L. & Sha'afi, R. I. (1985). Unique inhibitory profile of platelet activating factor induced calcium mobilization, polyphosphoinositide turnover and granule enzyme secretion in rabbit neutrophils towards pertussis toxin and phorbol esters. *Biochemical & Biophysical Research Communications*, **130**, 677–84.

Naccache, P. H., Faucher, N., Borgeat, P., Gasson, J. C. & DiPersio, J. F. (1988). Granulocyte-macrophage colony-stimulating factor modulates the excitation-response coupling sequence in human neutrophils. *Journal of Immunology*, **140**, 3541–6.

Naccache, P. H., McColl, S. R., Caon, A. C. & Borgeat, P. (1989). Arachidonic acid-induced mobilization of calcium in human neutrophils. Evidence for a multicomponent mechanism of action. *British Journal of Pharmacology*, **97**, 461–8.

Naccache, P. H., Sha'afi, R. I. & Borgeat, P. (1989). Mobilization, metabolism, and biological effects of eicosanoids in polymorphonuclear leukocytes. In: *The Neutrophil: Cellular Biochemistry and Physiology*, pp. 114–39, Ed. M. B. Hallett. CRC Press, Boca Raton, Florida.

Naccache, P. H., Therrien, S., Caon, A. C., Liao, N., Gilbert, C. & McColl, S. R. (1989). Chemoattractant-induced cytoplasmic pH changes and cytoskeletal reorganization in human neutrophils. Relationship to the stimulated calcium transients and oxidative burst. *Journal of Immunology*, **142**, 2438–44.

Nagata, K.-I., Nagao, S. & Nozawa, Y. (1989). Low Mr GTP-binding proteins in human platelets: cyclic AMP-dependent protein kinase phosphorylates m22KG(I) but not c22KG in cytosol. *Biochemical & Biophysical Research Communications*, **160**, 235–42.

Nasmith, P. E. & Grinstein, S. (1987). Are Ca^{2+} channels in neutrophils activated by a rise in cytosolic free Ca^{2+}. *FEBS Letters*, **221**, 95–100.

Nasmith, P. E., Mills, G. B. & Grinstein, S. (1989). Guanine nucleotides induce tyrosine phosphorylation and activation of the respiratory burst in neutrophils. *Biochemical Journal*, **257**, 893–7.

Nathan, C. F. (1987). Neutrophil activation on biological surfaces. Massive secretion of hydrogen peroxide in response to products of macrophages and lymphocytes. *Journal of Clinical Investigation*, **80**, 1550–60.

Nathan, C. F. (1989). Respiratory burst in adherent human neutrophils: triggering by colony-stimulating factors GM-CSF and CSF-G. *Blood*, **73**, 301–6.

Nathan, C., Srimal, S., Farber, C., Sanchez, E., Kabbash, L., Asch, A., Gailit, J. & Wright, S. D. (1989). Cytokine-induced respiratory burst of human neutrophils: dependence on extracellular matrix proteins and CD11/CD18 integrins. *Journal of Cell Biology*, **109**, 1341–9.

Ng, D. S. & Wong, K. (1989). Effect of platelet-activating factor (PAF) on cytosolic free calcium in human peripheral blood mononuclear leucocytes. *Research Communications in Chemical & Pathological Pharmacology*, **64**, 531–4.

Nigli, V. & Jenni, V. (1989). Actin-associated proteins in human neutrophils: identification and reorganization upon cell activation. *European Journal of Cell Biology*, **49**, 366–72.

Norgauer, J., Kownatzki, E., Seifert, R. & Aktories, A. (1988). Botulinum C2 toxin ADP-ribosylates actin and enhances O_2^- production and secretion but inhibits migration of activated human neutrophils. *Journal of Clinical Investigation*, **82**, 1376–82.

O'Flaherty, J. T., Jacobson, D. P. & Redman, J. F. (1989). Bidirectional effects of protein kinase C activators. Studies with human neutrophils and platelet-activating factor. *Journal of Biological Chemistry*, **264**, 6836–41.

Ohta, Y., Akiyama, T., Nishida, E. & Sakai, H. (1987). Protein kinase C and cAMP-dependent protein kinase induce opposite effects of actin polymerizability. *FEBS Letters*, **222**, 305–10.

Okajima, F., Katada, T. & Ui, M. (1985). Coupling of the guanine nucleotide regulatory protein to chemotactic peptide receptors in neutrophil membranes and its uncoupling by islet-activating protein, pertussis toxin. *Journal of Biological Chemistry*, **260**, 6761–8.

Omann, G. M., Allen, R. A., Bokoch, G. M., Painter, R. G., Traynor, A. E. & Sklar, L. A. (1987). Signal transduction and cytoskeletal activation in the neutrophil. *Physiological Reviews*, **67**, 285–322.

Omann, G. M., Porasik, M. M. & Sklar, L. A. (1989). Oscillating actin polymerization/depolymerization responses in human polymorphonuclear leukocytes. *Journal of Biological Chemistry*, **264**, 16355–8.

Pai, J.-K., Siegel, M. I., Egan, R. W. & Billah, M. M. (1988). Phospholipase

D catalyzes phospholipid metabolism in chemotactic peptide-stimulated HL-60 granulocytes. *Journal of Biological Chemistry*, **263**, 12472–7.

Pereira, H. A., Shafer, W. M., Pohl, J., Martin, L. E. & Spitznagel, J. K. (1990). CAP37, a human neutrophil-derived chemotactic factor with monocyte specific activity. *Journal of Clinical Investigation*, **85**, 1468–76.

Perlmutter, R. M., Marth, J. D., Ziegler, S. F., Garvin, A. M., Pawar, S., Cooke, M. P. & Abraham, K. M. (1988). Specialized protein tyrosine kinase proto-oncogenes in hematopoietic cells. *Biochimica Biophysica Acta*, in press.

Pike, M. C., Wicha, M. S., Yoon, P., Mayo, L. & Boxer, L. A. (1989). Laminin promotes the oxidative burst in human neutrophils via increased chemoattractant receptor expression. *Journal of Immunology*, **142**, 2004–11.

Pober, J. S., Gimbrone, M. A., Lapierre, L. A., Mendrick, D. L., Fiers, W., Rothlein, R. & Springer, T. A. (1986). Overlapping patterns of activation of human endothelial cells by interleukin 1, tumor necrosis factor, and immune interferon. *Journal of Immunology*, **137**, 1893–6.

Pohlman, T. H., Stanness, K. A., Beatty, P. G., Ochs, H. D. & Harlan, J. M. (1986). An endothelial cell surface factor(s) induced in vitro by lipopolysaccharide, interleukin 1, and tumor necrosis factor-α increases neutrophil adherence by a CDw18-dependent mechanism. *Journal of Immunology*, **136**, 4548–53.

Polakis, P. G., Snyderman, R. & Evans, T. (1989). Characterization of G25K, a GTP-binding protein containing a novel putative nucleotide binding domain. *Biochemical & Biophysical Research Communications*, **160**, 25–32.

Polakis, P. G., Weber, R. F., Nevins, B., Didsbury, J. R., Evans, T. & Snyderman, R. (1989). Identification of the ral and rac1 gene products, low molecular mass GTP-binding proteins from human platelets. *Journal of Biological Chemistry*, **264**, 16383–9.

Pommier, C. G., O'Shea, J., Chused, T., Yancey, K., Frank, M. J., Takahashi, T. & Brown, E. J. (1984). Studies on the fibronectin receptors of human peripheral blood leukocytes. Morphologic and functional characterization. *Journal of Experimental Medicine*, **159**, 137–51.

Pontremoli, S., Melloni, E., Sparatore, B., Michettii, M., Salamino, F. & Horecker, B. L. (1990). Isozymes of protein kinase C in human neutrophils and their modification by two endogenous proteinases. *Journal of Biological Chemistry*, **265**, 706–12.

Pozzan, T., Volpe, P., Zorzato, F., Bravin, M., Krause, K. H., Lew, D. P., Bruno, S., Hashimoto, B. & Meldolesi, J. (1988). The Ins(1,4,5)P$_3$-sensitive Ca^{2+} store of non-muscle cells: endoplasmic reticulum or calciosomes? *Journal of Experimental Biology*, **139**, 181–93.

Quinn, M. T., Parkos, C. A., Walker, L., Orkin, S. H., Dinauer, M. C. & Jesaitis, A. J. (1989). Association of a ras-related protein with cytochrome b of human neutrophils. *Nature, London*, **342**, 198–200.

Ray, L. B. & Sturgill, T. W. (1987). Rapid stimulation by insulin of a serine/threonine kinase in 3T3-L1 adipocytes that phosphorylates microtubule-associated protein-2 in vitro. *Proceedings of the National Academy of Sciences, USA*, **84**, 1502–6.

Reinhold, S. L., Prescott, S. M., Zimmerman, G. A. & McIntyre, T. M. (1990). Activation of human neutrophil phospholipase D by three separable mechanisms. *FASEB Journal*, **4**, 208–14.

Rhee, S. G., Suh, P.-G., Ryu, S.-H. & Lee, S. Y. (1989). Studies of inositol phospholipid-specific phospholipase C. *Science*, **244**, 546–50.

Roa, K. M. K. & Varani, J. (1982). Actin polymerization induced by chemotactic

peptide and concanavalin A in rat neutrophils. *Journal of Immunology*, **129**, 1605–7.
Rossi, A. G., McMillan, R. M. & MacIntyre, D. E. (1988). Agonist-induced calcium flux, phosphoinositide metabolism, aggregation and enzyme secretion in human neutrophils. *Agents and Actions*, **24**, 275–82.
Rossi, F., Grzeskowiak, M., Della Bianca, V., Calzetti, F. & Gandini, G. (1990). Phosphatidic acid and not diacylglycerol generated by phospholipase D is functionally linked to the activation of the NADPH oxidase by FMLP in human neutrophils. *Biochemical & Biophysical Research Communications*, **168**, 320–7.
Rothlein, R., Dustin, M. L., Marlin, S. D. & Springer, T. A. (1986). A human intercellular adhesion molecule (ICAM-1) distinct from LFA-1. *Journal of Immunology*, **137**, 1270–4.
Rotrosen, D., Gallin, J. I., Spiegel, A. M. & Malech, H. L. (1988). Subcellular localization of Gi α in human neutrophils. *Journal of Biological Chemistry*, **263**, 10958–64.
Rudolph, U., Koesling, D., Hinsch, K.-D., Seifert, R., Bigalke, M., Schultz, G. & Rosenthal, W. (1989a). G-protein α-subunits in cytosolic and membranous fractions of human neutrophils. *Molecular and Cellular Endocrinology*, **63**, 143–53.
Rudolph, U., Schultz, G. & Rosenthal, W. (1989b). The cytosolic G-protein α-subunit in human neutrophils responds to treatment with guanine nucleotides and magnesium. *FEBS Letters*, **251**, 137–42.
Ruoslahti, E. & Giancotti, F. G. (1989). Integrins and tumor cell dissemination. *Cancer Letters*, **1**, 119–26.
Selvaraj, P., Rosse, W. F., Silber, R. J. & Springer, T. A. (1988). The major Fc receptor in blood has a phosphatidylinositol anchor and is deficient in paroxysmal nocturnal haemaglobinuria. *Nature, London*, **333**, 565.
Sha'afi, R. I. & Molski, T. F. P. (1988). Activation of the neutrophil. *Progress in Allergy*, **42**, 1–64.
Sha'afi, R. I. & Naccache, P. H. (1981). Ionic events in neutrophil chemotaxis and secretion. *Advances in Inflammation Research*, **2**, 115–48.
Sha'afi, R. I., Shefcyk, J., Yassin, R., Molski, T. F. P., Volpi, M., Naccache, P. H., White, J. R., Feinstein, M. B. & Becker, E. L. (1986). Is a rise in intracellular concentration of free calcium necessary or sufficient for actin polymerization. *Journal of Cell Biology*, **102**, 1459–63.
Simchowitz, L., Spilberg, I. & Atkinson, J. P. (1983). Evidence that the functional responses of human neutrophils occur independently of transient elevations in cAMP levels. *Journal of Cyclic Nucleotides and Protein Phosphorylation Research*, **9**, 35–47.
Singer, I. I., Scott, S., Kawka, D. W. & Kazazis, D. M. (1989). Adhesomes: specific granules containing receptors for laminin, C3bi/fibrinogen, fibronectin, and vitronectin in human polymorphonuclear leukocytes and monocytes. *Journal of Cell Biology*, **109**, 3169–82.
Skubitz, K. M., Mendiola, J. R. & Collett, M. S. (1988). CD15 monoclonal antibodies react with a phosphotyrosine-containing protein on the surface of human neutrophils. *Journal of Immunology*, **141**, 4318–23.
Smith, C. D., Lane, B. C., Kusaka, I., Verghese, M. W. & Snyderman, R. (1985). Chemoattractant receptor induced hydrolysis of phosphatidylinositol 4,5-bisphosphate in human polymorphonuclear leucocyte membranes. Requirement for a guanine nucleotide regulatory protein. *Journal of Biological Chemistry*, **260**, 5875–8.
Smith, C. D., Uhing, R. J. & Snyderman, R. (1987). Nucleotide regulatory protein-

mediated activation of phospholipase C in human polymorphonuclear leukocytes is disrupted by phorbol esters. *Journal of Biological Chemistry*, **262**, 6121–7.

Smith, R. J., Justen, J. M. & Sam, L. M. (1990). Recombinant human granulocyte-macrophage colony-stimulating factor induces granule exocytosis from human polymorphonuclear neutrophils. *Inflammation*, **14**, 83–92.

Snyderman, R. & Uhing, R. J. (1988). Phagocytic cells: stimulus-response coupling mechanisms. In: *Inflammation: Basic Principles and Clinical Correlates*, pp. 309–24, eds. J. I. Gallin, I. M. Goldstein and R. Snyderman. Raven Press, New York.

Spät, A., Bradford, P. G., McKinney, J. S., Rubin, R. P. & Putney, J. W. (1986). A saturable receptor for ^{32}P-inositol-1,4,5-trisphosphate in hepatocytes and neutrophils. *Nature, London*, **319**, 514–16.

Stossel, T. P. (1989). From signal to pseudopod. *Journal of Biological Chemistry*, **264**, 18261–4.

Sullivan, R., Griffin, J. D., Simons, E. R., Schafer, A. I., Meshulam, T., Fredette, J. P., Maas, A. K. & Gadenne, A.-S. (1987). Effects of recombinant human granulocyte and macrophage colony-stimulating factors on signal transduction pathways in human granulocytes. *Journal of Immunology*, **139**, 3422–30.

Territo, M. C., Ganz, T., Selsted, M. E. & Lehrer, R. (1989). Monocyte- chemotactic activity of defensins from human neutrophils. *Journal of Clinical Investigation*, **84**, 2017–20.

Therrien, S. & Naccache, P. H. (1989). Guanine nucleotide-induced polymerization of actin in electropermeabilized human neutrophils. *Journal of Cell Biology*, **109**, 1125–32.

Tonks, N. K. & Charbonneau, H. (1989). Protein tyrosine dephosphorylation and signal transduction. *Trends in Biochemical Science*, **14**, 497–500.

Tonnesen, M. G., Anderson, D. C., Springer, T. A., Knedler, A., Avdi, N. & Henson, P. M. (1989). Adherence of neutrophils to cultured human microvascular endothelial cells. Stimulation by chemotactic peptides and lipid mediators and dependence upon the Mac-1, LFA-1, p150,95 glycoprotein family. *Journal of Clinical Investigation*, **83**, 637–46.

Traynor-Kaplan, A. E., Thompson, B. L., Harris, A. L., Taylor, P., Omann, G. M. & Sklar, L. A. (1989). Transient increase in phosphatidylinositol 3,4-bisphosphate and phosphatidylinositol trisphosphate during activation of human neutrophils. *Journal of Biological Chemistry*, **264**, 15668–73.

Tscharner, V. von, Prud'hom, B., Baggiolini, M. & Reuter, H. (1986). Ion channels in human neutrophils activated by a rise in free cytosolic calcium concentration. *Nature, London*, **324**, 369–72.

Tyers, M., Rachubinski, R. A., Stewart, M. I., Verrichio, A. M., Shorr, R. G., Haslam, R. J. & Harley, C. B. (1988). Molecular cloning and expression of the major protein kinase C substrate of platelets. *Nature, London*, **333**, 470–3.

Tzeng, D. Y., Duel, T. F., Huang, J. S., Senior, R. M., Boxer, L. A. & Baehner, R. L. (1984). Platelet-derived growth factor promotes polymorphonuclear leukocyte activation. *Blood*, **64**, 1123–8.

Veillette, A., Bookman, M. A., Horak, E. M., Samelson, L. E. & Bolen, J. B. (1989). Signal transduction through the CD4 receptor involves the activation of the internal membrane tyrosine-protein kinase p56^{lck}. *Nature, London*, **338**, 257–9.

Volpe, P., Krause, K.-H., Hashimoto, S., Zorzato, F., Pozzan, T., Meldolese, J. & Lew, D. P. (1988). 'Calciosome', a cytoplasmic organelle: The inositol 1,4, 5-trisphosphate-sensitive Ca^{2+} store of nonmuscle cells? *Proceedings of the National Academy of Sciences, USA*, **85**, 1091–5.

Volpi, M., Naccache, P. H. & Sha'afi, R. I. (1983). Calcium transport in inside-out membrane vesicles prepared from rabbit neutrophils. *Journal of Biological Chemistry*, **258**, 4153–8.

Volpi, M., Yassin, R., Tao, W., Molski, T. F. P., Naccache, P. H. & Sha'afi, R. I. (1984). Leukotriene B_4 mobilizes calcium without the breakdown of polyphosphoinositides and the production of phosphatidic acid in rabbit neutrophils. *Proceedings of the National Academy of Sciences, USA*, **81**, 5966–70.

Weisbart, R. H., Kwan, L., Golde, D. W. & Gasson, J. C. (1987). Human GM-CSF primes neutrophils for enhanced oxidative metabolism in response to the major physiological chemoattractants. *Blood*, **69**, 18–21.

White, J. R., Naccache, P. H. & Sha'afi, R. I. (1983). Stimulation by chemotactic factor of actin association with the cytoskeleton in rabbit neutrophils: effects of calcium and cytochalasin B. *Journal of Biological Chemistry*, **258**, 14041–7.

Wright, S. D. & Detmers, P. A. (1988). Adhesion-promoting receptors on phagocytes. In *Macrophage Plasma Membrane Receptors: Structure and Function*, pp. 99–120, ed. S. Gordon. The Company of Biologists, Cambridge.

Wright, S. D. & Meyer, B. C. (1986). Phorbol esters cause sequential activation and deactivation of complement receptors on polymorphonuclear leukocytes. *Journal of Immunology*, **136**, 1759–64.

Wright, S. D., Craigmyle, L. S. & Silverstein, S. C. (1983). Fibronectin and serum amyloid P component stimulate C3b- and C3bi-mediated phagocytosis in cultured human monocytes. *Journal of Experimental Medicine*, **158**, 1338–43.

Wright, S. D., Detmers, P. A., Jong, M. T. & Meyer, B. C. (1986). Interferon-gamma depresses binding of ligand by C3b and C3bi receptors on cultured human monocytes, an effect reversed by fibronectin. *Journal of Experimental Medicine*, **163**, 1245–59.

Wymann, M. P., Kernen, P., Bengtsson, T., Andersson, T., Baggiolini, M. & Deranleau, D. A. (1990). Corresponding oscillations in neutrophil shape and filamentous actin content. *Journal of Biological Chemistry*, **265**, 619–22.

Yamazaki, M., Gomez-Cambronero, J., Durstin, M., Molski, T. F. P., Becker, E. L. & Sha'afi, R. I. (1989). Phorbol 12-myristate 13-acetate inhibits binding of leukotriene B_4 and platelet-activating factor and the responses they induce in neutrophils: site of action. *Proceedings of the National Academy of Sciences, USA*, **86**, 5791–5.

Yang, D. C., Brown, A. B. & Chan, T. M. (1989). Stimulation of tyrosine-specific protein phosphorylation and phosphatidylinositol phosphorylation by orthovanadate in rat liver plasma membrane. *Archives in Biochemistry & Biophysics*, **274**, 659–62.

Yassin, R., Shefcyk, J., White, J. R., Tao, W., Volpi, M., Molski, T. F. P., Naccache, P. H., Becker, E. L. & Sha'afi, R. I. (1985). Effects of chemotactic factors and other agents on the amount of actin and a 65,000-mol-wt protein associated with the cytoskeleton of rabbit and human neutrophils. *Journal of Cell Biology*, **101**, 182–8.

Yin, H. L. & Hartwig, J. H. (1988). The structure of the macrophage actin skeleton. In *Macrophage Plasma Membrane Receptors: Structure and Function*, pp. 169–184, ed. S. Gordon. The Company of Biologists, Cambridge.

TISSUE-DERIVED LEUCOCYTE ATTRACTANT FACTORS IN INFLAMMATION

K. B. BACON, N. J. FINCHAM, D. G. QUINN & R. D. R. CAMP

Institute of Dermatology, St Thomas's Hospital, London SE1 7EH, UK

INTRODUCTION

Human inflammatory disease is common, often of uncertain cause and inadequately treated. One of the main histopathological features of both developing and established inflammatory lesions is leucocytic infiltration, which is likely to play a major pathogenic role. The mechanisms of induction of such infiltrates remain unclear, and have therefore been intensively studied with the aim not only of clarifying disease mechanisms, but also of identifying potential targets for anti-inflammatory drug therapy.

A major mechanism in the accumulation of leucocyte infiltrates in inflamed tissues is likely to be the local release of attractant compounds. Considerable effort has therefore been expended in the identification and biological characterization of such materials. If a particular leucocyte attractant mediator is to be proven to be important in the pathogenesis of an inflammatory process, evidence both for its release as well as for a mechanism for its release in the inflamed tissue, should be sought. It should also be shown to induce appropriate biological changes when introduced into the tissue *in vivo* in physiologically relevant concentrations, and inhibitors of its formation or effects, administered *in vivo*, should abolish that aspect of the inflammatory process for which it is thought responsible. Much of this work requires *in vivo* experimentation, which is either difficult or unethical to perform in detail in human volunteers, with the exception of studies involving the skin. We have therefore made extensive use of the skin of both normal volunteers and sufferers from the common inflammatory skin disease, psoriasis. This disease is of unknown cause, is frequently a therapeutic problem, and its lesions are readily sampled. It is also often possible to withhold treatment for an appropriate period prior to lesional sampling. The lesions of psoriasis are characterized by infiltrates of neutrophils (Chowaniec *et al.*, 1981), lymphocytes (Ragaz & Ackerman, 1979) and monocyte/macrophages (Braun-Falco & Schmoeckel, 1977), which are seen both in early developing lesions and in established, chronic disease. Psoriasis is therefore an ideal, accessible human disease for study of the pathogenic role of locally produced leucocyte attractants.

MEASUREMENT OF LEUCOCYTE MIGRATION

In order to detect and characterize leucocyte attractants in the relatively small samples obtainable from human skin, it has been necessary to develop biological assays of sufficient sensitivity, reproducibility and convenience.

Neutrophils

Mixed human leucocytes (70–80% neutrophils) from normal volunteers are prepared from heparinized venous blood by dextran sedimentation and hypotonic lysis of residual erythrocytes. These are used by us in the now well-documented agarose microdroplet assay (Smith & Walker, 1980; Camp et al., 1983) in which 2 μl droplets of agarose/leucocyte suspension are placed in the centre of microtitre plate wells, solidified by cooling and covered by medium containing test material. Neutrophil migration is quantified after a 2-hour incubation by measuring the distance randomly migrating cells have moved peripherally from the agarose droplet. The diameter of the 'flare' of migrating cells in each microtitre plate well is conveniently measured by use of a projecting microscope and digitizer tablet. Although this method does not detect directional migration, it is a quick, sensitive and reproducible assay of neutrophil locomotion. It was, for example, the method first used in experiments which led to the discovery of the potent neutrophil attractant properties of the eicosanoid, leukotriene B_4 (Ford-Hutchinson et al., 1980).

Lymphocytes

Mixed human peripheral blood lymphocytes have been isolated by density gradient centrifugation and plastic adherence to yield cells contaminated by less than 1% monocytes. Migration of these cells has been measured in a 48-well 'microchemotaxis' chamber (Neuro Probe, Cabin John, Maryland, USA) incorporating 8 μm pore-size polyvinylpyrrolidone-free polycarbonate membranes. Cells (50 μl of 2×10^6 ml^{-1} suspension) are placed in the upper wells of this Boyden chamber-type apparatus, and medium containing test material (25 μl) in the lower wells. Following a 60-minute incubation in the absence of serum, migration is quantified by determining the area of the lower surface of the filter occupied by adherent cells, by using image analysis. Image analysis is facilitated by the superior optical qualities of the thin polycarbonate membranes, and greatly increases the rapidity of the assay (Bacon et al., 1988). Several pieces of evidence indicate that this assay measures active lymphocyte migration rather than the simple adherence of passively falling cells. Firstly, concentration-related increases in the number of cells adherent to the lower surface of the filter are seen when agonist is placed in the lower wells only, whereas there is no increase in adherent cells above background when agonist is placed in both upper

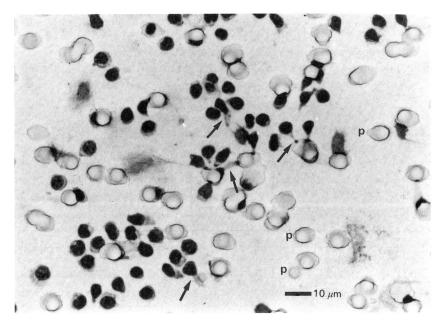

Fig. 1. Mixed human peripheral blood lymphocytes adherent to the underside of the polycarbonate filter. Arrows indicate cells polarised at the time of fixation. P represents the 8 μm pores.

and lower wells by using a modification of the checkerboard system of Zigmond & Hirsch (1973) (Bacon et al., 1988, 1990a). Secondly, although small numbers of cells do drop off the filter into the lower wells during the assay, the number does not reciprocally decrease in association with the increased cell numbers seen on the lower surface of the filter as agonist concentrations are increased (Bacon et al., 1988). The number of cells falling into the lower wells is also no different in the presence of inactive compounds or active attractants (Bacon et al., 1988). Furthermore, no cells adherent to the lower filter surface are seen when the assay is carried out at 4°C instead of at 37°C (Bacon et al., 1990b). Finally, polarization of lymphocytes on the filter undersurface is frequently seen (Fig. 1).

Responses are also completely inhibited by addition to the upper wells of $3\,\mu g\,ml^{-1}$ cytochalasin B, which interferes with cellular microfilament integrity (Bacon et al., 1990 b).

The responses seen in this assay system are also completely inhibited by addition to the lymphocyte suspensions of antibodies against the α and β subunits of the surface adhesion molecule, leucocyte function-associated antigen-1 (LFA-1; K. B. Bacon & R. D. R. Camp, submitted for publication), an observation which was not associated with an increase in the number of cells dropping from the lower surface of the filter into the lower wells. This finding, which has also been reported by others using a Boyden chamber system with nitrocellulose filters (van Epps et al., 1989) indicated that the translocation of lymphocytes to the lower surface of the polycarbonate filter was dependent on the integrity of surface LFA-1 molecules. Finally, responses in the assay appear to be calcium dependent as the calcium chelator, EGTA, and low concentrations of specific antagonists of voltage-dependent calcium channels, completely abolished agonist-induced activity (Bacon & Camp, 1989; Bacon et al., 1989).

A further methodological advantage of the 48-well lymphocyte migration assay is that it allows the phenotypic characterization of migrated cells on the undersurface of the polycarbonate filters by using the alkaline phosphatase anti-alkaline phosphatase (APAAP) immunocytochemical technique in conjunction with specific monoclonal antibodies (J. S. Ross, K. Mistry, K. B. Bacon & R. D. R. Camp, submitted for publication). This has allowed a more comprehensive characterization of the lymphocyte subsets responding to specific agonists.

Monocytes

Recently, peripheral blood monocytes have been prepared by density gradient centrifugation and plastic adherence, and used in the 48-well Neuro Probe chamber with 5 μm pore-size polyvinylpyrrolidone-free polycarbonate membranes and a 30-minute incubation. These purified cells have been used at a concentration of $10^6 \, ml^{-1}$, but mixed mononuclear cells (lymphocytes and monocytes) may also be used as lymphocytes do not migrate through the 5 μm diameter pores under these conditions.

The number of monocytes adherent to the undersurface of polycarbonate filters is again quantified by image analysis (Quinn & Camp, 1990). Under these conditions, migration of monocytes in response to a range of agonist preparations has been demonstrated, responses as high as 15-fold above the background occasionally being seen.

The low volumes and computerized analysis of responses in these assays of neutrophil, lymphocyte and monocyte migration have thus provided methods of sufficient sensitivity and convenience for the routine analysis of as many as 200 samples per day.

RESPONSES TO SPECIFIC AGONISTS

Dilution-related neutrophil migratory responses in the agarose microdroplet assay have been demonstrated with a range of agonists, including the eicosanoids leukotriene B_4 (LTB_4; Ford-Hutchinson et al., 1980; Brain et al., 1984a,b), 12[R]-hydroxyeicosatetraenoic acid (12[R]HETE; Cunningham et al., 1985), C5a des arg (Camp et al., 1986) and neutrophil activating peptide/IL-8 (Gearing et al., 1990). Material, similar or identical to all of these compounds, has been demonstrated in samples from psoriatic skin lesions (see below).

The lymphocyte migration assay has been found to give dose-related responses to a large range of standard agonists, including several recombinant cytokines. Active agonists have included zymosan-activated plasma, LTB$_4$, 12[R]-HETE, N-formylmethionyl-leucyl-phenylalanine (Bacon et al., 1988), recombinant interleukin (rIL)-1 alpha and beta (Bacon et al., 1989), rIL-2 (Ross et al., 1989), rIL-3, rIL-4, rIL-6 (Bacon et al., 1990 b) and rIL-8 (Bacon et al., 1989). The lymphocyte attractant effects of IL-1 (Miossec et al., 1984; Hunninghake et al., 1987), IL-2 (Kornfeld et al., 1985; Robbins et al., 1986), IL-8 (Larsen et al., 1989) and zymosan activated plasma/C5a (Russell et al., 1975; El Naggar et al., 1980, 1981; Van Epps, 1984) have also been reported by others. In contrast, we have found that recombinant granulocyte colony stimulating factor, macrophage colony stimulating factor, transforming growth factor beta (Bacon et al., 1990b) and the stereoisomer of 12[R]-HETE, 12[S]-HETE (Bacon et al., 1988), do not induce responses in the lymphocyte migration assay, and recombinant interferon gamma is only weakly active (Bacon et al., 1990b).

By comparison, preliminary results indicate that the responses of monocytes in the described 48-well assay are more selective for specific cytokines. Thus rIL-1, rIL-3, rIL-4, rIL-6 and rIL-7 were inactive, while recombinant interferon gamma was weakly active in some assays. However, recombinant transforming growth factor beta and monocyte chemotactic and activating factor-monocyte chemotactic peptide-1 (MCAF/MCP-1) gave good concentration-related responses (Quinn, D. G. & Camp, R. D. R., submitted for publication), as has been described by others (Wahl et al., 1987; Wiseman et al., 1988; Leonard & Yoshimura, 1990).

LEUCOCYTE ATTRACTANT ACTIVITY IN SAMPLES FROM HUMAN SKIN

Neutrophil attractants

Little or no neutrophil attractant activity is detectable in samples from normal skin, but a range of active compounds has been found in psoriatic lesions. These include the eicosanoids LTB$_4$ (Brain et al., 1984a,b; Ruzicka et al., 1986), 12[R]-HETE (Cunningham et al., 1985; Woollard et al., 1989, as well as C5a des arg (Takematsu et al., 1986a; Schröder & Christophers, 1986; Camp et al., 1986) and a group of polypeptides related to NAP/IL-8 (Schröder & Christophers, 1988; Gearing et al., 1990). The role of locally produced LTB$_4$ in the pathogenesis of psoriatic lesions has received much attention, but has been reviewed elsewhere and will not be considered in detail here. In brief, current evidence suggests that local release of LTB_4 in the skin is unlikely to be a primary pathogenic event in psoriasis. It may, nevertheless, play a role together with other events, and inhibitors of its formation or its effects may have a place in the treatment of this disease (Camp & Greaves, 1987; Camp, 1989).

In 1986, a novel neutrophil activating and attractant peptide distinct from C5a des arg, and chromatographically similar to IL-1, was described in aqueous extracts of the scale from the surface of psoriatic lesions (Schröder & Christophers, 1986; Camp et al., 1986). This was subsequently shown to be composed of several active components (Fincham et al., 1988). Recently, three components were resolved and shown to induce dilution-related locomotor activity in the agarose microdroplet assay, the activity of each component being neutralized by an anti-IL-8 antiserum (Gearing et al., 1990). Preliminary sequence information has confirmed the identity of one component as an IL-8 species, a second component showing homology with the amino-terminal sequence of melanoma growth-stimulating activity (Schröder et al., 1989).

It is of interest that the first reports of the isolation of the IL-8-like material from psoriatic lesions (Schröder & Christophers, 1986; Camp et al., 1986; Fincham et al., 1988) preceded the first descriptions of the *in vitro* production of IL-8 species by monocytes and other cells (Yoshimura et al., 1987a,b; Matsushima & Oppenheim, 1989). Furthermore, psoriatic skin lesions are characterized by rapid epidermal proliferation and the shedding on the surface of large amounts of incompletely differentiated stratum corneum containing biologically active compounds produced in deeper layers. It has been possible to recover sufficient quantities of this surface scale to allow detailed characterization of active material, to the level of amino acid sequencing, as described above. These facts highlight the importance of psoriasis as a model human inflammatory disease.

It appears that cytokine production in psoriasis is not a non-specific process. Analysis of aqueous extracts of lesional psoriatic stratum corneum and normal heel stratum corneum as control, for IL-1, IL-2, IL-4, IL-6, IL-8, granulocyte macrophage colony-stimulating factor, tumour necrosis factor and interferon activity showed that IL-8 was the only biologically active cytokine, of those tested, which was present in the lesional samples in increased levels (Gearing et al., 1990). Although immunoreactive amounts of interferons alpha and gamma were detected, no activity was found when the immunoreactive samples were tested in an interferon bioassay.

In vitro experiments have suggested that IL-8 may be produced locally in the skin not only by infiltrating leucocytes, but also by the epidermal keratinocytes (Barker et al., 1990a,b) and dermal fibroblasts (Schröder et al., 1990). These, and the above findings, point to IL-8 species as being of potential importance in the pathogenesis of psoriasis and possibly other skin diseases. However, the *in vivo* properties of IL-8 in the skin require further study.

Lymphocyte attractants

The evidence for the presence of activated T-cells in psoriatic lesions as well as the proven efficacy of cyclosporin A in the treatment of psoriasis, has heightened interest in lymphocyte-mediated pathogenic events in this disease (for review, see Bos, 1988). Other evidence for lymphocyte activation in psoriasis includes the finding of raised plasma levels of soluble IL-2 receptors and CD8 antigen (Kapp *et al.*, 1988; Kapp, 1990) and the report that therapeutic administration of the T-cell growth factor, IL-2, exacerbated the disease (Lee *et al.*, 1988). We have therefore considered that lymphocyte attractants, generated in the skin, may be of pathogenic importance in psoriasis. As described in previous sections, materials similar or identical to C5a and the eicosanoids LTB_4 and 12[R]-HETE, have been identified in extracts of psoriatic lesional samples, and have been shown to induce concentration-related lymphocyte migration *in vitro*. Using recombinant material, we have also confirmed the original observation (Larsen *et al.*, 1989) that IL-8 is a potent lymphocyte attractant *in vitro*, maximal responses in the migration assay being obtained with concentrations as low as 10^{-10} M (Bacon *et al.*, 1989). These various attractants may therefore all play a role in inducing the accumulation of lymphocytes in psoriatic lesions, although which of these compounds, if any, is of the greatest importance, is currently unclear. There is also limited evidence that reactive T-cell proliferation may contribute to lesional lymphocyte accumulation in psoriasis (Lange-Wantzin *et al.*, 1985; Morganroth *et al.*, 1990). (See also the chapter by Wilkinson, this volume, for a discussion of mitogens and activation of lymphocyte motility. *Eds*) Further studies of the role of lymphocyte attractants in psoriasis have been interrupted by our finding that normal skin samples contain substantial lymphocyte attractant activity. In view of the histological evidence that normal skin contains an extravascular population of activated lymphocytes (Smolle *et al.*, 1985; Bos *et al.*, 1987; Rowden *et al.*, 1988; Groh *et al.*, 1989), we have proposed that physiological attractants, which mediate the trafficking of lymphocytes in skin and other tissues, exist (Bacon *et al.*, 1990*a*). Aqueous extracts of normal heel stratum corneum, used as a convenient source of epidermal material, induce dilution-related responses in the lymphocyte migration assay. Active material in these extracts has been found in filtrates following ultrafiltration through YM10 and YM2 membranes, indicating the presence of low molecular weight material. Similar activity was found in suction blister and friction blister fluid from normal skin, as well as in exudate from abraded normal skin, recovered by a skin chamber method. The likely presence of plasma elements in the blister and chamber fluid samples prompted the testing of normal plasma, which was also found to contain dilution-related lymphocyte attractant activity which was exclusively less than 1 kD, as determined by ultrafiltration through YM2 membranes. The material in plasma and skin chamber fluid has been purified to homogeneity by gradient-elution reversed phase high performance liquid chromatography (RP-HPLC). The active material from the two sources is chromatographically identical on isocratic RP-HPLC and has the same UV absorbance spectrum (λ max 274 nm). It appears to be a polar but lipid extractable, non-peptide substance, the precise structure of which remains to be determined. Its recovery from plasma has led to the interim term 'plasma-associated lymphocyte chemoattractant', PALC (Bacon et al., 1990a). Apart from its structure, the source(s), in vivo properties, and role of PALC in the induction of pathological lymphocyte infiltrates remain to be further investigated. Its presence in plasma suggests that it may play a role in the physiological margination and adherence of trafficking lymphocytes to endothelium and it may also induce the migration of these cells into the skin if sufficient PALC accumulates in the extravascular compartment to create a concentration gradient. Analysis of the phenotypes of lymphocytes that migrated to the undersurface of polycarbonate filters in response to PALC indicates that $CD4^+$ helper T-cells are the main responders (K. B. Bacon, unpublished data). This corresponds well with the main T-cell phenotype reported to be present in normal human dermis (Bos et al., 1987).

Unexpectedly, aqueous extracts of normal human epidermal samples have been found to contain biologically active amounts of IL-1-like material (Gahring et al., 1985; Hauser et al., 1986; Schmitt et al., 1986; Camp et al., 1990). This cytokine, which possesses potent biological properties, may not be biologically available in human skin through intracellular retention in the epidermis, association with membranes or possibly through control by an inhibitor (Camp et al., 1990). Membrane perturbation or other events may allow its release and the induction of biological responses. IL-1 has also been found to be a potent lymphocyte attractant in vitro (Miossec et al., 1984; Hunninghake et al., 1987; Bacon et al., 1989). Its release from the epidermis in disease or following injury may therefore constitute an important mechanism for the induction of pathological lymphocyte infiltrates. Low level release of epidermal IL-1 under normal conditions may also, with PALC, be responsible for physiological trafficking of lymphocytes in normal skin. As for PALC, immunocytochemical analysis of migrated cells on the undersurface of polycarbonate filters showed that $CD4^+$ helper T-cells formed the major population responding to IL-1 (J. S. Ross, K. Mistry, K. B. Bacon & R. D. R. Camp, submitted for publication).

When injected intradermally in human volunteers, recombinant human IL-1 alpha (10–100 U) induced florid erythematous reactions lasting up to 48 hours, biopsies indicating the presence of leucocyte-rich infiltrates (Dowd et al., 1988). These findings confirmed previous evidence for the inflammatory properties of IL-1 in animal tissues, including mouse footpad (Granstein et al., 1986), rat ear (De Young et al., 1987) and rabbit skin (Rampart & Williams, 1988). The inflammatory properties of femtomole quantities of recombinant IL-1 led to the question of whether the human epidermal IL-1-like material, which has never been exactly identified by amino acid sequencing, might not be identical to authentic IL-1. Normal human heel stratum corneum-derived IL-1-like material has therefore been sequentially purified by reversed phase and anion exchange high performance liquid chromatography (HPLC). The retention time on anion exchange HPLC and the effects of neutralizing antibodies suggested the presence of IL-1 alpha-like material, but little or no IL-1 beta (Camp et al., 1990). Intradermal injection of purified autologous material in human volunteers induced persistent inflammatory reactions of a similar magnitude and duration to those seen with recombinant IL-1 alpha. Biopsies at 4 hours showed mixed infiltrates of neutrophils, monocytes and T-helper cells. As there is no consistent evidence that IL-1 affects neutrophil and monocyte migration *in vitro*, the infiltrates of the cells may have been due to the induced synthesis of other compounds (Camp et al., 1990). Preformed IL-1 alpha-like in the human epidermis also appears to be readily released, as shown by a skin chamber sampling method (Camp et al., 1990). Its release following membrane perturbation or a variety of other events may therefore constitute an important mechanism for the induction of inflammatory changes in human skin. As is the case in psoriatic lesions, cytokine production in normal skin is also not a non-specific process, as determined by our analysis of aqueous extracts of normal heel stratum corneum. Assay of these samples for IL-1, IL-2, IL-4, IL-6, IL-8, granulocyte–macrophage colony stimulating factor, tumour necrosis factor, and interferon activity showed that only IL-1 was present in biologically active amounts (Gearing et al., 1990). Extraction and bioassay of a range of organs in the rat has also shown that IL-1 production under normal conditions is specific to the skin – up to 900 times more IL-1 activity was found in extracts of epidermis than of any other organ examined, including lymph node, thymus and spleen, which appeared to contain essentially background levels (Schmitt et al., 1986). These findings provide further evidence for the importance of IL-1 as an initiator of inflammation in human skin.

Preliminary experiments have revealed decreased levels of IL-1 in samples from psoriatic lesions in comparison with levels in control samples (Takematsu et al., 1986b; Fincham et al., 1988; Hammerberg et al., 1989). This unexpected finding, which was not predicted by *in vitro* experiments, may be the result of decreased synthesis, increased breakdown, structural modification, receptor-mediated internalization or binding of IL-1 to an inhibitor in psoriatic lesions, but remains to be adequately explained. It suggests that, although the release of IL-1 from the epidermis may play a role in initiating inflammatory changes in psoriasis, IL-1 release may not be involved in maintaining the chronicity of lesions.

The skin lesions of the cutaneous T-cell lymphoma, mycosis fungoides, are characterized by infiltrates of both malignant and reactive T-cells (Rapaport & Thomas, 1974). Analysis of lesional and control skin for a range of cytokines, including IL-1, IL-2, IL-6, tumour necrosis factor, interferon gamma and granulocyte macrophage colony stimulating factor, has been carried out by using a skin chamber sampling method and either biological or immunoassays. Only IL-6 activity was found to be increased in lesional versus control samples, a result which supports the evidence that cytokine production in the skin is a selective process (Lawlor *et al.*, 1990). Recombinant IL-6 was shown to induce lymphocyte migration in the described 48-well assay (Bacon *et al.*, 1990*b*; Lawlor *et al.*, 1990), and it was calculated that sufficient IL-6 was likely to be released in lesions to influence lymphocyte migration *in vivo*. As for psoriasis, the lesional mycosis fungoides samples contained lower mean IL-1 activity than control samples.

Monocyte attractants

In spite of the evidence that monocytes are a prominent cell in developing psoriatic lesions (Braun-Falco & Schmoeckel, 1977; Schubert & Christophers, 1985), production of monocyte attractants in psoriasis has been the subject of only limited investigations. Aqueous extracts of lesional psoriatic stratum corneum have been found to contain increased monocyte attractant activity compared to samples from normal heel (a convenient source of control stratum corneum). Purification of lesional material by reversed phase HPLC has revealed at least two active, novel components, one of which possesses a molecular weight of about 10,000–13,000 as determined by ultrafiltration and size exclusion HPLC (Quinn & Camp, 1990). Whether this material is identical to the recently described cytokine, monocyte chemotactic and activating factor, MCAF (Matsushima & Oppenheim, 1989; Leonard & Yoshimura, 1990) is uncertain, although the production of MCAF mRNA and an MCAF-like peptide by cultured human epidermal keratinocytes has been demonstrated (Barker *et al.*, 1990*a,b*).

C5a (Yancey *et al.*, 1989) has been reported to induce monocyte migration *in vitro* and, as described above material similar or identical to this compound has been identified in samples from psoriatic lesions. However, the retention time of C5a/C5a des arg on size exclusion HPLC (Camp *et al.*, 1986) indicates that this compound cannot explain the monocyte attractant activity due to the two novel components described above (Quinn & Camp, 1990). Further identification of these components currently awaits the availability of MCAF-neutralizing antibodies or immunoassays.

CONCLUSIONS

Analysis of specific polypeptides, eicosanoids and other substances has revealed that a relatively wide range of compounds is capable of inducing *in vitro* leucocyte migration in the described assays. Preliminary evidence suggests that a particularly wide range of cytokines is able to induce *in vitro* lymphocyte migration, at least in our hands. By applying these, and other assays, to the analysis of samples from the skin of volunteers, it has been possible to focus attention on a more limited range of leucocyte attractants, which have been demonstrated to be produced *in vivo*. Compounds of current interest in relation to the pathophysiology of human skin include IL-1, IL-8, PALC and possibly MCAF. The production of agents which inhibit the formation or effects of these compounds is a major challenge, and may lead to the development of new anti-inflammatory drugs.

REFERENCES

Bacon, K. B., Camp, R. D. R., Cunningham, F. M. & Woollard, P. M. (1988). Contrasting *in vitro* chemotactic activity of the hydroxyl enantiomers of 12-hydroxy-5,8,10,14-eicosatetraenoic acid. *British Journal of Pharmacology*, **95**, 966–74.

Bacon, K. B. & Camp, R. D. R. (1989). Calcium channel antagonists inhibit leukotriene (LT)B_4 and interleukin (IL)−1α-induced lymphocyte migration. *Skin Pharmacology*, **2**, 47.

Bacon, K. B., Westwick, J. & Camp, R. D. R. (1989). Potent and specific inhibition of IL-8-, IL-1α- and IL-1β-induced lymphocyte migration by calcium channel antagonists. *Biochemical and Biophysical Research Communications*, **165**, 349–54.

Bacon, K., Fincham, N. & Camp, R. (1990*a*). Isolation of a novel low-molecular weight lymphocyte chemoattractant from normal human skin and plasma. *European Journal of Immunology*, **20**, 565–71.

Bacon, K., Gearing A. & Camp, R. (1990*b*). Induction of *in vitro* human lymphocyte migration by interleukin 3, interleukin 4 and interleukin 6. *Cytokine*, **2 (2)**, 1–6.

Barker, J. N. W. N., Sarma, V., Mitra, R. S., Dixit, V. M. & Nickoloff, B. J. (1990*a*). Marked synergism between tumour necrosis factor-α and interferon-γ in regulating keratinocyte-derived adhesion molecules and chemotactic factors. *Journal of Clinical Investigation*, **85**, 605–8.

Barker, J. N. W. N., Jones, M. L., Swenson, C., Mitra, R. S., Elder, J. T., Fantone, J. C., Ward, P. A., Dixit, V. M. & Nickoloff, B. J. (1990*b*). Keratinocyte (KC) production of a biologically active monocyte chemoattractant. *Journal of Investigative Dermatology*, **94**, 505.

Bos, J. D. (1988). The pathomechanisms of psoriasis: the skin immune system and cyclosporin. *British Journal of Dermatology*, **118**, 141–55.

Bos, J. D., Zonneveld, I., Das, P. K., Kreig, S. R., Van Der Loos, C. M. &

Kapsenberg, M. L. (1987). The skin immune system (SIS): Distribution and immunophenotype of lymphocyte populations in normal human skin. *Journal of Investigative Dermatology*, **88**, 569–73.

Brain, S. D., Camp, R. D. R., Cunningham, F. M., Dowd, P. M., Greaves, M. W. & Kobza Black, A. (1984a). Leukotriene B_4-like material in scale of psoriatic skin lesions. *British Journal of Pharmacology*, **83**, 313–17.

Brain, S. D., Camp, R. D. R., Dowd, P. M., Kobza Black, A. & Greaves, M. W. (1984b). The release of leukotriene B_4-like material in biologically active amounts from the lesional skin of patients with psoriasis. *Journal of Investigative Dermatology*, **83**, 70–3.

Braun-Falco, O. & Schmoekel, C. (1977). The dermal inflammatory reaction in initial psoriasis lesions. *Archives of Dermatological Research*, **258**, 9–16.

Camp, R. D. R. (1989). Cutaneous Pharmacology: Perspectives on the growth of investigation of mediators of inflammation. *Journal of Investigative Dermatology*, **92**, (Supplement), 785–835.

Camp, R. D. R., Mallet, A. I., Woollard, P. M., Brain, S. D., Kobza Black, A., & Greaves, M. W. (1983). The identification of hydroxy fatty acids in psoriatic skin. *Prostaglandins*, **26**, 431–47.

Camp, R. D. R., Fincham, N. J., Cunningham, F. M., Greaves, M. W., Morris, J. & Chu, A. C. (1986). Psoriasis skin lesions contain biologically active amounts of an interleukin 1-like compound. *Journal of Immunology*, **137**, 3469–74.

Camp, R., Fincham, N., Ross, J., Bird, C. & Gearing, A. (1990). Potent inflammatory properties in human skin of interleukin-1 alpha-like material isolated from normal skin. *Journal of Investigative Dermatology*, **94**, 735–41.

Camp, R. D. R. & Greaves, M. W. (1987). Inflammatory Mediators in the skin. *British Medical Bulletin*, **43**, 401–14.

Chowaniec, O., Jablonska, S., Beutner, E. H., Proniewska, M., Jarzabek-Corzelska, M. & Rzesa, G. (1981). Earliest clinical and histological changes in psoriasis. *Dermatologica*, **163**, 42–51.

Cunningham, F. M., Woollard, P. M. & Camp, R. D. R. (1985). Proinflammatory properties of unsaturated fatty acids and their monohydroxy metabolites. *Prostaglandins*, **30**, 497–509.

De Young, L. M., Spires, D. A., Kheifets, J. & Terrell, T. G. (1987). Biology and pharmacology of recombinant human interleukin-1β -induced rat ear inflammation. *Agents and Actions*, **21**, 325–7.

Dowd, P. M., Camp, R. D. R. & Greaves, M. W. (1988). Human recombinant interleukin-1α is proinflammatory in normal skin. *Skin Pharmacology*, **1**, 30–7.

El Naggar, A. K., Van Epps, D. E. & Williams, R. C. (1980). Human-B and T-lymphocyte locomotion in response to casein, C5a and fmet–leu–phe. *Cellular Immunology*, **56**, 365–73.

El Naggar, A. K., Van Epps, D. E. & Williams, R. C. (1981). Effects of cuturing on the human lymphocyte locomotion response to casein C5a and fmet–leu–phe. *Cellular Immunology*, **60**, 43–9.

Fincham, N. J., Camp, R. D. R., Gearing, A. J. H., Bird, C. R. & Cunningham, F. M. (1988). Neutrophil chemoattractant and IL-1-like activity in samples from psoriatic skin lesions. Further characterisation. *Journal of Immunology*, **140**, 4294–9.

Ford-Hutchinson, A. W., Bray, M. A., Doig, M. V., Shipley, M. E. & Smith, M. J. H. (1980). Leukotriene B, a potent chemokinetic and aggregating substance released from polymorphonuclear leucocytes. *Nature, London*, **286**, 264–5.

Gahring, L. C., Buckley, A. & Daynes, R. A. (1985). Presence of epidermal-derived thymocyte activating factor/interleukin 1 in normal human stratum corneum. *Journal of Clinical Investigation*, **76**, 1585–91.

Gearing, A. J. H., Fincham, N. J., Bird, C. R., Wadhwa, A. M., Meager, A., Cartwright, J. E. & Camp, R. D. R. (1990). Cytokines in skin lesions of psoriasis. *Cytokine*, **2(1)**, 68–75.

Granstein, R. D., Margolis, R. & Mizel, S. B. (1986). In vivo inflammatory activity of epidermal cell-derived thymocyte activating factor and recombinant interleukin 1 in the mouse. *Journal of Clinical Investigation*, **77**, 1020–7.

Groh, V., Porcelli, S., Fabbi, M., Lanier, L. L., Picker, L. J., Anderson, T., Warnke, R. A., Bhan, A. K., Strominger, J. L. & Brenner, M. B. (1989). Human lymphocytes bearing T-cell receptor γ/δ are phenotypically diverse and evenly distributed throughout the lymphoid system. *Journal of Experimental Medicine*, **169**, 1277–94.

Hammerberg, C., Fisher, G., Baadsgaard, O., Voorhees, J. J. & Cooper, K. D. (1989). Characterization of IL-1 and inhibitor in psoriatic and normal skin. *Journal of Investigative Dermatology*, **92**, 439.

Hauser, G., Saurat, J.-H., Schmitt, A., Jaunin, F. & Dayer, J. M. (1986). Interleukin 1 is present in normal human epidermis. *Journal of Immunology*, **136**, 3317–23.

Hunninghake, G. W., Glazier, A. J., Monick, M. M. & Dinarello, C. A. (1987). Interleukin 1 is a chemotactic factor for human T-lymphocytes. *American Review of Respiratory Diseases*, **135**, 66–71.

Kapp, A. (1990). Elevated levels, of soluble CD8 antigen in sera of patients with psoriasis – a possible sign of suppressor/cytotoxic T-cell activation. *Archives of Dermatological Research*, **282**, 6–7.

Kapp, A., Piskorski A. & Schöpf, E. (1988). Elevated levels of interleukin 2 receptor in patients with atopic dermatitis and psoriasis. *British Journal of Dermatology*, **119**, 707–10.

Kornfeld, H., Berman, J. S., Beer, D. J. & Centre, D. M. (1985). Induction of human T-lymphocyte motility by Interleukin 2. *Journal of Immunology*, **134**, 3887–90.

Lange-Wantzin, G., Larsen, J. K., Christensen, I. J., Ralfkiaer, E., Tjalve, M., Lundkofoed, M. & Thomsen, K. (1985). Activated and cycling lymphocytes in benign dermal lymphocytic infiltrates including psoriatic skin lesions. *Archives of Dermatological Research*, **278**, 92–6.

Larsen, C. G., Anderson, A. O., Appella, E., Oppenheim, J. J. & Matsushima, K. (1989). The neutrophil-activating protein (NAP-1) is also chemotactic for T-lymphocytes. *Science*, **243**, 1464–6.

Lawlor, F., Smith, N. P., Camp, R. D. R., Bacon, K. B., Kobza-Black, A., Greaves, M. W. & Gearing, A. J. H. (1990). Skin exudate levels of interleukin 6, interleukin-1, and other cytokines in mycosis fungoides. *British Journal of Dermatology, in press.*

Lee, R. E., Gaspari, A. A., Lotze, M. T., Chang, A. E. & Rosenberg, S. A. (1988). Interleukin 2 and psoriasis. *Archives of Dermatology*, **124**, 1811–15.

Leonard, E. J. & Yoshimura T. (1990). Human monocyte chemoattractant protein-1 (MCP-1). *Immunology Today*, **11**, 97–101.

Matsushima, K. & Oppenheim, J. J. (1989). Interleukin 8 and MCAF: Novel inflammatory cytokines inducible by IL1 and TNF. *Cytokine*, **1(1)**, 2–13.

Miossec, P., Yu, C. L. & Ziff, M. (1984). Lymphocyte chemotactic activity of human interleukin 1. *Journal of Immunology*, **133**, 2007–11.

Morganroth, G. S., Chan, L. S., Walter, J. D., Voorhees, J. J., Weinstein, J. D. & Cooper, K. D. (1990). Proliferating cells in psoriatic dermis are composed primarily of T-cells, endothelial cells and factor X111a$^+$ perivascular dendritic cells. *Journal of Investigative Dermatology*, **94**, 557.

Quinn, D. G. & Camp, R. D. R. (1990). Novel monocyte attractants in stratum corneum from psoriatic lesions. *Journal of Investigative Dermatology*, **94**, 569.

Ragaz, A. & Ackerman, A. B. (1979). Evolution, maturation and regression of lesions of psoriasis. *American Journal of Dermatopathology*, **1**, 199–214.

Rampart, M. & Williams, T. J. (1988). Evidence that neutrophil accumulation induced by interleukin-1 requires both local protein synthesis and neutrophil CD18 antigen expression *in vitro*. *British Journal of Pharmacology*, **94**, 1143–8.

Rapaport, H. & Thomas, L. B. (1974). Mycosis fungoides: The pathology of extra-cutaneous involvement. *Cancer*, **84**, 1198–229.

Robbins, R. A., Klassen, L., Rasmussen, J., Clayton, M. E. M. & Russ, W. D. (1986). Interleukin-2-induced chemotaxis of human T-lymphocytes. *Journal of Laboratory and Clinical Medicine*, **108**, 340–5.

Ross, J. S., Bacon, K. B. & Camp, R. D. R. (1989). Cyclosporin is a potent and selective inhibitor of *in vitro* human lymphocyte migration *British Journal of Pharmacology*, **96**, 290P.

Rowden, G., Davis, D., Luckett, D. & Poulter, L. (1988). Identification of CD4$^+$, 2H4$^+$ (T8 γ^+) suppressor-inducer cells in normal epidermis and superficial dermis. *British Journal of Dermatology*, **119**, 147–54.

Russell, R. J., Wilkinson, P. C., Sless, F. & Parrott, D. M. V. (1975). Chemotaxis of lymphoblasts. *Nature, London*, **256**, 646–8.

Ruzicka, T., Simmet, T., Peskar, B. A. & Ring, J. (1986). Skin levels of arachidonic acid-derived inflammatory mediators and histamine in atopic dermatitis and psoriasis. *Journal of Investigative Dermatology*, **86**, 105–8.

Schmitt, A., Hauser, C., Jaunin, F. & Dayer, J.-M. (1986). Normal epidermis contains high amounts of natural tissue IL-1. Biochemical analysis by HPLC identifies a MW $\approx$ 17 Kd form with a pI 5.7 and a MW $\approx$ 30 Kd form. *Lymphokine Research*, **5**, 105–18.

Schröder, J.-M. & Christophers, E. (1986). Identification of C5a des arg and an anionic neutrophil activating peptide (ANAP) in psoriatic scales. *Journal of Investigative Dermatology*, **87**, 53–8.

Schröder, J.-M. & Christophers, E. (1988). Identification of a novel family of highly potent neutrophil chemotactic peptides in psoriatic scales. *Journal of Investigative Dermatology*, **91**, 395.

Schröder, J.-M., Young, J., Gregory, H. & Christophers, E. (1989). Amino acid sequence characterization of two ultrastructurally related neutrophil activating peptides obtained from lesional psoriatic scales. *Journal of Investigative Dermatology*, **92**, 515.

Schröder, J.-M., Sticherling, M., Henneicke, H. H., Preissner, W. C. & Christophers, E. (1990). IL-1α or tumour necrosis factor-α stimulate release of three NAP-1/IL-8-related neutrophil chemotactic proteins in human dermal fibroblasts. *Journal of Immunology*, **144**, 2223–32.

Schubert, C. & Christophers, E. (1985). Mast cells and macrophages in early relapsing psoriasis. *Archives of Dermatological Research*, **277**, 352–8.

Smith, M. J. H. & Walker, J. R. (1980). The effect of some anti-rheumatic drugs in an *in vitro* model of human polymorphonuclear leukocyte chemotaxis. *British Journal of Pharmacology*, **69**, 473–8.

Smolle, J., Erhall, R. & Kerl, H. (1985). Inflammatory cell types in normal human epidermis an immunohistochemical and morphometric study. *Acta Dermatologica Venerologea (Stockh)*, **65**, 475–83.

Takematsu, H., Ohkohchi, K. & Tagami, H. (1986a). Demonstration of anaphylatoxins C3a, C4a and C5a in the scales of psoriasis and inflammatory pustular dermatoses. *British Journal of Dermatology*, **114**, 1–6.

Takematsu, Suzuki, R., Tagami, H. & Kumagai, K. (1986*b*). Interleukin 1-like activity in scale extracts of psoriasis and sterile pustular dermatoses. *Dermatologica*, **172**, 236–40.

Van Epps, D. E. (1984). Mediators and modulators of lymphocyte chemotaxis. *Agents and Actions (Supplement)*, **12**, 217–33.

Van Epps, D. E., Potter, J., Vachula, M., Smith, C. W. & Anderson, D. C. (1989). Suppression of human lymphocyte chemotaxis and transendothelial migration by anti-LFA-1 antibody. *Journal of Immunology*, **143**, 3207–10.

Wahl, S. M., Hunt, D. A., Wakefield, L. M., McCartney-Francis, N., Wahl, L. M., Roberts, A. B. & Sporn, M. B. (1987). Transforming growth factor type β induces monocyte chemotaxis and growth factor production. *Proceedings of the National Academy of Sciences (USA)*, **84**, 5788–99.

Wiseman, D. M., Polverini, P. J., Kamp, D. W. & Leibovich, S. J. (1988). Transforming growth factor-beta (TGFβ) is chemotactic for human monocytes and induces their expression of angiogenic activity. *Biochemical and Biophysical Research Communications*, **157**, 793–800.

Woollard, P. M., Cunningham, F. M., Murphy, G. M., Camp, R. D. R. & Greaves, M. W. (1989). A comparison of the proinflammatory effects of 12(R)-and 12(S)-hydroxy-5,8,10,14-eicosatetraenoic acid in human skin. *Prostaglandins*, **38**, 465–71.

Yancey, K. B., Lawley, T. J., Dersookian, M. & Harvath, L. (1989). Analysis of the interaction of human C5a and C5a des arg with human monocytes and neutrophils: Flow cytomatic and chemotaxis studies. *Journal of Clinical Dermatology*, **92**, 184–9.

Yoshimura, T., Matsushima, K., Oppenheim, J. J. & Leonard, E. J. (1987*a*). Neutrophil chemotactic factor produced by lipopolysaccharide (LPS)-stimulated human blood mononuclear leucocytes. Partial characterization and separation from interleukin 1 (IL-1). *Journal of Immunology*, **139**, 788–93.

Yoshimura, T., Matsushima, K., Tanaka, S., Robinson, E. A., Appella, E., Oppenheim, J. J. & Leonard, E. J. (1987*b*). Purification of a human monocyte-derived neutrophil chemotactic factor that shows sequence homology with other host defense cytokines. *Proceedings of the National Academy of Sciences (USA)*, **85**, 9233–7.

Zigmond, S. H. & Hirsch, J. G. (1973). Leukocyte locomotion and chemotaxis. *Journal of Experimental Medicine*, **137**, 387–410.

INDEX

'A' motility in *Myxococcus xanthus* 201–2
Acanthamoeba 257, 289
accumulated strain mechanism 98, 99
accumulation response, *Rhodobacter sphaeroides* 188–9
acetate 189, 303
acetate receptors 304
actin
 filamentous (F-actin); *Dictyostelium discoideum* 43, 244, 283–5; leucocytes 43, 328
 free monomeric (G-actin) 43, 283
 myosin I interactions 289–90
 polymerization; *Dictyostelium discoideum* 244, 261, 262, 263–4, 274, 283–5; leucocytes 43, 359–60
actin binding proteins 257–8, 360
adaptation
 Dictyostelium discoideum 262–3
 Escherichia coli 110, 129
 Halobacterium halobium 225
 leucocytes 38–9, 327
 Paramecium tetraurelia 303
 phosphotransferase carbohydrate (PTS) system 181
 phototaxis 192–3
 Rhodobacter sphaeroides 185
 role in klinokinesis 15–16
adaptive kinesis (kinesis-with-memory) 7–9
adenylate cyclase, olfactory receptors 298–9
adhesion
 cell–substrate 254, 333
 of leucocytes to endothelium 334–5, 361–3
aerotactic responses 189–91
agarose microdroplet assay 380
aggregation
 Dictyostelium discoideum 242–4, 253
 Myxococcus xanthus 202–4
Agrobacterium tumefaciens 115
alanine 169
alkaline phosphatase anti-alkaline phosphatase (APAAP) 382
allele-specific suppression 148–9
α-actinin 258, 360
ammonia (NH_3) 242
ammonium chloride (NH_4Cl) 310, 313–14
aphidicolin 231
arachidonic acid 352, 357
Arbacia spermatozoan 298, 301–2
aspartate 113, 121, 124
associations, multicellular, *Myxococcus xanthus* 199
ATP 109

Bacillus subtilis 115, 116, 225
 electron acceptor taxis 189–90
 RpoD sigma factor 208
bacteriorhodopsin (BR) 219–20
basal body, flagellar 80, 156–7
$\beta 1$ integrins 361
$\beta 2$ integrins 361, 362–3
$\beta 3$ integrins 361
β-spectrin 258
botulinum D toxin-substrate G proteins 350–1

C-factor, *Myxococcus xanthus* 204
C3bi 361, 362
C5a receptors 325
C5a/C5a des arg 324, 353, 382, 383, 385, 388
Ca–ATPase, *Paramecium tetraurelia* 312–14
calciosomes 356–7
calcium action potentials, *Paramecium tetraurelia* 309
calcium mobilization
 Dictyostelium discoideum 247, 252, 280–2, 283–4
 gustatory neurons 300–1
 leucocytes 284–5, 348, 355, 356–8
calcium pump, *Paramecium tetraurelia* 310, · 312–14
cAMP, *see* cyclic AMP
carbohydrates
 phosphotransferase, *see* phosphotransferase carbohydrates
 Rhodobacter sphaeroides responses 187–8
Caulobacter crescentus 115, 183, 206
 cell cycle 162
 flagella biogenesis 155–72
 heat shock proteins 168, 172
 protein localization 168–72
CD8 385
CD11/CD18 334, 361, 362
CD15 328
CD45 328, 335, 355
cell tracking, computerized 227–9
cGMP, *see* cyclic GMP
CheA protein 82, 83, 108, 112, 135–6, 193
 CheW interactions 136–7
 homologous proteins 116, 208, 209–10
 phosphotransferase system interactions 180, 181
$CheA_L/CheA_S/CheW$ complex 136–7, 151–2
 enhancement of $CheA_L$ autophosphorylation 138–42
 increased affinity for CheY 142–5

CheA$_L$ protein 135–6, 145
  autophosphorylation 136, 138–42
CheA$_S$/CheZ complex 145–7, 152
CheA$_S$ protein 135–6
cheB gene, *Caulobacter* 163, 167, 169
CheB protein 112, 135, 136
  function 108, 109
  similarity to FrzG 208, 210–11
chemokinesis 1, 6–7, 324
  leucocytes 327, 333
chemoreceptors, *see* receptors, chemosensory
chemotaxis
  bacterial: modelling 15–31; signalling complexes in 135–52; transducers 107–29
  definition 2
  Dictyostelium discoideum 244–7
  Halobacterium habobium 230–1
  leucocyte 323–40
  methylation-independent 178–89
  neutrophil 35–73
  Paramecium tetraurelia 302–4
chemotaxis genes
  Caulobacter 159–60; mapping 161; regulation of expression 160–4, 167
  Escherichia coli/Salmonella typhimurium 80, 81, 83
chemotaxis proteins
  Escherichia coli/Salmonella typhimurium 108, 135–6, 207–11: *see also* CheA protein; CheB protein; CheR protein etc
  methyl-accepting, *see* methyl-accepting chemotaxis proteins
  segregation in *Caulobacter crescentus* 169–70
cheR gene, *Caulobacter* 163, 167, 169
CheR protein
  function 108, 109, 135
  similarity to FrzF 208, 211
CheW protein 82, 83, 108, 112, 135–6
  CheA interactions 136–7
  phosphotransferase system interactions 180, 181
  similarity to FrzA 207, 208
CheY protein 88, 108, 135, 136, 193
  CheA$_L$/CheA$_S$/CheW complex interactions 142–5
  homologous proteins 116, 208, 209–10
  motor (FliG) interactions 148–9, 150, 151
  phosphotransferase system interactions 180, 181
CheZ protein 108, 135, 136, 149, 151
  binding to CheY 143–5
  CheA$_S$ interaction 145–7, 152
  phosphotransferase system interactions 180, 181

colchicine 332
computerized cell tracking 227–9
contact guidance, leucocytes 333–4
csgA mutants 204
cyclic AMP
  Dictyostelium discoideum: actin response 283, 284, 290; chemotactic responses 244–7; role in aggregation 242, 273; transduction pathways 250, 251, 259–60
  Paramecium tetraurelia responses 302, 303–4, 310–12
  regulation of flagellar gene expression 85
  role in olfaction 299
cyclic AMP receptors
  Dictyostelium discoideum 42, 247, 259–60, 273–4, 298, 301–2: binding affinities 248; structure 248, 275
  Paramecium tetraurelia 298, 304–9
cyclic GMP
  Dictyostelium discoideum 247, 248, 274, 275: intracellular messenger function 252–3, 282–3; myosin II interaction 286, 288–9
  Halobacterium halobium 227
cyclic nucleotides, in leucocytes 348
cyclosporin 338
cytoadhesins 361
cytochalasin B 381
cytokines 326, 354, 389
  leucocyte migratory responses 383
  mediation of leucocyte adhesion to endothelium 362
  modulation of leucocyte responsiveness 363–4
  in mycosis fungoides 388
  in psoriatic skin 384, 387
Cytophaga johnsonae 200
cytoskeletal mutants, *Dictyostelium discoideum* 253–4

defensins 347–8
demethylation
  Escherichia coli 108–9, 110
  Halobacterium halobium 225–6
Dendrocoelum lacteum 30
diacylglycerol 252, 261
dibutyryl-cGMP 227
Dictyostelium discoideum 43, 241–64, 298, 301
  calcium mobilization 247, 252, 280–2, 283–4
  cyclic AMP receptors, *see* cyclic AMP receptors, *Dictyostelium discoideum*
  cyclic GMP, *see* cyclic GMP, *Dictyostelium discoideum*
  F-actin association with cytoskeleton 261, 262, 263–4, 274, 283–5
  folate receptors 248–50

G proteins 250–2, 274, 277–8
inositol lipids 278–9, 280
inositol polyphosphates 275–7
life cycle 241–4
model for chemotactic movement 290
motility and chemotactic behaviour 244–7
mutant studies 247–59, 282–3: actin binding proteins 257–8; intracellular messengers 252–3; motility and cytoskeletal mutants 253–4; myosin II 254–7; proteins interacting with receptors 250–2; receptors 248–50
myosin I 43, 257, 289–90
myosin II, see myosin II, *Dictyostelium discoideum*
ras genes 252, 279–80
signal transduction pathways 259–63, 273–90
Dictyostelium lacteum 273, 283
diffusion coefficients, variations in 16–20, 21, 30
diglyceride (DG) 348, 353
dimethylsulphoxide (DMSO) 191
dipeptide/dipeptide-binding protein 113
directional persistence 37
discoidin 253–4
discoidin I 254
DnaK protein 168, 172
Dras gene 252
DrasG gene 252

E-ring 156, 157
EI kinase 179, 180–1
elasticotaxis 200
electron acceptor taxis 189–91
endothelial venules, high (HEV) 334–5, 336
endothelium, adhesion of leucocytes to 334–5, 361–3
Enterobacter aerogenes 115
enzyme II (EII) 178–81
enzyme III (EIII) 178, 179
epidermal growth factor receptors 117
equilibrium, approach to 25
equilibrium distributions 22–5
erbstatin 354, 355, 356
Escherichia coli 107–29, 225
  chemotactic behaviour 107–8
  components of chemotactic system 108–9
  electron acceptor taxis 189–90
  flagellum 77–101, 108
  methylation-independent chemotaxis 178–81, 189
  motility compared to that of *Myxococcus xanthus* 206, 207, 214–15
  signalling complexes 135–52
  strategies for chemotaxis 15, 20, 31
  transducers, see methyl-accepting chemotaxis proteins, *Escherichia coli*

ethionine 225
excitation response, *Rhodobacter sphaeroides* 185–8
export, flagellum-specific 91–4
  control of order 96
  substrate recognition 94–5
export apparatus, flagellum-specific 79, 88, 93–4
extracellular matrix components, modulation of leucocyte responsiveness 361–3

Fc receptors 328, 333
FcRII receptors 328
fgdA mutants 247, 250–2, 259, 260, 277–8
fibronectin 362
Fick's First Equation 17, 20, 21
filament, flagellar
  Caulobacter 156, 158–9
  Escherichia coli/Salmonella typhimurium 79, 88, 91, 92, 96–7
  Halobacterium halobium 222
filament-capping protein, see hook-associated protein 2
filamin 258
filopods 244
filter assay 323–4
first-signal model 329
flaD gene 157
flaE gene 167
flagella
  Caulobacter 155–72, 183
  Escherichia coli/Salmonella typhimurium 77–101, 108
  Halobacterium halobium 220–1
  Rhodobacter sphaeroides 183–5
flagellar assembly
  Caulobacter 160–72: positional control 168, 170–2; regulation of gene expression 160–8
  Escherichia coli/Salmonella typhimurium 88–101: control of order of export 96; flagellum-specific export 91–4; hook length control 97–101; morphogenetic pathway 88–90; stoichiometry of flagellar components 96–7; substrate recognition for export 94–5
flagellar genes
  Caulobacter 159–60: genomic mapping 160, 161; regulation of expression 160–8
  Escherichia coli/Salmonella typhimurium 80–8; region I genes 80–2; region II genes 81, 82–4; region IIIa genes 80, 81, 83, 84; region IIIb genes 80, 81, 83, 84–5; regulation of expression 85–8
flagellar motor switching
  Escherichia coli/Salmonella typhimurium 78–9, 136–52, 193–4

flagellar motor switching (cont.)
   Halobacterium halobium 225, 226–7, 229–32
flagellins
   Caulobacter 156, 158–60, 163–4, 170–2
   Escherichia coli/Salmonella typhimurium 80, 84, 86, 92–3, 94–5, 96
   Halobacterium halobium 222
flagellum-specific export, *see* export, flagellum-specific
flaK gene 156, 157–8, 163, 165
flaO gene 163, 165
flaP gene 156, 157
FLAP protein 352
flaS gene 163, 164–5, 168
flaY gene 163, 167
flbD gene/FlbD protein 166, 167
flbN gene 156, 157, 163–4, 165, 166, 168
flgA operon 80, 82
flgB operon 80–1, 82
flgD gene/FlgD protein 81, 99
flgJ gene
   Caulobacter 156, 158, 163, 170–1
   Escherichia coli/Salmonella typhimurium 81
flgk gene, *Caulobacter* 156, 158, 160, 166, 171
flgk operon, *Escherichia coli/Salmonella typhimurium* 81, 82, 86
FlgK protein, *see* hook-associated protein 1
flgL gene, *Caulobacter* 156, 158, 163, 166
FlgL protein, *see* hook-associated protein 3
flhA gene 83, 94
flhB operon 82–3
FlhC protein 85, 86–7
flhD operon 82, 85, 86, 87
flhD protein 85, 86–7
flhE gene 81
fliA operon 83, 84, 85, 86, 87
fliB operon 81, 83, 84
fliC operon 81, 83, 84
FliC protein 79
fliD operon 83, 84, 85, 86, 87–8, 96
FliD protein, *see* hook-associated protein 2
fliE operon 83, 84
fliF operon 83, 84
fliG gene 149, 150
FliG protein 79, 83, 136, 149
fliH gene 84, 94
fliI gene 84, 94
fliJ gene 84
fliK gene 84, 97
FliK protein, control of hook length 97–100
fliL operon 83, 84–5
fliM gene 85, 149
FliM protein 79, 149
fliN gene 85, 94
fliS gene 81, 84, 87

fliT gene 81, 84, 87
fluoroaluminate (AlF$_4$) 227
fodrin 355
folA mutants 247, 248–50, 259
folate
   Dictyostelium discoideum responses 241–2, 251–2
   Paramecium tetraurelia responses 304, 310
folate receptors 248–150, 304, 305–6
formyl-methionyl-leucyl-phenylalanine (fMet-Leu-Phe, FMLP) 325, 383
   polarization response 327
   transduction pathways 284–5, 349, 350, 352, 353, 357
formyl-norleucyl-leucyl-phenylalanine (FNLLP) receptors 42, 46–9
formyl peptide receptors 325, 328
formyl peptides 325
frizzy genes, *see frz* genes
fructose-protein (FPr) 181
fruiting body formation
   Dictyostelium discoideum 242–4, 256–7
   Myxococcus xanthus 202–4
frz (frizzy) genes 205–15
   function of products 211–13
   isolation and genetic analysis 205–6
   motility patterns of mutants 206–17
   regulation of expression 213–14, 215
   sequence similarities to chemotaxis genes of enteric bacteria 207–11
frzA gene 205, 206, 213–14, 215
FrzA protein 207, 208, 212
frzB gene 205, 206, 213–14, 215
frzC gene 205, 206
frzCD, regulation of expression 213–14, 215
FrzCD protein 214, 215
   methylation 211–12
   similarity to methyl-accepting chemotaxis proteins (MCPs) 208–9, 211
frzD mutants 206, 207
frzE gene 205, 206, 213–14, 215
FrzE protein 208, 209–10, 212, 213
frzF gene 205, 206
FrzF protein 208, 211, 212
frzG mutants 206
FrzG protein 208, 210–11, 212
ftr 166
fumaric acid 226–7, 232

G proteins
   Dictyostelium discoideum 250–2, 274, 277–8
   Halobacterium halobium 227
   leucocytes 42–3, 325, 338, 349–51, 364–6:
      actin polymerization and 360;
      phospholipase A$_2$ activation 352;
      phospholipase D activation 353;
      tyrosine phosphorylation and 354

olfactory receptors 298–9
ras-related small molecular weight 350–1
Gα1 protein 251, 252
Gα2 protein 251, 252, 260, 277–8
G_{i2} protein, alpha subunit 350, 360
galactose 113, 169
galactose-binding protein 113, 123, 124
gelation factor, 120 kD 258
genetic suppression, allele-specific 148–9
gliding motility, *Myxococcus xanthus* 200–1
glorin 273, 283
glucose 169
glutamine 169
glycerol 114
gradient perception
  Escherichia coli 108, 110–11
  leucocytes/neutrophils 35–73, 326–7
  Rhodobacter sphaeroides 185–9
granulocyte colony-stimulating factor (G-CSF) 354, 383
granulocyte-macrophage colony-stimulating factor (GM-CSF) 354, 363–4, 365
growth factor receptors 117
GTP binding proteins, *see* G proteins
GTPγS 227, 277, 278, 350, 352
guanine nucleotide-binding proteins, *see* G proteins
gustation 300–1

Halobacterium halobium 115, 219–35
measurement of behaviour 227–9
signal transduction 219–35: diffusible signal inducing motor switching 231–2; integration of photosensory and chemosensory signals 230–1; kinetics of near UV reception by SR-I 234–5; molecular components 221–7; photocatalytic signal formation 232–3; Poisson stimulus–response curves 233–4
spontaneous motor switching 229–30
swimming behaviour 220–1
halorhodopsin (HR) 220
HB5 mutants 247, 259
heat-denatured proteins, neutrophil responses 324
heat shock proteins, *Caulobacter crescentus* 168, 172
hexose protein (HPr) 178, 179, 180–1, 190
high endothelial venules (HEV) 334–5, 336
hook
  Caulobacter 156, 157–8, 166–7, 168
  Escherichia coli/Salmonella typhimurium 79, 157–8: assembly 88, 89, 91, 92, 93, 94–5, 96–7; control of length 97–101
hook-associated protein 1 (HAP1, FlgK) 79, 89, 92, 93, 95, 100
hook-associated protein 2 (HAP2, FliD, filament-capping protein) 79, 84, 87, 88, 92–3, 95
hook-associated protein 3 (HAP3, FlgL) 79, 89, 92, 95
hook-associated proteins (HAPs) 79–80, 86
  assembly 92–3, 94–5, 100
  Caulobacter 158
  stoichiometry 96–7
12[R]-hydroxyeicosatetraenoic acid (12[R]-HETE) 382, 383, 385
12[S]-hydroxyeicosatetraenoic acid(12[S]-HETE) 383

IBMX (isobutyl methylxanthine) 227, 311–12
immune complexes, adhesion of neutrophils to 333
immunosuppressant drugs 338
IMP 303–4
inflammation
  mechanisms of leucocyte accumulation 332–9
  tissue-derived leucocyte attractant factors 379–87
inositol 1,4-bisphosphate 275
inositol 4,5-bisphosphate (Ins(4,5)P_2) 277
inositol hexakisphosphate (InsP_6) 277
inositol lipids 278–9, 280, 348–9, 366
inositol phosphate cycle, *Dictyostelium discoideum* 277
inositol polyphosphates
  Dictyostelium discoideum 275–7
  Paramecium tetraurelia 310
inositol 1,3,4,5-tetrakisphosphate (Ins(1,3,4,5)P_4) 349
inositol 1,3,4,-trisphosphate (Ins(1,3,4)P_3) 349
inositol 1,4,5,-trisphosphate (Ins(1,4,5)P_3)
  Dictyostelium discoideum 252, 260–1, 274, 275–7, 278, 280, 284
  leucocytes 348, 356–7
  role in olfaction 299
inositol 1,4,5-trisphosphate (Ins(1,4,5)P_3) binding protein 356
insulin receptors 117
integration host factor (IHF) 166–7
integrin receptors 361–2
integrins 361–3
intercellular adhesion molecules (ICAMs) 361, 362
interferon-alpha 384
interferon-gamma (IFNγ) 354, 362, 383, 384
interleukin 1 (IL-1) 362, 383, 386–8, 389
interleukin 1 (IL-1)-like material 386
interleukin 2 (IL-2) 383
interleukin 2 (IL-2) receptors 385
interleukin 3 (IL-3) 383
interleukin 4 (IL-4) 383
interleukin 6 (IL-6) 383, 388

interleukin 7 (IL-7) 383
interleukin 8 (IL-8) 389
  lymphocyte attractant activity 337, 383, 385
  neutrophil attractant activity 382, 384–5
ionic conductances, *Paramecium tetraurelia* 309–10

kinesis
  abstraction of concepts 2–6
  alternative concepts 10–12
  definitions 1–2
  distinction from taxis 6–10
kinesis-with-memory (adaptive kinesis) 7–9
Klebsiella pneumoniae 115, 180
klinokinesis 2, 3, 30
  with adaptation 15–16
  neutrophils 38
  without adaptation 15

L-ring
  Caulobacter 156, 157
  Escherichia coli/Salmonella typhimurium 79, 88, 90, 96
lactate, *Paramecium tetraurelia* responses 303
lamellipod extension and retraction 36, 65, 325
  molecular basis 43
  stochastic modelling 65–73
  thinking models 44
  in uniform chemoattractant concentrations 37
  see also pseudopod formation
laminin (LM) 361, 362
leucocyte adhesion deficiency (LAD) 334
leucocyte cell-adhesion molecules (Leu-CAM, $\beta 2$ integrins) 361, 362–3
leucocyte function-associated antigen-1 (LFA-1) 381–2
leucocytes 323–40, 379–87
  attractants in human skin 383–9
  gradient detection 35–73, 326–7
  initial studies of chemotaxis 323–6
  mechanism of polarization 327–8
  mechanisms of tissue accumulation 332–9: adhesion to endothelium 334–5; non-chemotactic responses 333–4
  migration assays 380–2
  signal transduction 347–66: activation pathways 348–56; agonists and their receptors 347–8; cytoskeleton 359–60; modulation of responsiveness by cytokines 363–4; modulation of responsiveness by extracellular matrix components 361–3; second messengers and effector systems 284–5, 356–9
  stochastic models of movement 53–73,
  328–32
  see also lymphocytes; monocytes; neutrophils
leucotriene B_4 (LTB$_4$) 64, 349, 352
  calcium mobilization 357
  leucocyte attractant activity 382, 383, 385
  in psoriatic skin 383–4
light
  aerotactic responses and 191
  Halobacterium halobium responses 230–1: biochemical processes 219–20, 222–4; transduction 231–5
  measurement of behavioural responses 228–9
  see also phototaxis
lipocortins 352, 359
lipopolysaccharide (LPS) 361, 362
5-lipoxygenase 352
lobopod formation, *see* pseudopod formation
Lon protein 172
lymphocyte chemotactic factor (LCF) 337
lymphocytes 328, 379
  activation and locomotor activity 335–9
  attractants 383, 385–8
  chemotactic responses to macrophages 339, 340
  migration assay 380–2
  migratory pathways 334–5
  signal transduction pathways 355
lymphoid tissue, migration of lymphocytes in 334–5, 336
lymphoma, cutaneous T-cell 388

M-ring
  Caulobacter 156–7
  Escherichia coli/Salmonella typhimurium 79, 88, 90, 94, 96
macrophage colony-stimulating factor (M-CSF) 383
macrophages 379
  effects of cytokines 363–4
  lymphocyte chemotactic responses 339, 340
magnesium ions (Mg^{2+}) 312
magnetotaxis 5–6
maltose/maltose-binding protein 113, 121, 123–4
mating type receptors, yeast 298, 301–2
meche (*tar*) operon 82, 83
melanoma growth-stimulating activity 384
membrane potential, *Paramecium tetraurelia* 309–12
membrane proteins, redistribution in polarized leucocytes 327–8
methyl-accepting chemotaxis proteins (MCPs) 177
  Caulobacter crescentus 167, 169–70

Escherichia coli 112: intramolecular transduction 117–20; ligand interaction 120–4; membrane-spanning region 124–7; a model for structural organization 127–9; related receptors in other systems 115–17; role of 109–11; structure 111–15
Escherichia coli/Salmonella typhimurium; CheA$_L$/CheA$_S$/CheW complex formation and 141–2, 151–2; gene localization 80, 82, 83–4; similarity to FrzCD 208–9, 211
Halobacterium halobium 224–6
see also Tap protein; Tar protein; Trg protein; Tsr protein
methylation-dependent transduction
 Escherichia coli 108–9, 110–11, 129, 135
 Halobacterium halobium 224–6
 Myxococcus xanthus 211–13
phosphotransferase carbohydrate system interactions 180, 181
methylation-independent taxis 177–94
 chemotaxis 178–89
 electron acceptor taxis 189–91
 phototaxis 192–3
mgl gene 201
microbeam irradiation 231–2
microtubule-associated protein (MAP) kinase 359
microtubule-associated protein-2 (MAP-2) 359
microtubule-disaggregating drugs 332
microtubules 244, 261
mitogens, lymphocyte 336
molecular ruler mechanism 97–9
monapterin 283
monocyte chemotactic and activating factor–monocyte chemotactic peptide-1 (MCAF/MCP-1) 383, 388, 389
monocytes 325, 328, 379
 attractants, 348, 383, 388–9
 effects of cytokines 363–4
 migration assay 382
 signal transduction pathways 354–5
Montecarlo simulations 15, 22–30
Mot (A and B) proteins 79, 83, 88
mot mutants, *Caulobacter* 159
motA (mocha) operon 82, 83
motility mutants, *Dictyostelium discoideum* 247, 253–4
motor switching, flagellar, *see* flagellar motor switching
mycosis fungoides 388
myosin I 43, 257, 289–90, 328
myosin II
 Dictyostelium discoideum 244, 247, 274: mutant studies 254–7, 285–6; role in chemotactic responses 261, 262, 263–4,

285–9, 290
 leucocytes 43, 359
Myxococcus xanthus 115, 199–215
 'A' and 'S' motility 201–2
 cell motility and vegetative growth 199–200
 directional control of motility: frizzy genes 205–15
 fruiting body formation 202–4
 gliding motility 200–1
 motility compared to that of enteric bacteria 214–15

N-ethylmaleimide (NEM) 125, 126
neu oncogene product 117
neutrophil activating peptide/interleukin 8 (NAP/IL-8) 348, 382, 383
neutrophils 328, 379
 accumulation in tissues 333, 334
 attractants 324, 325, 347–8, 382–3
 in human skin 383–5
 cytokines modulating responsiveness of 363–4
 extracellular matrix components modulating function 362–3
 gradient perception 35–73: continuous spatial chemoattractant gradients 39–40; continuous temporal chemoattractant gradients 38–9; discontinuous temporal chemoattractant gradients 38; first generation probabilistic receptor models 50–2; receptor binding and dynamics 41–2; receptor signal transduction and motility activation 42–3; second generation stochastic receptor-cell turning models 53–65; theoretical concepts in receptor-sensing 44–50; thinking models 44; third generation stochastic receptor–lamellipodial activity model 65–73; uniform chemoattractant concentrations 37–8
 migration assay 380
 signal transduction pathways 42–3, 350, 352, 353, 354–60
 see also leucocytes
nitrate 190
nocodazole 332

odorant binding proteins (OBP) 299–300, 301
olfaction 298–300
orientation 3
 neutrophils 39, 40
orthokinesis 1–2, 16, 30
 neutrophils 37
oxygen responses 189–91

P-ring
  Caulobacter 156, 157
  Escherichia coli/Salmonella typhimurium
    79, 88, 90, 96
p55$^{c\text{-fgr}}$ 353-4
p56$^{c\text{-lck1}}$ 353
p59$^{c\text{-hck}}$ 353
p60$^{c\text{-src}}$ 353
p93$^{c\text{-fes}}$ 353
P$_{480}$, *see* sensory rhodopsin-II
Paramecium tetraurelia 298, 302–15
  behaviour 302–4
  Ca-ATPase 312–14
  receptors 304–9
  transduction 309–12
pertussis toxin 338, 349, 355
pH
  extracellular, *Dictyostelium discoideum*
    247
  Paramecium tetraurelia responses 309, 310
pheromone binding protein (PBP) 300
pheromones, insect 298, 299
phoborhodopsin, *see* sensory rhodopsin-II
phorbol esters 358–9
phosphatidic acid 353, 364
phosphatidylinositol 4-phosphate (PtdIns4P)
    280
phosphatidylinositol 4,5-bisphosphate
    (PtdIns4,5-P$_2$) 278–9, 280, 348
phosphatidylinositol trisphosphate (PIP$_3$)
    349
phosphodiesterase, cGMP-specific 227,
    252–3, 261–2, 282
phosphoenolpyruvate (PEP) 178
phosphoinositide cycle
  Dictyostelium discoideum 278–9
  leucocytes 348–9, 366
phospholipase A$_2$ 353, 366
phospholipase A$_2$-activating protein 352
phospholipase C, phosphoinositide-specific
    (PIC)
  Dictyostelium discoideum 279
  leucocytes 348, 349, 351–2, 363–4
phospholipase D (PLD) 348, 353, 364, 366
phosphoryl-chemotaxis-protein (PCP) 180,
    181
phosphorylation
  cAMP receptors 247, 275
  CheA$_L$ protein 136, 138–42
  by Frz proteins in *Myxococcus xanthus*
    211–13
  integrated responses to sensory stimuli and
    193–4
  myosin II 247, 286–9
  phosphotransferase system 178–9, 180–1
  protein kinase C-mediated 358–9
  tyrosine, in leucocytes 353–6
phosphotransferase carbohydrates
    chemotaxis towards 178–81
  Rhodobacter sphaeroides responses 187–8
phosphotransferase (PTS) system 178, 190
photochemical processes, *Halobacterium halobium* 219–20, 222–4
phototaxis 2
  Halobacterium halobium 230–1
  photosynthetic bacteria 192–3
  see also light
pili 202
plasma-associated lymphocyte
    chemoattractant (PALC) 386, 389
platelet activating factor (PAF) 357
platelet-derived growth factor (PDGF)
    354–5
platelet-derived growth factor (PDGF)
    receptors 117
ple mutants 159
pod mutants 159
Poisson stimulus–response curves,
    Halobacterium halobium 233–4
polarization 327–8
  lymphocytes 337
  neutrophils 37, 327
polarization assay 325–6
Polysphondylium 273
Polysphondylium violaceum 283
position, concept of 2–3
potassium
  extracellular, *Dictyostelium discoideum*
    247
  Rhodobacter sphaeroides responses 186,
    187
probabilistic receptor models 50–2
profilin 359, 360
proline 169, 181–2
propionate, *Rhodobacter sphaeroides*
    responses 186, 187
protein kinase C 348, 358–9, 366
protein kinases
  calcium–calmodulin-dependent 359
  cyclic AMP-dependent 359
  see also tyrosine kinases
proton-driven flagellar motors 78
PS-3 thermophilic bacterium 115
pseudochemotaxis 21, 27, 30
Pseudomonas aeruginosa 115
Pseudomonas pudita 115
pseudoplasmodium (slug) formation,
  Dictyostelium discoideum 242, 256–7
pseudopod formation
  Dictyostelium discoideum 244–7, 261–2:
    actin and 289–90; mechanism of
    chemotactic response 259–63, 290;
    myosin I and 257, 289–90; myosin II and
    256, 257
  leucocytes 36, 326
  see also lamellipod extension and retraction

pseudospatial mechanism of gradient sensing 44, 53, 326–7
  modelling 53–65
psoriasis 379
  lymphocyte attractants 385, 387–8
  monocyte attractants 388–9
  neutrophil attractants 383–5
pteridines 273, 283
pyruvate, *Rhodobacter sphaeroides* responses 187

rac1 350
rac2 350
random walks
  Escherichia coli 107
  neutrophils 37
  one-dimensional model 16–20
  other models 20–1
  persistent 327
rap1 350–1
ras genes, *Dictyostelium discoideum* 252, 279–80
ras-related small molecular weight G proteins 350–1
reaction, *see* response
receptor noise 45–50, 52
receptors, chemosensory
  Dictyostelium discoideum 248–50
  eukaryotes 297–8, 301–2
  leucocytes 347–8
  neutrophil: binding and dynamics 41–2; probabilistic models 50–2; signal transduction and motility activation 42–3; stochastic models 53–73; theoretical concepts in receptor-sensing 44–50
  Paramecium tetraurelia 304–9
  see also cyclic AMP receptors; methyl-accepting chemotaxis proteins
resact receptor, *Arbacia* 298, 301–2
response (reaction)
  alternative concepts 10–12
  concepts of 2–4, 5–6
  distinction of a kinesis from a taxis and 6–8
retinal proteins 219–20
RF-1 protein 166, 167
RF-2 protein 165–6
rflA activity 85, 86, 87–8
Rhizobium meliloti 115
Rhodobacter sphaeroides
  chemotaxis 181–9: accumulation response 188–9; temporal responses 185–8
  electron acceptor taxis 189–91
  motility 182–5
  phototaxis 192–3
rhodopsin, bovine 275
Rhodospirillum rubrum 115, 182, 232
  phototaxis 192, 193

ribose 113, 169
ribose-binding protein 113, 124
rod, flagellar
  Caulobacter 156–7
  Escherichia coli/Salmonella typhimurium 79, 80, 88: assembly 92, 94, 95, 96–7

S-adenosylhomocysteine (SAH) 109
S-adenosylmethionine (SAM) 108–9, 135
'S' motility in *Myxococcus xanthus* 201–2
S-ring 79, 156
Saccharomyces cerevisiae (yeast) 298, 301–2
Salmonella typhimurium 115, 206
  chemotaxis proteins, *see* chemotaxis proteins, *Escherichia coli/Salmonella typhimurium*
  electron acceptor taxis 189–90
  flagellum 77–101
  methyl-accepting chemotaxis proteins, *see* methyl-accepting chemotaxis proteins, *Escherichia coli/Salmonella typhimurium*
  methylation-independent chemotaxis 179
  signalling complexes 135–52
Sec system 90
sensory rhodopsin-I (SR-I) 220, 222–4, 231
  kinetics of near UV reception 234–5
  metastable intermediate (SR-I$_{373}$) 223, 224, 233, 234–5
  signal transduction 225, 226, 232
sensory rhodopsin-II (SR-II) 220, 222–4, 226, 231, 232, 234
serine 113, 121, 124
sevenless gene 117
severin 258
sigma factors
  flagellum-specific 86–7, 164, 165–7
  FrzCD protein and 208
  heat shock 172
signal relay, cyclic AMP 274
signal threshold (S/Δ) ratio 51
signal to noise (S/N) ratio 50, 51
signal transduction 297–315
  aerotactic responses 191
  Dictyostelium discoideum 259–63, 273–90
  Escherichia coli 150–2, 189: intermolecular 110; intramolecular 109–10, 117–20
  gustatory receptor neurons 300–1
  Halobacterium halobium 219–35
  leucocytes 42–3, 347–66
  methylation-independent chemotaxis 179–81
  Myxococcus xanthus 211–13
  olfactory receptor neurons 298–300
  Paramecium tetraurelia 309–12, 314–15
  unicellular eukaryotes 301–15
signalling complexes 135–52
slime trails, *Myxococcus xanthus* 199–200

slug formation, *Dictyostelium discoideum* 242, 256–7
sodium-driven flagellar motors 78
spatial chemoattractant gradients
  Dictyostelium discoideum responses 244–6, 263–4
  E. coli responses 108
  neutrophil/leucocyte responses 39–40, 326
  Rhodobacter sphaeroides responses 185, 188–9
spatial mechanism of gradient sensing 44, 50–2
Spirillum voluntans 30
Spirochaeta aurantia 115
ST638 354
staurosporine 358
stimulus
  alternative concepts 10–12
  concepts of 4–6
  distinction of a kinesis from a taxis and 8–10
stmF (streamer F) mutants 247, 253, 282–3, 286–9, 290
stochastic models of leucocyte movement 328–32
  second generation 53–65
  third generation 65–73
streamer F mutants, *see stmF* mutants
substrate adhesion
  Dictyostelium discoideum 254
  neutrophils 333
sucrose 180
superflagella 222
superoxide anions 353, 364
swarms, *Myxococcus xanthus* 199
switch factor, *Halobacterium halobium* 226–7, 232
switching, flagellar motor, *see* flagellar motor switching

T-cell lymphoma, cutaneous 388
T-cell receptor (TCR) 347, 353, 355
T-cells 335, 339
  $CD4^+$ helper subtype 386
  in psoriasis 385
T-maze assay 302–3
talin 360
Tap protein 108, 111, 113
Tar–EnvZ hybrid protein 116–17
tar gene 115
tar (*meche*) operon 82, 83
Tar protein 108, 117
  intramolecular transduction 117–18, 119–20
  ligand interaction 121, 124
  similarity to FrzCD 208
  structure 111, 112, 113, 114–15
taste receptor cells 300–1

τ-proteins, microtubule-related 359
taxis
  abstraction of concepts 2–6
  alternative concepts 10–12
  definitions 1, 2
  distinction from kinesis 6–10
temperature-sensitive chemotactic mutants, *Dictyostelium discoideum* 259
temporal chemoattractant gradients
  Dictyostelium discoideum responses 246–7, 263–4
  Escherichia coli responses 108, 110
  neutrophil/leucocyte responses 38–9, 326
  Rhodobacter sphaeroides responses 185–8
temporal mechanism of gradient sensing 44, 50–2
thinking models, neutrophil gradient detection 44
Thy-1 328
time-averaging hypothesis 49–50, 52
torso gene 117
trans-acting factors, *Caulobacter* flagellum synthesis 164, 165–7
transducers, *see* methyl-accepting chemotaxis proteins; receptors, chemosensory
transduction, *see* signal transduction
transforming growth factor-beta 383
Trg–EnvZ hybrid protein 117
Trg protein 108
  homologues 115–16
  intramolecular transduction 117–18, 120
  ligand interaction 121–2, 123–4
  membrane-spanning region 125
  structure 111, 112, 113, 114
tsr gene 115
Tsr protein 108
  intramolecular transduction 117–18, 120
  ligand interaction 121, 122, 124
  structure 111, 112, 113–14
tumour necrosis factor-α (TNFα) 354, 362, 363–4
turning, defining 3–4
tyrosine kinases 117
  leucocytes 353–6, 364, 366
tyrosine phosphorylation, leucocytes 353–6

vanadate 355
very late antigens (VLA) 361
vimentin 359
voting hypothesis 150–1

Wiener process 47–8

xylose 169

zymosan-activated plasma 383